Probability and Mathematical Statistics (C ▮▮▮▮▮▮▮▮▮▮

D0793527

ROHATGI • An Introduction to Pr
ematical Statistics
SCHEFFE • The Analysis of Varian
SEBER • Linear Regression Analysis
SERFLING • Approximation Theorems of Mathematical Statistics
TJUR • Probability Based on Radon Measures
WILKS • Mathematical Statistics
WILLIAMS • Diffusions, Markov Processes, and Martingales, Volume I: Foundations
ZACKS • The Theory of Statistical Inference

Applied Probability and Statistics

ANDERSON, AUQUIER, HAUCK, OAKES, VANDAELE, and WEISBERG • Statistical Methods in Comparative Studies
BAILEY • The Elements of Stochastic Processes with Applications to the Natural Sciences
BAILEY • Mathematics, Statistics and Systems for Health
BARNETT and LEWIS • Outliers in Statistical Data
BARTHOLEMEW • Stochastic Models for Social Processes, *Second Edition*
BARTHOLOMEW and FORBES • Statistical Techniques for Manpower Planning
BECK and ARNOLD • Parameter Estimation in Engineering and Science
BELSLEY, KUH, and WELSCH • Regression Diagnostics: Identifying Influential Data and Sources of Collinearity
BENNETT and FRANKLIN • Statistical Analysis in Chemistry and the Chemical Industry *Book Perrad VMI*
BHAT • Elements of Applied Stochastic Processes
BLOOMFIELD • Fourier Analysis of Time Series: An Introduction
BOX • R. A. Fisher, The Life of a Scientist
BOX and DRAPER • Evolutionary Operation: A Statistical Method for Process Improvement
BOX, HUNTER, and HUNTER • Statistics for Experimenters: An Introduction to Design, Data Analysis, and Model Building *1978*
BROWN and HOLLANDER • Statistics: A Biomedical Introduction
BROWNLEE • Statistical Theory and Methodology in Science and Engineering, *Second Edition*
BURY • Statistical Models in Applied Science *'86*
CHAMBERS • Computational Methods for Data Analysis
CHATTERJEE and PRICE • Regression Analysis by Example
CHERNOFF and MOSES • Elementary Decision Theory
CHOW • Analysis and Control of Dynamic Economic Systems
CLELLAND, deCANI, and BROWN • Basic Statistics with Business Applications, *Second Edition*
COCHRAN • Sampling Techniques, *Third Edition*
COCHRAN and COX • Experimental Designs, *Second Edition*
CONOVER • Practical Nonparametric Statistics, *Second Edition*
COX • Planning of Experiments
DANIEL • Biostatistics: A Foundation for Analysis in the Health Sciences, *Second Edition*
DANIEL • Applications of Statistics to Industrial Experimentation
DANIEL and WOOD • Fitting Equations to Data: Computer Analysis of Multifactor Data, *Second Editon*
DAVID • Order Statistics
DEMING • Sample Design in Business Research

continued on back

Regression Diagnostics

Regression Diagnostics:
Identifying Influential Data and Sources of Collinearity

DAVID A. BELSLEY

Professor of Economics, Department of Economics,
Boston College, and

Visiting Professor, MIT Center for Computational
Research in Economics and Management Science

EDWIN KUH

Professor of Economics, Sloan School of Management and
Department of Economics, Massachusetts Institute of
Technology, and

Director, MIT Center for Computational Research in
Economics and Management Science

ROY E. WELSCH

Professor of Management Science and Statistics, Sloan
School of Management, Massachusetts Institute of
Technology, and

Associate Director, MIT Center for Computational
Research in Economics and Management Science.

JOHN WILEY & SONS

New York • Chichester • Brisbane • Toronto • Singapore

Library of Congress Cataloging in Publication Data:

Belsley, David A
    Regression diagnostics.

(Wiley series in probability and mathematical
statistics)
    Bibliography: p.
    Includes index.
    1. Regression analysis.   I. Kuh, Edwin, joint
author. II. Welsch, Roy E., joint author.
III. Title.   IV. Title: Collinearity.

QA278.2.B44   519.5'36   79-19876
ISBN 0-471-05856-4

Printed in the United States of America

10 9

To

G. L. B.
D. F. B.

C. G. K.
B. K. K.

K. B. W.
K. W. W.

Preface

This book deals with an important but relatively neglected aspect of regression theory: exploring the characteristics of a given data set for a particular regression application. Two diagnostic techniques are presented and examined. The first identifies influential subsets of data points, and the second identifies sources of collinearity, or ill conditioning, among the regression variates. Both of these techniques can be used to assess the potential damage their respective conditions may cause least-squares estimates, and each technique allows the user to examine more fully the suitability of his data for use in estimation by linear regression. We also show that there is often a natural hierarchy to the use of the two diagnostic procedures; it is usually desirable to assess the conditioning of the data and to take any possible remedial action in this regard prior to subsequent estimation and further diagnosis for influential data.

Although this book has been written by two economists and a statistician and the examples are econometric in orientation, the techniques described here are of equal value to all users of linear regression. The book serves both as a useful reference and as a collateral text for courses in applied econometrics, data analysis, and applied statistical techniques. These diagnostic procedures provide all users of linear regression with a greater arsenal of tools for learning about their data than have hitherto been available.

This study combines results from four disciplines: econometrics, statistics, data analysis, and numerical analysis. This fact causes problems of inclusion. While some effort has been expended to make this book self-contained, complete success here would unduly expand the length of the text, particularly since the needed concepts are well developed in readily available texts that are cited where appropriate. Some notational problems arise, since econometrics, statistics, data analysis, and numerical analysis employ widely different conventions in use of symbols and terminology. No choice of notation will satisfy any one reader, and so the indulgence of all is requested.

This book has been written with both the theorist and the practitioner in mind. Included are the theoretical bases for the diagnostics as well as straightforward means for implementing them. The practitioner who chooses to ignore some of the theoretical complexities can do so without jeopardizing the usefulness of the diagnostics.

The division of labor lending to this book is reasonably straightforward. The diagnostics for influential data presented in Chapter 2 and its related aspects are an outgrowth of Roy E. Welsch and Edwin Kuh (1977), while the collinearity diagnostics of Chapter 3 and its related aspects stem from David A. Belsley (1976). While both chapters are written with the other in view, the blending of the two techniques occurs most directly in the examples of Chapter 4. It is therefore possible for the reader interested primarily in influential data to omit Chapter 3 and for the reader primarily interested in collinearity to omit Chapter 2.

Research opportunities at the Massachusetts Institute of Technology Center for Computational Research in Economics and Management Science (previously under the aegis of the National Bureau of Economic Research) expedited this research in a number of important ways. The TROLL interactive econometric and statistical analysis system is a highly effective and adaptable research environment. In particular, a large experimental subsystem, SENSSYS, with over 50 operations for manipulating and analyzing data and for creating graphical or stored output, has been created by Stephen C. Peters for applying the diagnostic techniques developed here. The analytical and programming skills of Mark Gelfand, David Jones, Richard Wilk, and later, Robert Cumby and John Neese, have also been essential in this endeavor.

Economists, statisticians, and numerical analysts at the Center and elsewhere have made many helpful comments and suggestions. The authors are indebted to the following: John Dennis, Harry Eisenpress, David Gay, Gene Golub, Richard Hill, Paul Holland, and Virginia Klema. Special thanks go to Ernst Berndt, Paolo Corsi, and David C. Hoaglin for their careful reading of earlier drafts and to John R. Meyer who, as president of the NBER, strongly supported the early stages of this research. It is not possible to provide adequate superlatives to describe the typing efforts of Karen Glennon through the various drafts of this manuscript.

David A. Belsley would like to acknowledge the Center for Advanced Study in the Behavioral Sciences at Stanford where he began his inquiry into collinearity during his tenure there as Fellow in 1970–1971. Roy E.

Welsch would like to acknowledge many helpful conversations with members of the research staff at the Bell Telephone Laboratories at Murray Hill, New Jersey. Grants from the National Science Foundation (GJ-1154X3, SOC-75-13626, SOC-76-14311, and SOC-77-07412) and from IBM helped to make this work possible, and we wish to express our deep appreciation for their support.

DAVID A. BELSLEY
EDWIN KUH
ROY E. WELSCH

Cambridge, Massachusetts
February 1980

Contents

Regression Diagnostics

CHAPTER 1

Introduction and Overview

Over the last several decades the linear regression model and its more sophisticated offshoots, such as two- and three-stage least squares, have surely become among the most widely employed quantitative tools of the applied social sciences and many of the physical sciences. The popularity of ordinary least squares is attributable to its low computational costs, its intuitive plausibility in a wide variety of circumstances, and its support by a broad and sophisticated body of statistical inference. *Given the data*, the tool of least squares can be employed on at least three separate conceptual levels. First, it can be applied mechanically, or descriptively, merely as a means of curve fitting. Second, it provides a vehicle for hypothesis testing. Third, and most generally, it provides an environment in which statistical theory, discipline-specific theory, and data may be brought together to increase our understanding of complex physical and social phenomena. From each of these perspectives, it is often the case that the relevant statistical theory has been quite well developed and that practical guidelines have arisen that make the use and interpretation of least squares straightforward.

When it comes to examining and assessing the quality and potential influence of the data that are assumed "given," however, the same degree of understanding, theoretical support, and practical experience cannot be said to exist. The thrust of standard regression theory is based on sampling fluctuations, reflected in the coefficient variance-covariance matrix and associated statistics (t-tests, F-tests, prediction intervals). The explanatory variables are treated as "given," either as fixed numbers, or, in elaboration of the basic regression model, as random variables correlated with an otherwise independently distributed error term (as with estimators of simultaneous equations or errors-in-variables models). In reality, however, we know that data and model often can be in conflict in ways not readily analyzed by standard procedures. Thus, after all the t-tests have been

1

examined and all the model variants have been compared, the practitioner is frequently left with the uneasy feeling that his regression results are less meaningful and less trustworthy than might otherwise be the case because of possible problems with the data—problems that are typically ignored in practice. The researcher, for example, may notice that regressions based on different subsets of the data produce very different results, raising questions of model stability. A related problem occurs when the practitioner knows that certain observations pertain to unusual circumstances, such as strikes or war years, but he is unsure of the extent to which the results depend, for good or ill, on these few data points. An even more insidious situation arises when an unknown error in data collecting creates an anomalous data point that cannot be suspected on prior grounds. In another vein, the researcher may have a vague feeling that collinearity is causing troubles, possibly rendering insignificant estimates that were thought to be important on the basis of theoretical considerations.

In years past, when multivariate research was conducted on small models using desk calculators and scatter diagrams, unusual data points and some obvious forms of collinearity could often be detected in the process of "handling the data," in what was surely an informal procedure. With the introduction of high-speed computers and the frequent use of large-scale models, however, the researcher has become ever more de-tached from intimate knowledge of his data. It is increasingly the case that the data employed in regression analysis, and on which the results are conditioned, are given only the most cursory examination for their suitabil-ity. In the absence of any appealing alternative strategies, data-related problems are frequently brushed aside, all data being included without question on the basis of an appeal to a law of large numbers. But this is, of course, absurd if some of the data are in error, or they come from a different regime. And even if all the data are found to be correct and relevant, such a strategy does nothing to increase the researcher's understanding of the degree to which his regression results depend on the specific data sample he has employed. Such a strategy also leaves the researcher ignorant of the properties that additionally collected data could have, either to reduce the sensitivity of the estimated model to some parts of the data, or to relieve ill-conditioning of the data that may be preventing meaningful estimation of some parameters altogether.

The role of the data in regression analysis, therefore, remains an important but unsettled problem area, and one that we begin to address in this book. It is clear that strides made in this integral but neglected aspect of regression analysis can have great potential for making regression an even more useful and meaningful statistical tool. Such considerations have led us to examine new ways for analyzing regression models with an emphasis on diagnosing potential data problems rather than on inference or curve fitting.

This book provides the practicing statistician and econometrician with new tools for assessing the quality and reliability of their regression estimates. Diagnostic techniques are developed that (1) aid in the systematic location of data points that are either unusual or inordinately influential and (2) measure the presence and intensity of collinear relations among the regression data, help to identify the variables involved in each, and pinpoint the estimated coefficients that are potentially most adversely affected.

Although the primary emphasis of these contributions is on diagnostics, remedial action is called for once a source of trouble has been isolated. Various strategies for dealing with highly influential data and for ill-conditioned data are therefore also discussed and exemplified. Whereas the list of possible legitimate remedies will undoubtedly grow in time, it is hoped that the procedures suggested here will forestall indiscriminate use of the frequently employed, and equally frequently inappropriate, remedy: throw out the outliers (many of which, incidentally, may not be influential) and drop the collinear variates. While the efforts of this book are directed toward single-equation ordinary least squares, some possible extensions of these analytical tools to simultaneous-equations models and to nonlinear models are discussed in the final chapter.

Chapter 2 is devoted to a theoretical development, with an illustrative example, of diagnostic techniques that systematically search for unusual or influential data, that is, observations that lie outside patterns set by other data, or those that strongly influence the regression results. The impact of such data points is rarely apparent from even a close inspection of the raw-data series, and yet such points clearly deserve further investigation either because they may be in error or precisely because of their differences from the rest of the data.

Unusual or influential data points, of course, are not necessarily bad data points; they may contain some of the most interesting sample information. They may also, however, be in error or result from circumstances different from those common to the remaining data. Only after such data points have been identified can their quality be assessed and appropriate action taken. Such an analysis must invariably produce regression results in which the investigator has increased confidence. Indeed, this will be the case even if it is determined that no corrective action is required, for then the investigator will at least know that the data showing the greatest influence are legitimate.

The basis of this diagnostic technique is an analysis of the response of various regression model outputs to controlled perturbations of the model inputs. We view model inputs broadly to include data, parameters-to-be-estimated, error and model specifications, estimation assumptions, and the ordering of the data in time, space, or other characteristics. Outputs include fitted values of the response variable, estimated parameter values, residuals, and functions of them (R^2, standard errors, autocorrelations, etc.). Specific perturbations of model inputs are developed that reveal where model outputs are particularly sensitive. The perturbations take various forms including differentiation or differencing, deletion of data, or a change in model or error specification. These diagnostic techniques prove to be quite successful in highlighting unusual data, and an example is provided using typical economic cross-sectional data.

Chapter 3 is devoted to the diagnosis of collinearity among the variables comprising a regression data matrix. Collinear (ill-conditioned) data are a frequent, if often unanalyzed, component of statistical studies, and their presence, whether exposed or not, renders ordinary least-squares estimates less precise and less useful than would otherwise be the case. The ability to diagnose collinearity is therefore important to users of least-squares regression, and it consists of two related but separable elements: (1) detecting the presence of collinear relationships among the variates, and (2) assessing the extent to which these relationships have degraded regression parameter estimates. Such diagnostic information would aid the investigator in determining whether and where corrective action is necessary and worthwhile. Until now, attempts at diagnosing collinearity have not been wholly successful. The diagnostic technique presented here, however, provides a procedure that deals succesfully with both diagnostic elements. First, it provides numerical indexes whose magnitudes signify the presence of one or more near dependencies among the columns of a data matrix. Second, it provides a means for determining, within the linear regression model, the extent to which each such near dependency is degrading the least-squares estimate of each regression coefficient. In most instances this latter information also enables the investigator to determine specifically which columns of the data matrix are involved in each near dependency, that is, it isolates the variates involved and the specific relationships in which they are included. Chapter 3 begins with a development of the necessary theoretical basis for the collinearity analysis and then provides empirical verification of the efficacy of the process.

Simple rules and guidelines are stipulated that aid the user, and examples are provided based on actual economic data series.

Chapter 4 provides extended and detailed application to statistical models (drawn from economics) of both sets of diagnostic techniques and examines their interrelationship. Material is also presented here on corrective actions. Mixed estimation is employed to correct the strong collinearity that besets standard consumption-function data, and both sets of diagnostic methods are given further verification in this context. A monetary equation is analyzed for influential observations, and the use of ridge regession is examined as a means for reducing ill-conditioning in the data. A housing-price model, based on a large cross-sectional sample, shows the merits of robust estimation for diagnosis of the error structure and improved parameter estimates.

The book concludes with a summary chapter in which we discuss important considerations regarding the use of the diagnostics and their possible extensions to analytic frameworks outside linear least squares, including simultaneous-equations models and nonlinear models.

CHAPTER 2

Detecting Influential Observations and Outliers

In this chapter we identify subsets of the data that appear to have a disproportionate influence on the estimated model and ascertain which parts of the estimated model are most affected by these subsets. The focus is on methods that involve both the response (dependent) and the explanatory (independent) variables, since techniques not using both of these can fail to detect multivariate influential observations.

The sources of influential subsets are diverse. First, there is the inevitable occurrence of improperly recorded data, either at their source or in their transcription to computer-readable form. Second, observational errors are often inherent in the data. Although procedures more appropriate for estimation than ordinary least squares exist for this situation, the diagnostics we propose below may reveal the unsuspected existence and severity of observational errors. Third, outlying data points may be legitimately occurring extreme observations. Such data often contain valuable information that improves estimation efficiency by its presence. Even in this beneficial situation, however, it is constructive to isolate extreme points and to determine the extent to which the parameter estimates depend on these desirable data. Fourth, since the data could have been generated by a model(s) other than that specified, diagnostics may reveal patterns suggestive of these alternatives.

The fact that a small subset of the data can have a disproportionate influence on the estimated parameters or predictions is of concern to users of regression analysis, for, if this is the case, it is quite possible that the model-estimates are based primarily on this data subset rather than on the majority of the data. If, for example, the task at hand is the estimation of the mean and standard deviation of a univariate distribution, exploration

6

of the data will often reveal outliers, skewness, or multimodal distributions. Any one of these might cast suspicion on the data or the appropriateness of the mean and standard deviation as measures of location and variability, respectively. The original model may also be questioned, and transformations of the data consistent with an alternative model may be suggested. Before performing a multiple regression, it is common practice to look at the univariate distribution of each variate to see if any oddities (outliers or gaps) strike the eye. Scatter diagrams are also examined. While there are clear benefits from sorting out peculiar observations in this way, diagnostics of this type cannot detect multivariate discrepant observations, nor can they tell us in what ways such data influence the estimated model.

After the multiple regression has been performed, most detection procedures focus on the residuals, the fitted (predicted) values, and the explanatory variables. Although much can be learned through such methods, they nevertheless fail to show us directly what the estimated model would be if a subset of the data were modified or set aside. Even if we are able to detect suspicious observations by these methods, we still will not know the extent to which their presence has affected the estimated coefficients, standard errors, and test statistics. In this chapter we develop techniques for diagnosing influential data points that avoid some of these weaknesses. In Section 2.1 the theoretical development is undertaken. Here new techniques are developed and traditional procedures are suitably modified and reinterpreted. In Section 2.2 the diagnostic procedures are exemplified through their use on an intercountry life-cycle savings function employing cross-sectional data. Further examples of these techniques and their interrelation with the collinearity diagnostics that are the subject of the next chapter are found in Chapter 4.

Before describing multivariate diagnostics, we present a brief two-dimensional graphic preview that indicates what sort of interesting situations might be subject to detection. We begin with an examination of Exhibit 2.1a which portrays a case that we might call (to avoid statistical connotations) evenly distributed. If the variance of the explanatory variable is small, slope estimates will often be unreliable, but in these circumstances standard test statistics contain the necessary information.

In Exhibit 2.1b, the point ∘ is anomalous, but since it occurs near the mean of the explanatory variable, no adverse effects are inflicted on the slope estimate. The intercept estimate, however, will be affected. The source of this discrepant observation might be in the response variable, or the error term. If it is the last, it could be indicative of heteroscedasticity or thick-tailed error distributions. Clearly, more such points are needed to analyze those problems fully, but isolating the single point is instructive.

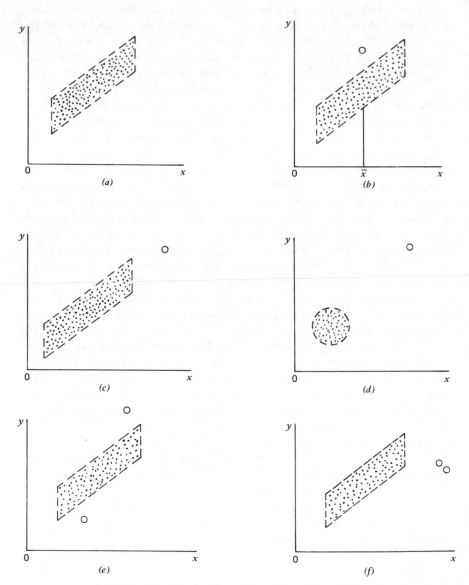

Exhibit 2.1 Plots for alternative configurations of data.

Exhibit 2.1c illustrates the case in which a gap separates the discrepant point from the main body of data. Since this potential outlier is consistent with the slope information contained in the rest of the data, this situation may exemplify the benevolent third source of influence mentioned above in which the outlying point supplies crucially useful information—in this case, a reduction in variance. Exhibit 2.1d is a more troublesome configuration that can arise frequently in practice. In this situation, the estimated regression slope is almost wholly determined by the extreme point. In its absence, the slope might be almost anything. Unless the extreme point is a crucial and valid piece of evidence (which, of course, depends on the research context), the researcher is likely to be highly suspicious of the estimate. Given the gap and configuration of the main body of data, the estimated slope surely has fewer than the usual degrees of freedom; in fact, it might appear that there are effectively only two data points.

The situation displayed in Exhibit 2.1e is a potential source of concern since either or both o's will heavily influence the slope estimate, but differently from the remaining data. Here is a case where some corrective action is clearly indicated—either data deletion or, less drastically, a downweighting of the suspicious observations or possibly even a model reformulation.

Finally, Exhibit 2.1f presents an interesting case in which deletion of either o by itself will have little effect on the regression outcome. The potential effect of one outlying observation is clearly being masked by the presence of the other. This example serves as simple evidence for the need to examine the effects of more general subsets of the data.

2.1 THEORETICAL FOUNDATIONS

In this section we present the technical background for diagnosing influential data points. Our discussion begins with a description of the technique of row deletion, at first limited to deleting one row (observation) at a time. This procedure is easy to understand and to compute. Here we examine in turn how the deletion of a single row affects the estimated coefficients, the predicted (fitted) values, the residuals, and the estimated covariance structure of the coefficients. These four outputs of the estimation process are, of course, most familiar to users of multiple regression and provide a basic core of diagnostic tools.

The second diagnostic procedure is based on derivatives of various regression outputs with respect to selected regression inputs. In particular, it proves useful to examine the sensitivity of the regression output to small perturbations away from the usual regression assumption of homoscedasticity. Elements of the theory of robust estimation can then be used to convert these derivatives into diagnostic measures.

The third diagnostic technique moves away from the traditional regression framework and focuses on a geometric approach. The y vector is adjoined to the X matrix to form n data points in a $p+1$ dimensional space. It then becomes possible for multivariate methods, such as ratios of determinants, to be used to diagnose discrepant points. The emphasis here is on locating outliers in a geometric sense.

Our attention then turns to more comprehensive diagnostic techniques that involve the deletion or perturbation of more than one row at a time. Such added complications prove necessary, for, in removing only one row at a time, the influence of a group of influential observations may not be adequately revealed. Similarly, an influential data point that coexists with others may have its influence masked by their presence, and thus remain hidden from detection by single-point (one-at-a-time) diagnostic techniques. The first multiple-point (more-than-one-at-a-time) procedures we examine involve the deletion of subsets of data, with particular emphasis on the resulting change in coefficients and fitted values. Since multiple deletion is relatively expensive, lower-cost stepwise[1] methods are also introduced.

The next class of procedures adjoins to the X matrix a set of dummy variables, one for each row under consideration. Each dummy variate consists of all zeros except for a one in the appropriate row position. Variable-selection techniques, such as stepwise regression or regressions with all possible subsets removed, can be used to select the discrepant rows by noting which dummy variables remain in the regression. The derivative approaches can also be generalized to multiple rows. The emphasis is placed both on procedures that perturb the homoscedasticity assumption in exactly the same way for all rows in a subset and on low-cost stepwise methods.

Next we examine the usefulness of Wilks' Λ statistic applied to the matrix Z, formed by adjoining y to X, as a means for diagnosing groups of outlying observations. This turns out to be especially useful either when there is no natural way to form groups, as with most cross-sectional data, or when unexpected groupings occur, such as might be the case in census tract data. We also examine the Andrews-Pregibon (1978) statistic.

---

[1]The use of the term *stepwise* in this context should not be confused with the concept of stepwise regression, which is not being indicated. The term *sequential* was considered but not adopted because of its established statistical connotations.

Finally we consider generalized distance measures (like the Mahalanobis distance) applied to the Z matrix. These distances are computed in a stepwise manner, thus allowing more than one row at a time to be considered.

A useful summary of the notation employed is given in Exhibit 2.2.

Single-Row Effects

We develop techniques here for discovering influential observations.[2] Each observation, of course, is closely associated with a single row of the data matrix X and the corresponding element of y.[3] An influential observation is one which, either individually or together with several other observations, has a demonstrably larger impact on the calculated values of various estimates (coefficients, standard errors, t-values, etc.) than is the case for most of the other observations. One obvious means for examining such an impact is to delete each row, one at a time, and note the resultant effect on the various calculated values.[4] Rows whose deletion produces relatively large changes in the calculated values are deemed influential. We begin, then, with an examination of this procedure of row deletion, looking in turn at the impact of each row on the estimated coefficients and the predicted (fitted) values ($\hat{y}$'s), the residuals, and the estimated parameter variance-covariance matrix. We then turn to other means of locating single data points with high impact: differentiation of the various calculated values with respect to the weight attached to an observation, and a geometrical view based on distance measures. Generalizations of some of these procedures to the problem of assessing the impact of deleting more than one row at a time are then examined.

Deletion.

Coefficients and Fitted Values. Since the estimated coefficients are often of primary interest to users of regression models, we look first at the change in the estimated regression coefficients that would occur if the ith row were deleted. Denoting the coefficients estimated with the ith row

---

[2]A number of the concepts employed in this section have been drawn from the existing literature. Relevant citations accompany the derivation of these formulae in Appendix 2A.

[3]Observations and rows need not be uniquely paired, for in time-series models with lagged variables, the data relevant to a given observation could occur in several neighboring rows. We defer further discussion of this aspect of time-series data until Chapters 4 and 5, and continue here to use these two terms interchangeably.

[4]The term *row deletion* is used generally to indicate the deletion of a row from both the X matrix and the y vector.

Exhibit 2.2 Notational conventions

Population Regression $y = X\beta + \epsilon$		Estimated Regression $y = Xb + e$	
y:	$n \times 1$ column vector for response variable	same	
X:	$n \times p$ matrix of explanatory variables*	same	
β:	$p \times 1$ column vector of regression parameters	b:	estimate of β
ϵ:	$n \times 1$ column vector of errors	e:	residual vector
σ^2:	error variance	s^2:	estimated error variance

Additional Notation

x_i:	ith row of X matrix	$b(i)$:	estimate of β when ith row of X and y have been deleted.
X_j:	jth column of X matrix	$s^2(i)$:	estimated error variance when ith row of X and y have been deleted.
$X(i)$:	X matrix with ith row deleted.		

Matrices are transposed with a superscript T, as in $X^T X$. Mention should also be made of a convention that is adopted in the reporting of regression results. Estimated standard errors of the regression coefficients are always reported in parentheses beneath the corresponding coefficient. In those cases where emphasis is on specific tests of significance, the t's are reported instead, and are always placed in square brackets. Other notation is either obvious or is introduced in its specific context.

*We typically assume X to contain a column of ones, corresponding to the constant term. In the event that X contains no such column, certain of the formulas must have their degrees of freedom altered accordingly. In particular, at a later stage we introduce the notation $\tilde{X}$ to indicate the matrix formed by centering the columns of X about their respective column means. If the $n \times p$ matrix X contains a constant column of ones, $\tilde{X}$ is assumed to be of size $n \times (p-1)$, the column of zeros being removed. The formulas as written take into account this change in degrees of freedom. Should X contain no constant column, however, all formulas dealing with centered matrices must have their degrees of freedom increased by one.

deleted by $\mathbf{b}(i)$, this change is easily computed from the formula

$$\text{DFBETA}_i \equiv \mathbf{b} - \mathbf{b}(i) = \frac{(\mathbf{X}^T\mathbf{X})^{-1}\mathbf{x}_i^T e_i}{1 - h_i}, \tag{2.1}$$

where

$$h_i = \mathbf{x}_i(\mathbf{X}^T\mathbf{X})^{-1}\mathbf{x}_i^T, \tag{2.2}$$

and the reader is reminded that $\mathbf{x}_i$ is a *row* vector. The quantity h_i occurs frequently in the diagnostics developed in this chapter and it is discussed more below.[5]

Whether the change in b_j, the jth component of $\mathbf{b}$, that results from the deletion of the ith row is large or small is often most usefully assessed relative to the variance of b_j, that is, $\sigma^2(\mathbf{X}^T\mathbf{X})_{jj}^{-1}$. If we let

$$\mathbf{C} = (\mathbf{X}^T\mathbf{X})^{-1}\mathbf{X}^T, \tag{2.3}$$

then

$$b_j - b_j(i) = \frac{c_{ji} e_i}{1 - h_i}. \tag{2.4}$$

Since

$$\sum_{i=1}^{n} (\mathbf{X}^T\mathbf{X})^{-1}\mathbf{x}_i^T\mathbf{x}_i(\mathbf{X}^T\mathbf{X})^{-1} = (\mathbf{X}^T\mathbf{X})^{-1}, \tag{2.5}$$

it follows that [see Mosteller and Tukey (1977)]

$$\text{var}(b_j) = \sigma^2 \sum_{k=1}^{n} c_{jk}^2. \tag{2.6}$$

Thus a scaled measure of change can be defined as

$$\text{DFBETAS}_{ij} \equiv \frac{b_j - b_j(i)}{s(i)\sqrt{(\mathbf{X}^T\mathbf{X})_{jj}^{-1}}} = \frac{c_{ji}}{\sqrt{\sum_{k=1}^{n} c_{jk}^2}} \frac{e_i}{s(i)(1 - h_i)}, \tag{2.7}$$

[5]See Appendixes 2A and 2B for details on the computation of the h_i.

where we have replaced s^2, the usual estimate of σ^2, by

$$s^2(i) = \frac{1}{n-p-1} \sum_{k \neq i} \left[y_k - \mathbf{x}_k \mathbf{b}(i) \right]^2$$

in order to make the denominator stochastically independent of the numerator in the Gaussian (normal) case. A simple formula for $s(i)$ results from

$$(n-p-1)s^2(i) = (n-p)s^2 - \frac{e_i^2}{1-h_i}. \tag{2.8}$$

In the special case of location,

$$\text{DFBETA}_i = \frac{e_i}{n-1}$$

and

$$\text{DFBETAS}_i = \frac{\sqrt{n}\, e_i}{(n-1)s(i)}. \tag{2.9}$$

As we might expect, the chance of getting a large DFBETA is reduced in direct proportion to the increase in sample size. Deleting one observation should have less effect as the sample size grows. Even though scaled by a measure of the standard error of b, DFBETAS_i decreases in proportion to $\sqrt{n}$.

Returning to the general case, large values of $|\text{DFBETAS}_{ij}|$ indicate observations that are influential in the determination of the jth coefficient, b_j.[6] The nature of "large" in relation to the sample size, n, is discussed below.

Another way to summarize coefficient changes and, at the same time, to gain insight into forecasting effects when an observation is deleted is by

---

[6] When the Gaussian assumption holds, it can also be useful to look at the change in t-statistics as a means for assessing the sensitivity of the regression output to the deletion of the ith row, that is, to examine

$$\text{DFTSTAT}_{ij} \equiv \frac{b_j}{s\sqrt{(\mathbf{X}^T\mathbf{X})_{jj}^{-1}}} - \frac{b_j(i)}{s(i)\sqrt{[\mathbf{X}^T(i)\mathbf{X}(i)]_{jj}^{-1}}}.$$

Studying the changes in regression statistics is a good second-order diagnostic tool because, if a row appears to be overly influential on other grounds, an examination of the regression statistics will show whether the conclusions of hypothesis testing would be affected.

the change in fit, defined as

$$\text{DFFIT}_i \equiv \hat{y}_i - \hat{y}_i(i) = \mathbf{x}_i[\mathbf{b} - \mathbf{b}(i)] = \frac{h_i e_i}{1 - h_i}. \tag{2.10}$$

This diagnostic measure has the advantage that it does not depend on the particular coordinate system used to form the regression model. For scaling purposes, it is natural to divide by $\sigma \sqrt{h_i}$, the standard deviation of the fit, $\hat{y}_i = \mathbf{x}_i \mathbf{b}$, giving

$$\text{DFFITS}_i \equiv \left[\frac{h_i}{1 - h_i} \right]^{1/2} \frac{e_i}{s(i)\sqrt{1 - h_i}}, \tag{2.11}$$

where σ has been estimated by $s(i)$. A measure similar to (2.11) has been suggested by Cook (1977).

It is natural to ask about the scaled changes in fit for other than the ith row; that is,

$$\frac{\mathbf{x}_k(\mathbf{b} - \mathbf{b}(i))}{s(i)\sqrt{h_k}} = \frac{h_{ik} e_i}{s(i)\sqrt{h_k}\,(1 - h_i)}, \tag{2.12}$$

where $h_{ik} \equiv \mathbf{x}_i(\mathbf{X}^T\mathbf{X})^{-1}\mathbf{x}_k^T$. Since

$$\sup_{\boldsymbol{\lambda}} \frac{\left| \boldsymbol{\lambda}^T[\mathbf{b} - \mathbf{b}(i)] \right|}{s(i)\left[\boldsymbol{\lambda}^T(\mathbf{X}^T\mathbf{X})^{-1}\boldsymbol{\lambda} \right]^{1/2}} = \frac{\left\{ [\mathbf{b} - \mathbf{b}(i)]^T(\mathbf{X}^T\mathbf{X})[\mathbf{b} - \mathbf{b}(i)] \right\}^{1/2}}{s(i)}$$

$$\equiv |\text{DFFITS}_i|, \tag{2.13}$$

it follows that

$$\left| \frac{\mathbf{x}_k[\mathbf{b} - \mathbf{b}(i)]}{s(i)\sqrt{h_k}} \right| \leqslant |\text{DFFITS}_i|. \tag{2.14}$$

Thus $|\text{DFFITS}_i|$ dominates the expression in (2.12) for all k and these latter measures need only be investigated when $|\text{DFFITS}_i|$ is large.

A word of warning is in order here, for it is obvious that there is room for misuse of the above procedures. High-influence data points could conceivably be removed solely to effect a desired change in a particular estimated coefficient, its t-value, or some other regression output. While this danger surely exists, it is an unavoidable consequence of a procedure that successfully highlights such points. It should be obvious that an influential point is legitimately deleted altogether only if, once identified, it can be shown to be uncorrectably in error. Often no action is warranted, and when it is, the appropriate action is usually more subtle than simple deletion. Examples of corrective action are given in Section 2.2 and in Chapter 4. These examples show that the benefits obtained from information on influential points far outweigh any potential danger.

The Hat Matrix.   Returning now to our discussion of deletion diagnostics, we can see from (2.1) to (2.11) that h_i and e_i are fundamental components. Some special properties of h_i are discussed in the remainder of this section and we study special types of residuals (like $e_i/s(i)\sqrt{1-h_i}$) in the next section.[7]

The h_i are the diagonal elements of the least-squares projection matrix, also called the hat matrix,

$$\mathbf{H} = \mathbf{X}(\mathbf{X}^T\mathbf{X})^{-1}\mathbf{X}^T, \tag{2.15}$$

which determines the fitted or predicted values, since

$$\hat{\mathbf{y}} \equiv \mathbf{X}\mathbf{b} = \mathbf{H}\mathbf{y}. \tag{2.16}$$

The influence of the response value, y_i, on the fit is most directly reflected in its impact on the corresponding fitted value, $\hat{y}_i$, and this information is seen from (2.16) to be contained in h_i. The diagonal elements of $\mathbf{H}$ can also be related to the distance between $\mathbf{x}_i$ and $\bar{\mathbf{x}}$ (the row vector of explanatory variable means). Denoting by tilde data that have been centered, we show in Appendix 2A that

$$h_i - \frac{1}{n} = \tilde{h}_i = \tilde{\mathbf{x}}_i(\tilde{\mathbf{X}}^T\tilde{\mathbf{X}})^{-1}\tilde{\mathbf{x}}_i^T. \tag{2.17}$$

We see from (2.17) that $\tilde{h}_i$ is a positive-definite quadratic form and thus possesses an appropriate distance interpretation.[8]

Where there are two or fewer explanatory variables, scatter plots will quickly reveal any x-outliers, and it is not hard to verify that they have

---

[7]The immediately following material closely follows Hoaglin and Welsch (1978).

[8]As is well known [Rao (1973), Section 1c.1], any $n \times n$ positive-definite matrix $\mathbf{A}$ may be decomposed as $\mathbf{A} = \mathbf{P}^T\mathbf{P}$ for some non-singular matrix $\mathbf{P}$. Hence the positive-definite quadratic form $\mathbf{x}^T\mathbf{A}\mathbf{x}$ ($\mathbf{x}$ an n-vector) is equivalent to the sum of squares $\mathbf{z}^T\mathbf{z}$ (the squared Euclidean length of the n-vector $\mathbf{z}$), where $\mathbf{z} = \mathbf{P}\mathbf{x}$.

relatively large h_i values. When $p > 2$, scatter plots may not reveal "multivariate outliers," which are separated from the bulk of the x-points but do not appear as outliers in a plot of any single explanatory variable or pair of them. Since, as we have seen, the diagonal elements of the hat matrix **H** have a distance interpretation, they provide a basic starting point for revealing such "multivariate outliers." These diagonals of the hat matrix, the h_i, are diagnostic tools in their own right as well as being fundamental parts of other such tools.

H is a projection matrix and hence, as is shown in Appendix 2A,

$$0 \leqslant h_i \leqslant 1. \tag{2.18}$$

Further, since **X** is of full rank,

$$\sum_{i=1}^{n} h_i = p. \tag{2.19}$$

The average size of a diagonal element, then, is p/n. Now if we were designing an experiment, it would be desirable to use data that were roughly equally influential, that is, each observation having an h_i near to the average p/n. But since the **X** data are usually given to us, we need some criterion to decide when a value of h_i is large enough (far enough away from the average) to warrant attention.

When the explanatory variables are independently distributed as the multivariate Gaussian, it is possible to compute the exact distribution of certain functions of the h_i's. We use these results for guidance only, realizing that independence and the Gaussian assumption are often not valid in practice. In Appendix 2A, $(n-p)[h_i - (1/n)]/(1-h_i)(p-1)$ is shown to be distributed as F with $p-1$ and $n-p$ degrees of freedom. For $p > 10$ and $n-p > 50$ the 95% value for F is less than 2 and hence $2p/n$ (twice the balanced average h_i) is a good rough cutoff. When $p/n > 0.4$, there are so few degrees of freedom per parameter that all observations become suspect. For small p, $2p/n$ tends to call a few too many points to our attention, but it is simple to remember and easy to use. In what follows, then, we call the ith observation a *leverage point* when h_i exceeds $2p/n$. The term *leverage* is reserved for use in this context.

Note that when $h_i = 1$, we have $\hat{y}_i = y_i$; that is, $e_i = 0$. This is equivalent to saying that, in some coordinate system, one parameter is determined completely by y_i or, in effect, dedicated to one data point. A proof of this result is given in Appendix 2A where it is also shown that

$$\det\left[\mathbf{X}^T(i)\mathbf{X}(i)\right] = (1 - h_i)\det(\mathbf{X}^T\mathbf{X}). \tag{2.20}$$

Clearly when $h_i = 1$ the new matrix $\mathbf{X}(i)$, formed by deleting the ith row, is singular and we cannot obtain the usual least-squares estimates. This is extreme leverage and does not often occur in practice.

We complete our discussion of the hat matrix with a few simple examples. For the sample mean, all elements of $\mathbf{H}$ are $1/n$. Here $p = 1$ and each $h_i = p/n$, the perfectly balanced case.

For a straight line through the origin,

$$h_{ij} = \frac{x_i x_j}{\sum_{k=1}^{n} x_k^2}, \tag{2.21}$$

and

$$\sum_{i=1}^{n} h_i = p = 1.$$

Simple linear regression is slightly more complicated, but a few steps of algebra give

$$h_{ij} = \frac{1}{n} + \frac{(x_i - \bar{x})(x_j - \bar{x})}{\sum_{k=1}^{n}(x_k - \bar{x})^2} \tag{2.22}$$

Residuals. We turn now to an examination of the diagnostic value of the effects that deleting rows can have on the regression residuals. The use of the regression residuals in a diagnostic context is, of course, not new. Looking at regression residuals, $e_i = y_i - \hat{y}_i$, and especially large residuals, has traditionally been used to highlight data points suspected of unduly affecting regression results. The residuals have also been employed to detect departures from the Gauss-Markov assumptions on which the desirable properties of least squares rest. As is well known, the residuals can be used to detect some forms of heteroscedasticity and auto-correlation, and can provide the basis for mitigating these problems. The residuals can also be used to test for the approximate normality of the disturbance term. Since the least-squares estimates retain their property of best-linear-unbiasedness even in the absence of normality of the disturbances, such tests are often overlooked in econometric practice, but even moderate departures from normality can noticeably impair estimation efficiency[9] and the meaningfulness of standard tests of hypotheses. Harmful departures from normality include pronounced skewness, multiple modes, and thick-tailed distributions. In all these uses of residuals,

---

[9] The term efficiency is used here in a broad sense to indicate minimum mean-squared error.

one should bear in mind that large outliers among the true errors, ϵ_i, can often be reflected in only modest-sized least-squares residuals, since the squared-error criterion weights extreme values heavily.

Three diagnostic measures based on regression residuals are presented here; two deal directly with the estimated residuals and the third results from a change in the assumption on the error distribution. The first is simply a frequency distribution of the residuals. If there is evident visual skewness, multiple modes, or a heavy-tailed distribution, a graph of the frequency distribution will prove informative. It is worth noting that economists often look at time plots of residuals but seldom at their frequency or cumulative distribution.

The second is the normal probability plot, which displays the cumulative normal distribution as a straight line whose slope measures the standard deviation and whose intercept reflects the mean. Thus a failure of the residuals to be normally distributed will often reveal itself as a departure of the cumulative residual plot from a straight line. Outliers often appear at either end of the cumulative distribution.

Finally, Denby and Mallows (1977) and Welsch (1976) have suggested plotting the estimated coefficients and residuals as the error likelihood, or, equivalently, as the criterion function (negative logarithm of the likelihood) is changed. One such family of criterion functions has been suggested by Huber (1973); namely,

$$
\rho_c(t) = \begin{cases} \dfrac{t^2}{2} & |t| \leqslant c \\[2ex] c|t| - \dfrac{c^2}{2} & |t| > c, \end{cases} \tag{2.23}
$$

which goes from least squares ($c = \infty$) to least absolute residuals ($c \to 0$). This approach is attractive because of its relation to robust estimation, but it requires considerable computation.

For diagnostic use the residuals can be modified in ways that will enhance our ability to detect problem data. It is well known [Theil (1971)] that

$$
\text{var}(e_i) = \sigma^2(1 - h_i). \tag{2.24}
$$

Consequently, many authors have suggested that, instead of studying e_i, we should use the *standardized residuals*

$$
e_{si} \equiv \frac{e_i}{s\sqrt{1 - h_i}}. \tag{2.25}
$$

We prefer instead to estimate σ by $s(i)$ [cf. (2.8)]. The result is a *studentized residual* (RSTUDENT),

$$e_i^* \equiv \frac{e_i}{s(i)\sqrt{1-h_i}}, \qquad (2.26)$$

which, in a number of practical situations, is distributed closely to the t-distribution with $n-p-1$ degrees of freedom. Thus, if the Gaussian assumption holds, we can readily assess the significance of any single studentized residual. Of course, the e_i^* will not be independent.

The studentized residuals have another interesting interpretation. If we were to add to the data a dummy variable consisting of a column with all zeros except for a one in the ith row (the new model), then e_i^* is the t-statistic that tests for the significance of the coefficient of this new variable. To prove this, let SSR stand for sum of squared residuals and note that

$$\frac{\left[\,\text{SSR(old model)} - \text{SSR(new model)}\,\right]/1}{\text{SSR(new model)}/(n-p-1)} \qquad (2.27)$$

$$= \frac{(n-p)s^2 - (n-p-1)s^2(i)}{s^2(i)} = \frac{e_i^2}{s^2(i)(1-h_i)}. \qquad (2.28)$$

Under the Gaussian assumption, (2.27) is distributed as $F_{1,n-p-1}$, and the result follows by taking the square root of (2.28). Some additional details are contained in Appendix 2A.

The studentized residuals thus provide a better way to examine the information in the residuals, both because they have equal variances and because they are easily related to the t-distribution in many situations. However, this does not tell the whole story, since some of the most influential data points can have relatively small studentized residuals (and very small e_i).

To illustrate with the simplest case, regression through the origin, note that

$$b - b(i) = \frac{x_i e_i}{\sum_{j\neq i} x_j^2}. \qquad (2.29)$$

Equation (2.29) shows that the residuals are related to the change in the least-squares estimate caused by deleting one row, but each contains different information, since large values of $|b - b(i)|$ can be associated with small $|e_i|$ and vice versa. Hence row deletion and the analysis of residuals need to be treated together and on an equal footing.

When the index of observations is time, the studentized residuals can be related to the recursive residuals proposed by Brown, Durbin, and Evans (1975). If $\mathbf{b}(t)$ is the least-squares estimate based on the first $t-1$ observations, then the recursive residuals are defined to be

$$q_t = \frac{y_t - \mathbf{x}_t \mathbf{b}(t)}{\left\{1 + \mathbf{x}_t \left[\mathbf{X}^T(t)\mathbf{X}(t)\right]^{-1}\mathbf{x}_t^T\right\}^{1/2}}, \qquad t = p+1,\ldots,T. \qquad (2.30)$$

which by simple algebra (see Appendix 2A) can be written as

$$\frac{y_t - \mathbf{x}_t \mathbf{b}}{\sqrt{1 - h_t}}, \qquad (2.31)$$

where h_t and $\mathbf{b}$ are computed from the first t observations. For a related interpretation see a discussion of the PRESS residual by Allen (1971).

When we set

$$S_t \equiv \sum_{i=1}^{t} (y_i - \mathbf{x}_i \mathbf{b})^2, \qquad (2.32)$$

(2.8) gives

$$S_t = S_{t-1} + q_t^2. \qquad (2.33)$$

Brown, Durbin, and Evans propose two tests for studying the constancy of regression relationships over time. The first uses the cusum

$$W_t = \frac{T-p}{S_T} \sum_{j=p+1}^{t} q_j, \qquad t = p+1,\ldots,T, \qquad (2.34)$$

and the second the cusum-of-squares

$$c_t = \frac{S_t}{S_T}, \qquad t = p+1,\ldots,T. \qquad (2.35)$$

Schweder (1976) has shown that certain modifications of these tests, obtained by summing from $j = T$ to $t \geqslant p+1$ (backward cusum, etc.) have greater average power. The reader is referred to that paper for further details. An example of the use of these tests is given in Section 4.3.

Covariance Matrix. So far we have focused on coefficients, predicted (fitted) values of y, and residuals. Another major aspect of regression is the covariance matrix of the estimated coefficients.[10] We again consider the diagnostic technique of row deletion, this time in a comparison of the covariance matrix using all the data, $\sigma^2(X^TX)^{-1}$, with the covariance matrix that results when the ith row has been deleted, $\sigma^2[X^T(i)X(i)]^{-1}$. Of the various alternative means for comparing two such positive-definite symmetric matrices, the ratio of their determinants $\det[X^T(i)X(i)]^{-1}/\det(X^TX)^{-1}$ is one of the simplest and, in the present application, is quite appealing. Since these two matrices differ only by the inclusion of the ith row in the sum of squares and cross products, values of this ratio near unity can be taken to indicate that the two covariance matrices are close, or that the covariance matrix is insensitive to the deletion of row i. Of course, the preceding analysis is based on information from the X matrix alone and ignores the fact that the estimator s^2 of σ^2 also changes with the deletion of the ith observation. We can bring the y data into consideration by comparing the two matrices $s^2(X^TX)^{-1}$ and $s^2(i)[X^T(i)X(i)]^{-1}$ in the determinantal ratio,

$$\text{COVRATIO} \equiv \frac{\det\left\{s^2(i)\left[X^T(i)X(i)\right]^{-1}\right\}}{\det\left[s^2(X^TX)^{-1}\right]}$$

$$= \frac{s^{2p}(i)}{s^{2p}}\left\{\frac{\det\left[X^T(i)X(i)\right]^{-1}}{\det(X^TX)^{-1}}\right\}. \tag{2.36}$$

Equation (2.36) may be given a more useful formulation by applying (2.20) to show

$$\frac{\det\left[X^T(i)X(i)\right]^{-1}}{\det(X^TX)^{-1}} = \frac{1}{1-h_i}. \tag{2.37}$$

Hence, using (2.8) and (2.26) we have

$$\text{COVRATIO} = \frac{1}{\left[\dfrac{n-p-1}{n-p}+\dfrac{e_i^{*2}}{n-p}\right]^p(1-h_i)}. \tag{2.38}$$

---

[10] A diagnostic based on the diagonal elements of the covariance matrix can be obtained from the expression (2.6). By noting which c_{ji}^2 appear to be excessively large for a given j, we determine those observations that influence the variance of the jth coefficient. This diagnostic, however, has two weaknesses. First, it ignores the off-diagonal elements of the covariance matrix and second, emphasis on the c_{ji}^2 ignores s^2.

As a diagnostic tool, then, we are interested in observations that result in values of COVRATIO from (2.38) that are not near unity, for these observations are possibly influential and warrant further investigation.

In order to provide a rough guide to the magnitude of such variations from unity, we consider the two extreme cases $|e_i^*| \geqslant 2$ with h_i at its minimum $(1/n)$ and $h_i \geqslant 2p/n$ with $e_i^* = 0$. In the first case we get

$$\text{COVRATIO} \approx \frac{1}{\left(1 + \dfrac{e_i^{*2} - 1}{n-p}\right)^p} \leqslant \frac{1}{\left(1 + \dfrac{3}{n-p}\right)^p}.$$

Further approximation leads to

$$\frac{1}{\left(1 + \dfrac{3}{n-p}\right)^p} \approx \left(1 + \frac{3p}{n}\right)^{-1} \approx 1 - \frac{3p}{n}, \tag{2.39}$$

where $n - p$ has been replaced by n for simplicity. The latter bounds are, of course, not useful when $n \leqslant 3p$. For the second case

$$\text{COVRATIO} \approx \frac{1}{\left(1 - \dfrac{1}{n-p}\right)^p} \frac{1}{(1 - h_i)} \geqslant \frac{1}{\left(1 - \dfrac{1}{n-p}\right)^p \left(1 - \dfrac{2p}{n}\right)}.$$

A cruder but simpler bound follows from

$$\frac{1}{\left(1 - \dfrac{1}{n-p}\right)^p \left(1 - \dfrac{2p}{n}\right)} \approx \frac{1}{\left(1 - \dfrac{p}{n-p}\right)\left(1 - \dfrac{2p}{n}\right)}$$

$$\approx \left(1 - \frac{3p}{n}\right)^{-1} \approx 1 + \frac{3p}{n}. \tag{2.40}$$

Therefore we investigate points with $|\text{COVRATIO} - 1|$ near to or larger than $3p/n$.[11]

The formula in (2.38) is a function of basic building blocks, such as h_i and the studentized residuals. Roughly speaking (2.38) will be large when h_i is large and small when the studentized residual is large. Clearly those

---

[11] Some prefer to normalize expressions like (2.36) for model size by raising them to the $1/p$th power. Had such normalization been done here, the approximations corresponding to (2.39) and (2.40) would be $1 - (3/n)$ and $1 + (3/n)$ respectively.

two factors can offset each other and that is why it is useful to look at them separately and in combinations as in (2.38).

We are also interested in how the variance of $\hat{y}_i$ changes when an observation is deleted. To do this we compute

$$\text{var}(\hat{y}_i) = s^2 h_i$$

$$\text{var}(\hat{y}_i(i)) = \text{var}(\mathbf{x}_i \mathbf{b}(i)) = s^2(i)\left[\frac{h_i}{1 - h_i}\right],$$

and form the ratio

$$\text{FVARATIO} \equiv \frac{s^2(i)}{s^2(1 - h_i)}.$$

This expression is similar to COVRATIO except that $s^2(i)/s^2$ is not raised to the pth power. As a diagnostic measure it will exhibit the same patterns of behavior with respect to different configurations of h_i and the studentized residual as described above for COVRATIO.

Differentiation. We examine now a second means for identifying influential observations, differentiation of regression outputs with respect to specific model parameters. In particular, we can alter the weight attached to the ith observation if, in the assumptions of the standard linear regression model, we replace $\text{var}(\epsilon_i) = \sigma^2$ with $\text{var}(\epsilon_i) = \sigma^2/w_i$, for the specific i only. Differentiation of the regression coefficients with respect to w_i, evaluated at $w_i = 1$, provides a means for examining the sensitivity of the regression coefficients to a slight change in the weight given to the ith observation. Large values of this derivative indicate observations that have large influence on the calculated coefficients. This derivative, as is shown in Appendix 2A, is

$$\frac{\partial \mathbf{b}(w_i)}{\partial w_i} = \frac{(\mathbf{X}^T\mathbf{X})^{-1}\mathbf{x}_i^T e_i}{\left[1 - (1 - w_i)h_i\right]^2}, \tag{2.41}$$

and it follows that

$$\left.\frac{\partial \mathbf{b}(w_i)}{\partial w_i}\right|_{w_i = 1} = (\mathbf{X}^T\mathbf{X})^{-1}\mathbf{x}_i^T e_i. \tag{2.42}$$

This last formula is often viewed as the influence of the ith observation on the estimated coefficients. Its relationship to the formula (2.1) for DFBETA is obvious and it could be used as an alternative to that statistic.

The theory of robust estimation [cf. Huber (1973)] implies that influence functions such as (2.42) can be used to approximate the covariance matrix of **b** by forming

$$\sum_{i=1}^{n} (\mathbf{X}^T\mathbf{X})^{-1}\mathbf{x}_i^T e_i e_i \mathbf{x}_i (\mathbf{X}^T\mathbf{X})^{-1} = \sum_{i=1}^{n} e_i^2 (\mathbf{X}^T\mathbf{X})^{-1}\mathbf{x}_i^T \mathbf{x}_i (\mathbf{X}^T\mathbf{X})^{-1}. \quad (2.43)$$

This is not quite the usual covariance matrix, but if e_i^2 is replaced by the average value, $\sum_{k=1}^{n} e_k^2 / n$, we get

$$\frac{\sum_{k=1}^{n} e_k^2}{n} \sum_{i=1}^{n} (\mathbf{X}^T\mathbf{X})^{-1}\mathbf{x}_i^T\mathbf{x}_i(\mathbf{X}^T\mathbf{X})^{-1} = \frac{\sum_{k=1}^{n} e_k^2}{n}(\mathbf{X}^T\mathbf{X})^{-1}, \quad (2.44)$$

which, except for degrees of freedom, is the estimated least-squares covariance matrix.

To assess the influence of an individual observation, we compare

$$\sum_{k \neq i} e_k^2 (\mathbf{X}^T\mathbf{X})^{-1}\mathbf{x}_k^T\mathbf{x}_k(\mathbf{X}^T\mathbf{X})^{-1} \quad (2.45)$$

with

$$s^2(\mathbf{X}^T\mathbf{X})^{-1}. \quad (2.46)$$

The use of determinants with the sums in (2.45) is difficult, so we replace e_k^2 for $k \neq i$ by $s^2(i)$, leaving

$$s^2(i) \sum_{k \neq i} (\mathbf{X}^T\mathbf{X})^{-1}\mathbf{x}_k^T\mathbf{x}_k(\mathbf{X}^T\mathbf{X})^{-1} = s^2(i)(\mathbf{X}^T\mathbf{X})^{-1}[\mathbf{X}^T(i)\mathbf{X}(i)](\mathbf{X}^T\mathbf{X})^{-1}.$$

$$(2.47)$$

Forming the ratio of the determinant of (2.47) to that of (2.46) we get

$$\frac{s^{2p}(i)}{s^{2p}} \cdot \frac{\det[\mathbf{X}^T(i)\mathbf{X}(i)]}{\det(\mathbf{X}^T\mathbf{X})} = \frac{(1-h_i)}{\{[(n-p-1)/(n-p)] + [e_i^{*2}/(n-p)]\}^p},$$

$$(2.48)$$

which is just (2.38) multiplied by $(1-h_i)^2$. We prefer (2.38) because no substitution for e_i^2 is required.

A similar result for the variances of the fit, $\hat{y}_i$, compares the ratio of $\sum_{k \neq i} e_k^2 h_{ik}^2$ and $s^2 h_i$ giving, after some manipulation,

$$\frac{s^2(i)(1-h_i)}{s^2} = \frac{1-h_i}{\left(\dfrac{n-p-1}{n-p} + \dfrac{e_i^{*2}}{n-p}\right)}, \tag{2.49}$$

which we note to be FVARATIO multiplied by $(1-h_i)^2$. This ratio can be related to some of the geometric procedures discussed below.

A Geometric View. In the previous sections we have examined several techniques for diagnosing those observations that are influential in the determination of various regression outputs. We have seen that key roles are played in these diagnostic techniques by the elements of the hat matrix **H**, especially its diagonal elements, the h_i, and by the residuals, the e_i. The former elements convey information from the **X** matrix, while the latter also introduce information from the response vector, **y**. A geometric way of viewing this interrelationship is offered by adjoining the **y** vector to the **X** matrix to form a matrix $\mathbf{Z} \equiv [\mathbf{X}\,\mathbf{y}]$, consisting of $p+1$ columns. We can think of each row of **Z** as an observation in a $p+1$ dimensional space and search for "outlying" observations.

In this situation, it is natural to think of Wilks' Λ statistic [Rao (1973)] for testing the differences in mean between two populations. Here one such population is represented by the ith observation and the second by the rest of the data. If we let $\tilde{\mathbf{Z}}$ denote the centered (by $\bar{z}$) **Z** matrix, then the statistic is

$$\Lambda(\tilde{\mathbf{z}}_i) = \frac{\det\left(\tilde{\mathbf{Z}}^T\tilde{\mathbf{Z}} - (n-1)\bar{\tilde{\mathbf{z}}}^T(i)\bar{\tilde{\mathbf{z}}}(i) - \tilde{\mathbf{z}}_i^T\tilde{\mathbf{z}}_i\right)}{\det(\tilde{\mathbf{Z}}^T\tilde{\mathbf{Z}})},$$

where $\bar{\tilde{\mathbf{z}}}(i)$ is the p-vector (row) of column means of $\tilde{\mathbf{Z}}(i)$.

As part of our discussion of the hat matrix in Appendix 2A, we show that

$$\Lambda(\tilde{\mathbf{x}}_i) = \frac{n}{n-1}(1-h_i), \tag{2.50}$$

and a simple application of the formulas for adding a column to a matrix [Rao (1973), p. 33] shows that

$$\Lambda(\tilde{\mathbf{z}}_i) = \left(\frac{n}{n-1}\right)(1-h_i)\left[1 + \frac{e_i^{*2}}{(n-p-1)}\right]^{-1}. \tag{2.51}$$

This index is again seen to be composed of the basic building blocks, h_i, and the studentized residuals, e_i^*, and is similar (in the case of a single observation in one group) to (2.49). Small values of (2.51) would indicate possible discrepant observations.

If we are willing to assume, for purposes of guidance, that $\tilde{\mathbf{Z}}$ consists of n independent samples from a p-dimensional Gaussian distribution, then $\Lambda(\tilde{\mathbf{z}}_i)$ can be easily related to the F-statistic by

$$\left(\frac{n-p-1}{p}\right)\frac{1-\Lambda(\tilde{\mathbf{z}}_i)}{\Lambda(\tilde{\mathbf{z}}_i)} \sim F_{p,n-p-1}. \tag{2.52}$$

In place of $\Lambda(\tilde{\mathbf{z}}_i)$ we could have used the Mahalanobis distance between one row and the mean of the rest; that is,

$$M(\tilde{\mathbf{z}}_i) = (n-2)\left(\tilde{\mathbf{z}}_i - \bar{\tilde{\mathbf{z}}}(i)\right)\left(\tilde{\mathbf{Z}}^T(i)\tilde{\mathbf{Z}}(i)\right)^{-1}\left(\tilde{\mathbf{z}}_i - \bar{\tilde{\mathbf{z}}}(i)\right)^T, \tag{2.53}$$

where $\tilde{\mathbf{Z}}(i)$ is $\tilde{\mathbf{Z}}(i)$ centered by $\bar{\tilde{\mathbf{z}}}(i)$. This is seen by noting that Λ and M are simply related by

$$\frac{1-\Lambda}{\Lambda} = \frac{(n-1)M}{(n-2)n}. \tag{2.54}$$

However, $\Lambda(\tilde{\mathbf{x}}_i)$ has a more direct relationship to h_i and its computation is somewhat easier when, later on, we consider removing more than one observation at a time.

The major disadvantage of diagnostic approaches based on $\mathbf{Z}$ is that the special nature of $\mathbf{y}$ in the regression context is ignored (except when $\mathbf{X}$ is considered as fixed in the distribution of diagnostics based on $\mathbf{Z}$). The close parallel of this approach to that of the covariance comparisons as given in (2.48) and (2.49) suggests, however, that computations based on $\mathbf{Z}$ will prove useful as well.

Criteria for Influential Observations. In interpreting the results of each of the previously described diagnostic techniques, a problem naturally arises in determining when a particular measure of leverage or influence is large enough to be worthy of further notice. When, for example, is a hat-matrix diagonal large enough to indicate a point of leverage, or a DFBETA an influential point? As with all empirical procedures, this question is ultimately answered by judgment and intuition in choosing reasonable cutoffs most suitable for the problem at hand, guided wherever possible by statistical theory. There are at least three sources of information for determining such cutoffs that seem useful: external scaling, internal scaling, and gaps. Elasticities, such as $(\partial b_j(w_i)/\partial w_i)(w_i/b_j)$, and approximations to them like $(b_j - b_j(i))/b_j$, may also be useful in specific applications, but will not be pursued here.

External Scaling. External scaling denotes cutoff values determined by recourse to statistical theory. Each of the t-like diagnostics RSTUDENT, DFBETAS, and DFFITS, for example, has been scaled by an appropriate estimated standard error, which, under the Gaussian assumption, is stochastically independent of the given diagnostic. As such, it is natural to say, at least to a first approximation, that any of the diagnostic measures is large if its value exceeds two in magnitude. Such a procedure defines what we call an *absolute cutoff*, and it is most useful in determining cutoff values for RSTUDENT, since this diagnostic is less directly dependent on the sample size. Absolute cutoffs, however, are also relevant to determining extreme values for the diagnostics DFBETAS and DFFITS, even though these measures do depend directly on the sample size, since it would be most unusual for the removal of a single observation from a sample of 100 or more to result in a change in any estimated statistic by two or more standard errors. By way of contrast, there can be no absolute cutoffs for the hat-matrix diagonals h_i or for COVRATIO, since there is no natural standard-error scaling for these diagnostics.

While the preceding absolute cutoffs are of use in providing a stringent criterion that does not depend directly on the sample size n, there are many cases in which it is useful to have a cutoff that would tend to expose approximately the same proportion of potentially influential observations, regardless of sample size. Such a measure defines what we call a *size-adjusted cutoff*. In view of (2.7) and (2.9) a size-adjusted cutoff for DFBETAS is readily calculated as $2/\sqrt{n}$. Similarly, a size-adjusted cutoff for DFFITS is possible, for we recall from (2.19) that a perfectly balanced design matrix $\mathbf{X}$ would have $h_i = p/n$ for all i, and hence [see (2.11)],

$$\text{DFFITS}_i = \left(\frac{p}{n-p}\right)^{1/2} e_i^*.$$

A convenient size-adjusted cutoff in this case would be $2\sqrt{p/n}$, which accounts both for the sample size n and the fact that DFFITS$_i$ increases as p does. In effect, then, the perfectly balanced case acts as a standard from which this measure indicates sizable departures. As we have noted above, the only cutoffs relevant to the hat-matrix diagonals h_i and COVRATIO are the size-adjusted cutoffs $2p/n$ and $1 \pm 3(p/n)$, respectively.

Both absolute and size-adjusted cutoffs have practical value, but the relation between them becomes particularly important for large data sets.

In this case, it is unlikely that the deletion of any single observation can result in large values for |DFBETAS| or |DFFITS|; that is, when n is large there are not likely to be any observations that are influential in the absolute sense. However, it is still extremely useful to discover those observations that are most strongly influential in relation to the others, and the size-adjusted cutoffs provide a convenient means for doing this.

Internal Scaling.   Internal scaling defines extreme values of a diagnostic measure relative to the "weight of the evidence" provided by the given diagnostic series itself. The calculation of each deletion diagnostic results in a series of n values. The hat-matrix diagonals, for example, form a set of size n, as do DFFIT and the p series of DFBETA. Following Tukey (1977) we compute the interquartile range $\tilde{s}$ for each series and indicate as extreme those values that exceed $(7/2)\tilde{s}$. If these diagnostics were Gaussian this would occur less than 0.1% of the time. Thus, these limits can be viewed as a convenient point of departure in the absence of a more exact distribution theory. The use of an interquartile range in this context provides a more robust estimate of spread than would the standard deviation when the series are non-Gaussian, particularly in instances where the underlying distribution is heavy tailed.[12]

Gaps.   With either internal or external scaling, we are always alerted when a noticeable gap appears in the series of a diagnostic measure; that is, when one or more values of the diagnostic measure show themselves to be singularly different from the rest. The question of deciding when a gap is worthy of notice is even more difficult than deriving the previous cutoffs. Our experience with the many data sets examined in the course of our research, however, shows that in nearly every instance a large majority of the elements of a diagnostic series bunches in the middle, while the tails frequently contain small fractions of observations clearly detached from the remainder.

It is important to note that, in any of these approaches to scaling, we face the problems associated with extreme values, multiple tests, and multiple comparisons. Bonferroni-type bounds can be useful for small data sets or for situations where only few diagnostics need to be examined because the rest have been excluded on other grounds. Until more is known about the issue, we suggest a cautious approach to the use of the

---

[12] For further discussion of appropriate measures of spread for non-Gaussian data, see Mosteller and Tukey (1977).

diagnostics, but not so cautious that we remain ignorant of the potentially damaging effects of highly influential data.

Partial-Regression Leverage Plots. Simple two-variable regression scatter-plots (like the stylized examples in Exhibit 2.1*e* and *f*) contain much diagnostic information about residuals and leverage and, in addition, provide guidance about influential subsets of data that might escape detection through the use of single-row techniques.

It is natural to ask if a similar tool exists for multiple regression, and this leads to the *partial-regression leverage plot.* This graphical device can be motivated as follows. Let $X[k]$ be the $n \times (p-1)$ matrix formed from the data matrix, X, by removing its kth column, X_k. Further let u_k and v_k, respectively, be the residuals that result from regressing y and X_k on $X[k]$. As is well known, the kth regression coefficient of a multiple regression of y on X can be determined from the simple two-variate regression of u_k on v_k. We define, then, the partial-regression leverage plot for b_k as a scatter plot of the u_k against the v_k along with their simple linear-regression line. The residuals from this regression line are, of course, just the residuals from the multiple regression of y on X, and the slope is b_k, the multiple-regression estimate of β_k. Also, the simple correlation between u_k and v_k is equal to the partial correlation between y and X_k in the multiple regression.

We feel that these plots are an important part of regression diagnostics and that they should supplant the traditional plots of residuals against explanatory variables. Needless to say, however, partial-regression leverage plots cannot tell us everything. Certain types of multivariate influential data can be overlooked and the influence of the leverage points detected in the plot will sometimes be difficult to quantify. The computational details for these plots are discussed by Mosteller and Tukey (1977) who show that the u_k are equal to $b_k v_k + e$, where e is the vector of residuals from the multiple regression. This fact saves considerable computational effort.

The v_k have another interesting interpretation. Let $h_{ij}[k]$ denote the elements of the hat matrix for the regression of y on all of the explanatory variables except X_k. Then the elements of the hat matrix for the full regression are

$$h_{ij} = h_{ij}[k] + \frac{v_{k,i} v_{k,j}}{\sum_{l=1}^{n} v_{k,l}^2}, \tag{2.55}$$

where $v_{k,i}$ denotes the ith component of the vector v_k. This expression can be usefully compared with (2.21) for regression through the origin. Thus the v_k are closely related to the partial leverage added to $h_{ij}[k]$ by the addition of X_k to the regression.

Multiple-Row Effects

In the preceding discussion, we have presented various diagnostic techniques for identifying influential observations that have been based on the deletion or alteration of a single row. While such techniques can satisfactorily identify influential observations much of the time, they will not always be successful. We have already seen, for example, in the simple case presented in Exhibit 2.1*f* that one outlier can mask the effect of another. It is necessary, therefore, to develop techniques that examine the potentially influential effects of subsets or groups of observations. We turn shortly to several multiple-row techniques that tend to avoid the effects of masking and that have a better chance of isolating influential subsets in the data.

Before doing this, however, we must mention an inherent problem in delimiting influential subsets of the data, namely, when to stop—with subsets of size two, three, or more? Clearly, unusual observations can only be recognized relative to the bulk of the remaining data that are considered to be typical, and we must select an initial base subset of observations to serve this purpose. But how is this subset to be found? One straightforward approach would be to consider those observations that do not appear exceptional by any of the single-row measures discussed above. Of course, we could always be fooled, as in the example of Exhibit 2.1*f*, into including some discrepant observations in this base subset, but this would be minimized if we used low cutoffs, such as relaxing our size-adjusted cutoff levels to 90% or less instead of holding to the more conventional 95% level. We could also remove exceptional observations noticed in the partial-regression leverage plots. Some of the following procedures are less dependent on a base subset than others, but it cannot be avoided entirely, for the boundary between the typical and the unusual is inherently vague. We denote by B^* (of size m^*) the largest subset of potentially influential observations that we wish to consider. The complement of B^* is the base subset of observations defined to be typical.

We follow the same general outline as before and discuss deletion, residuals, differentiation, and geometric approaches in the multiple-row context.

Deletion. A natural multiple-row generalization of (2.4) would be to examine the larger values of

$$\frac{|b_j - b_j(D_m)|}{\text{scale}},\tag{2.56}$$

for $j = 1, \ldots, p$ and $m = 2, 3, 4$, and so on, and where "scale" indicates some appropriate measure of standard error. Here D_m is a set (of size m) of indexes of the rows to be deleted. If fitted values are of interest, then the appropriate measure becomes

$$\frac{|\mathbf{x}_k[\mathbf{b} - \mathbf{b}(D_m)]|}{\text{scale}}, \tag{2.57}$$

for $k = 1, \ldots, n$. Although computational formulas exist for these quantities [Bingham (1977)], the cost is great and we feel most of the benefits can be obtained more simply.

To avoid the consideration of p quantities in (2.56) or n quantities in (2.57), squared norms, such as

$$[\mathbf{b} - \mathbf{b}(D_m)]^T[\mathbf{b} - \mathbf{b}(D_m)] \tag{2.58}$$

or

$$[\mathbf{b} - \mathbf{b}(D_m)]^T\mathbf{X}^T\mathbf{X}[\mathbf{b} - \mathbf{b}(D_m)] \tag{2.59}$$

can be considered as summary measures. Since we are often most interested in changes in fit that occur for the data points remaining after deletion, (2.59) can be modified to

$$\text{MDFFIT} \equiv [\mathbf{b} - \mathbf{b}(D_m)]^T\mathbf{X}^T(D_m)\mathbf{X}(D_m)[\mathbf{b} - \mathbf{b}(D_m)]. \tag{2.60}$$

Bingham (1977) shows (2.60) can also be expressed as

$$\mathbf{e}_{D_m}^T\mathbf{X}_{D_m}[\mathbf{X}^T(D_m)\mathbf{X}(D_m)]^{-1}\mathbf{X}_{D_m}^T\mathbf{e}_{D_m}, \tag{2.61}$$

where $\mathbf{e}$ is the column vector of least-squares residuals and where D_m, used as a subscript, denotes a matrix or vector with rows whose indexes are contained in D_m. Because of (2.61) MDFFIT can be computed at lower cost than (2.59). Unfortunately, even (2.61) is computationally expensive when m exceeds about 20 observations. Some inequalities, however, are available for MDFFIT which may ease these computational problems. More details are provided at the end of Appendix 2B.

For larger data sets, a stepwise approach is available that can provide useful information at low cost. This method begins for $m = 2$ by using the two largest $|\text{DFFIT}|$ (or $|\text{DFFITS}|$) to form $D_2^{(1)}$. If the two largest values of

$$|\mathbf{x}_k[\mathbf{b} - \mathbf{b}(D_2^{(1)})]| \tag{2.62}$$

do not have their indexes k contained in $D_2^{(1)}$, a set $D_2^{(2)}$ is formed consisting of the indexes for the two largest. This procedure is iterated until a set D_2 is found with indexes coinciding with the two largest values of (2.62). The resulting statistic is designated SMDFFIT.

For $m = 3$, a starting set $D_3^{(1)}$ is found by using the three largest values of (2.62) from the final iteration for $m = 2$. Once the starting set is found the iteration proceeds as for $m = 2$. The overall process continues for $m = 4$, 5, and so on. An alternative approach is to use the m largest values of |DFFIT| to start the iterations for each value of m. Different starting sets can lead to different final results.

This stepwise approach is motivated by the idea that the fitted values most sensitive to deletion should be those which correspond to the deleted observations because no attempt is being made to fit these points. Since (2.14) does not hold in general when two or more points are deleted, the stepwise process attempts to find a specific set for each m where it does hold.

We conclude our study of multiple-row deletion by generalizing the covariance ratio to a deletion set D_m; namely,

$$\text{COVRATIO}(D_m) \equiv \frac{\det s^2(D_m)[\mathbf{X}^T(D_m)\mathbf{X}(D_m)]^{-1}}{\det s^2(\mathbf{X}^T\mathbf{X})^{-1}}. \tag{2.63}$$

Computation of this ratio is facilitated by the fact that

$$\frac{\det[\mathbf{X}^T(D_m)\mathbf{X}(D_m)]}{\det(\mathbf{X}^T\mathbf{X})} = \det(\mathbf{I} - \mathbf{H})_{D_m}, \tag{2.64}$$

where $(\mathbf{I} - \mathbf{H})_{D_m}$ stands for the submatrix formed by considering only the rows and columns of $\mathbf{I} - \mathbf{H}$ that are contained in D_m. FVARATIO also can be generalized.[13]

Studentized Residuals and Dummy Variables.   The single-row studentized residual given in (2.26) is readily extended to deletions of more than one row at a time. Instead of adding just one dummy variate with a unity in row i and zeros elsewhere, we add many such dummies, each with its unity

---

[13] It is easily seen that equations (2.48) and (2.49) can also be generalized.

only in the row to be deleted. In the extreme, we could add n such variables, one for each row. This leads to a singular problem which can, in fact, be studied. However, we assume that no more than $n-p$ columns of dummy variates are to be added.

Once the subset of dummy columns to be added has been decided on, a problem that we turn to below, it is natural to make use of standard regression selection techniques to decide which, if any, of these dummy variables should be retained. Each dummy variable that is retained indicates that its corresponding row warrants special attention, just as we saw that the studentized residual calls attention to a single observation. The advantage here is that several rows can be considered simultaneously and we have a chance to overcome the masking situation in Exhibit 2.1f.

There are no clear-cut means for selecting the set of dummy variables to be added. As already noted, we could use the previously described single-row techniques along with partial-regression leverage plots to determine a starting subset of potentially influential observations. Rather generally, however, the computational efficiency of some of these selection algorithms allows this starting subset to be chosen quite large.

To test any particular subset D_m of dummy variables a generalization of (2.27) is available. For example, we could consider

$$\text{RESRATIO} \equiv \frac{\left[\text{SSR(no dummies)} - \text{SSR}(D_m \text{ dummies used}) \right]/m}{\text{SSR}(D_m \text{ dummies used})/(n-p-m)},$$

$$(2.65)$$

which is distributed as $F_{m,\,n-p-m}$ if the appropriate probability assumptions hold. For further details see Gentleman and Wilk (1975).

The use of stepwise regression has been considered as a solution to this problem by Mickey, Dunn, and Clark (1967). The well-known difficulties of stepwise regression arise in this context, and, in general, it is best to avoid attempting to discover the model (i.e., explanatory variables) and influential points at the same time. Thus one must first choose a set of explanatory variables and stay with them while the dummy variables are selected. Of course, this process may be iterated and, if some observations are deleted, a new stepwise regression on the explanatory variable set should be performed. Stepwise regression also clearly fails to consider all possible combinations of the dummy variables and can therefore miss influential points when more than one is present.

A natural alternative to stepwise regression is to consider all-possible-subsets regression.[14] The computational costs are higher and

[14] See Furnival and Wilson (1974).

more care must be taken in choosing the starting subset of dummy variables. Wood (1973) has suggested using partial-residual plots[15] to find an initial subset which is subjected in turn to the C_p selection technique developed in Mallows (1973b) in order to find which dummy variables are to be retained. We think this method is appealing, especially if partial-regression leverage plots are combined with the methods discussed earlier in this chapter as an aid to finding the initial subset of dummies. Computational costs will tend to limit the practical size of this subset to 20 or fewer dummy variates.

The use of dummy variates has considerable appeal but the single-row analogue, the studentized residual, is, as we have seen, clearly not adequate for finding influential data points. This criticism extends to the dummy-variable approach because the use of sums of squares of residuals fails to give adequate weight to the structure and leverage of the explanatory-variable data.

The deletion methods discussed above provide one way to deal with this failure. Another is to realize that $\mathbf{I} - \mathbf{H}$ is proportional to the covariance matrix of the least-squares residuals. A straightforward argument using (2.64) shows that

$$1 - h_i(D_m) = \frac{\det(\mathbf{I} - \mathbf{H})_{D_m, i}}{\det(\mathbf{I} - \mathbf{H})_{D_m}}, \qquad (2.66)$$

where the numerator submatrix of $(\mathbf{I} - \mathbf{H})$ contains the ith row and column of $\mathbf{I} - \mathbf{H}$ in addition to the rows and columns in D_m. When this is specialized to a single deleted row, $k \neq i$, we obtain

$$1 - h_i(k) = \frac{(1 - h_i)(1 - h_k) - h_{ik}^2}{1 - h_k}$$
$$= (1 - h_i)\left[1 - \mathrm{cor}^2(e_i, e_k) \right]. \qquad (2.67)$$

This means that $h_i(k)$ can be large when the magnitude of the correlation between e_i and e_k is large. Thus useful clues about subsets of leverage points can be provided by looking at large diagonal elements of $\mathbf{H}$ and at the large residual correlations. This is an example of the direct use of the off-diagonal elements of $\mathbf{H}$, elements implicitly involved in most multiple-row procedures. This is further exemplified in the next two sections.

Differentiation. Generalizing the single-row differentiation techniques to multiple-row deletion is straightforward. Instead of altering the weight, w_i, attached to only one observation, we now consider a diagonal weight

---

[15] On these plots, see Larson and McCleary (1972).

matrix $\mathbf{W}$, with diagonal elements $(w_1, w_2, \ldots, w_n) \equiv \mathbf{w}$, and define $\mathbf{b}(\mathbf{w}) \equiv (\mathbf{X}^T\mathbf{W}\mathbf{X})^{-1}\mathbf{X}^T\mathbf{W}\mathbf{y}$. This $\mathbf{b}(\mathbf{w})$ is a vector-valued function of $\mathbf{w}$ whose first partial derivatives evaluated at $\mathbf{w} = \iota$ (the vector of ones) are

$$\nabla\mathbf{b}(\iota) = \mathbf{CE}, \tag{2.68}$$

where ∇ is the standard gradient operator, $\mathbf{C}$ is defined in (2.3), and $\mathbf{E} = \mathrm{diag}(e_1, \ldots, e_n)$. If we are interested in fitted values, this becomes

$$\mathbf{X}\nabla\mathbf{b}(\iota) = \mathbf{HE}. \tag{2.69}$$

Our concern is with subsets of observations that have a large influence. One way to identify such subsets is to consider the directional derivatives $\nabla\mathbf{b}(\iota)\mathit{l}$ where l is a column vector of unit length with nonzero entries in rows with indexes in D_m, that is, the rows to be perturbed. For a fixed m, the indexes corresponding to the nonzero entries in those l which give large values of

$$\mathit{l}^T\nabla\mathbf{b}^T(\iota)\mathbf{A}\nabla\mathbf{b}(\iota)\mathit{l} \tag{2.70}$$

would be of interest. The matrix $\mathbf{A}$ is generally $\mathbf{I}$, $\mathbf{X}^T\mathbf{X}$, or $\mathbf{X}^T(D_m)\mathbf{X}(D_m)$.

These l vectors are just the eigenvectors corresponding to largest eigenvalues of the matrix

$$\left[\nabla\mathbf{b}^T(\iota)\mathbf{A}\nabla\mathbf{b}(\iota)\right]_{D_m}. \tag{2.71}$$

When $\mathbf{A} = \mathbf{X}^T\mathbf{X}$, (2.71) is just the matrix whose elements are $h_{ij}e_ie_j$, with $i, j \in D_m$. While the foregoing procedure is conceptually straightforward, it has the practical drawback that, computationally, finding these eigenvectors is expensive. We therefore explore two less costly simplifications.

In the first simplification we place equal weight on all the rows of interest, and consider the effect of an infinitesimal perturbation of that single weight. This is equivalent to using a particular directional derivative, l^*, that has all of its nonzero entries equal. When $\mathbf{A}$ is $\mathbf{X}^T\mathbf{X}$, this gives

$$\mathit{l}^{*T}\nabla\mathbf{b}^T(\iota)\mathbf{X}^T\mathbf{X}\nabla\mathbf{b}(\iota)\mathit{l}^* = \sum_{i,j \in D_m} h_{ij}e_ie_j. \tag{2.72}$$

Much less computational effort is required to find the large values of (2.72) than to compute the eigenvectors for (2.71). A more complete discussion of this issue is contained in Appendix 2B. The expression $\mathbf{b} - \mathbf{b}(i)$ is an approximation to the ith column of $\nabla\mathbf{b}(\iota)$, and could be used instead in the preceding discussion. In this case (2.72) becomes

$$\sum_{i,j \in D_m} h_{ij} \frac{e_i e_i}{(1-h_i)(1-h_j)} \equiv \text{MEWDFFIT}. \qquad (2.73)$$

In the second simplification, we use a stepwise approach for large data sets, employed in the same manner as (2.62), using the statistic

$$\left| \mathbf{x}_k \sum_{i \in D_m} (\mathbf{X}^T\mathbf{X})^{-1} \mathbf{x}_i^T e_i \right| = \left| \sum_{i \in D_m} h_{ki} e_i \right|. \qquad (2.74)$$

Geometric Approaches. Wilks' Λ statistic generalizes to the multiple-row situation quite readily and is useful for discovering groups of outliers. This is particularly interesting when the observations cannot be grouped on the basis of prior knowledge (e.g., time) or when there is prior knowledge but unexpected groupings occur.

The generalization goes as follows. Let l_1 be an $n \times 1$ vector consisting of ones for rows contained in D_m and zeros elsewhere and $l_2 = \iota - l_1$. The relevant Λ statistic for this case is [Rao (1973), p. 570]

$$\Lambda(D_m) = \frac{\det\left[\tilde{\mathbf{Z}}^T\tilde{\mathbf{Z}} - (1/m)\tilde{\mathbf{Z}}^T l_1 l_1^T\tilde{\mathbf{Z}} - (1/(n-m))\tilde{\mathbf{Z}}^T l_2 l_2^T\tilde{\mathbf{Z}} \right]}{\det(\tilde{\mathbf{Z}}^T\tilde{\mathbf{Z}})}.$$

Using an argument similar to that in Appendix 2A, this statistic reduces to

$$\Lambda(D_m) = 1 - \frac{n}{m(n-m)} (l_1^T \tilde{\mathbf{P}} l_1), \qquad (2.75)$$

where $\tilde{\mathbf{P}} \equiv \tilde{\mathbf{Z}}(\tilde{\mathbf{Z}}^T\tilde{\mathbf{Z}})^{-1}\tilde{\mathbf{Z}}^T$. Thus $\Lambda(D_m)$ is directly related to sums of elements of a matrix, $\tilde{\mathbf{P}}$, ($\tilde{\mathbf{H}}$ if $\tilde{\mathbf{Z}}$ is replaced by $\tilde{\mathbf{X}}$) and, as we show in Appendix 2B, this greatly simplifies computation.

To use Λ we examine the smaller values for each $m = 1, 2,$ and so on. If we assume for guidance that the rows of $\tilde{\mathbf{Z}}$ are independent samples from a p-variate Gaussian distribution, then

$$\left(\frac{n-p-1}{p} \right)\left[\frac{1-\Lambda(D_m)}{\Lambda(D_m)} \right] \sim F_{p, n-p-1}. \qquad (2.76)$$

This is only approximate, since we are interested in extreme values. It would be even better to know the distribution of Λ conditional on $\mathbf{X}$, but this remains an open problem. More than just the smallest value of Λ should be examined for each m, since there may be several significant groups. Gaps in the values of Λ are also usually worth noting.

Andrews and Pregibon (1978) have proposed another method based on $\mathbf{Z}$. They consider the statistic

$$Q(D_m) = \frac{\det\left[\mathbf{Z}^T(D_m)\mathbf{Z}(D_m)\right]}{\det(\mathbf{Z}^T\mathbf{Z})}$$

$$= \frac{(n-p-m)s^2(D_m)\det\left[\mathbf{X}^T(D_m)\mathbf{X}(D_m)\right]}{(n-p)s^2\det(\mathbf{X}^T\mathbf{X})}, \qquad (2.77)$$

which relates to (2.49) and (2.51) for $m=1$. The idea is to ascertain the change in volume (measured by the determinant of $\mathbf{Z}^T\mathbf{Z}$) caused by the deletion of the rows in D_m. If $\check{\mathbf{Z}}$ instead of $\mathbf{Z}$ had been used, Q becomes another form of Wilks' Λ statistic where there are $m+1$ groups: one for each row in D_m and one group for all the remaining rows.

Computationally, Q is about the same order as MDFFIT and considerably more complicated than Λ. However, Andrews and Pregibon have succeeded in developing a distribution theory for Q when $\mathbf{y}$ is Gaussian and $\mathbf{X}$ is fixed. While useful only for n of modest size, it does provide some significance levels for finding sets of outliers.

Both Λ and Q are computationally feasible for $m \leqslant 20$. A stepwise approach based on the Mahalanobis distance and the ideas of robust covariance [Devlin, Gnanadesikan, and Kettenring (1975)] can be used for larger subsets. The philosophy is similar to that developed for (2.62) and (2.74). If we think the points in D are outliers, it is reasonable to remove them from our estimate of the covariance and means of the columns of $\tilde{\mathbf{Z}}$ by computing $\tilde{\mathbf{Z}}^T(D)\tilde{\mathbf{Z}}(D)$ and $\bar{\tilde{\mathbf{z}}}(D)$. The distance from any row $\tilde{\mathbf{z}}_i$ to $\bar{\tilde{\mathbf{z}}}(D)$ is then measured by

$$M(i,D) = (n-2)\left[\tilde{\mathbf{z}}_i - \bar{\tilde{\mathbf{z}}}(D)\right]\left[\tilde{\mathbf{Z}}^T(D)\tilde{\mathbf{Z}}(D)\right]^{-1}\left[\tilde{\mathbf{z}}_i - \bar{\tilde{\mathbf{z}}}(D)\right]^T. \qquad (2.78)$$

The starting set $D_2^{(1)}$ consists of the rows corresponding to the two largest values of the single-row Mahalanobis distance $M(\tilde{\mathbf{z}}_i)$. $D_2^{(2)}$ consists of the indexes of the two largest values of $M(i, D_2^{(1)})$. If $D_2^{(2)} = D_2^{(1)}$, we stop. If not we iterate with $D_2^{(k+1)}$ consisting of the two largest values of $M(i, D_2^{(k)})$, and the process stops when $D_2^{(k+1)} = D_2^{(k)}$.

Final Comments

The multiple-row techniques presented here form a subset of the possible procedures that could be devised. Our choices have been made on the basis of limited experience and theory. We originally hoped that multiple-row methods could be avoided because of their computational cost. A number of examples has shown that a single-row analysis alone is not enough, but that such an analysis coupled with the partial-regression leverage plots and stepwise multiple-row methods is quite often adequate and of modest cost.

On small data sets ($n < 100$), the fact that it is expensive to have m^* exceed 20 is not so serious since this includes over 20 percent of the data. However, B^* should still be chosen carefully, using relaxed cutoffs and partial-regression leverage plots. The inequalities developed at the end of Appendix 2B can also provide additional information. Of the measures discussed, we have had best results with MDFFIT and Λ.

For large data sets, the restriction on m^* is serious because potentially masked observations may be omitted from B^*, and we recommend partial-regression leverage plots and stepwise techniques such as SMDFFIT.

A somewhat less expensive alternative than full deletion is to use the inequalities at the end of Appendix 2B to find potentially influential subsets based on the values of h_i and e_i^2. This provides no rank ordering among subsets, and we lose the ability to look for specific groupings of influential observations. Computing and storage costs will probably determine whether stepwise, inequality, or full-deletion methods are used.

2.2   APPLICATION: AN INTERCOUNTRY LIFE-CYCLE SAVINGS FUNCTION

We now exemplify some of the more important aspects of the various diagnostic procedures presented above in their application to a body of data relevant to a cross-sectional study of an intercountry life-cycle savings function. Not all of the potential battery of diagnostic techniques are applied, since some of them contain redundant information and others, while providing interesting insights, are not essential to an understanding of the analysis at hand.

A Diagnostic Analysis of the Model

Arlie Sterling (1977) of MIT has provided us with data he has collected on 50 countries relevant to a cross-sectional study of the life-cycle savings hypothesis as developed by Franco Modigliani (1975). In this model the savings ratio (aggregate personal saving divided by disposable income) is explained by per-capita disposable income, the percentage rate of change in per-capita disposable income, and two demographic variables: the percentage of population less than 15 years old and the percentage of population over 75 years old. The data are averaged over the decade 1960–1970 to remove the business cycle or other short-term fluctuations. We examine first the intercountry life-cycle savings model and its basic regression results. We then examine in turn the application of the single- and multiple-row diagnostics.

The Model and Regression Results. According to the life-cycle hypothesis, savings rates should be smaller if nonmembers of the labor force constitute a large part of the population. Income is not expected to be important, since age distribution and the rate of income growth constitute the core of life-cycle savings behavior. The regression equation and variable definitions are then

$$\text{SR}_i = \beta_1 + \beta_2\text{POP15}_i + \beta_3\text{POP75}_i + \beta_4\text{DPI}_i + \beta_5\Delta\text{DPI}_i + \epsilon_i, \quad (2.79)$$

where

SR_i = the average aggregate personal savings rate in country i over the period 1960–1970,

POP15_i = the average percentage of the population under 15 years of age over the period 1960–1970,

POP75_i = the average percentage of the population over 75 years of age over the period 1960–1970,

DPI_i = the average level of real per-capita disposable income in country i over the period 1960–1970 measured in U.S. dollars,

ΔDPI_i = the average percentage growth rate of DPI_i over the period 1960–1970.

A full list of countries and data appears in Exhibit 2.3. It is evident that a wide geographic area and span of economic development have been included. It is also plausible to suppose that the quality of the underlying data is uneven. With these obvious caveats, the ordinary least-squares estimates are given in (2.80), for which standard errors appear in parentheses. To comment briefly on the results, the R^2 is not uncharacteristically low for cross sections, and the population variables have correct negative signs. Testing at the 5% level, we note that $b_3(\text{POP75})$ has a small t-statistic but $b_2(\text{POP15})$ does not—b_4 (income) is

Exhibit 2.3 Intercountry life-cycle savings data

Index	Country	SR	POP15	POP75	DPI	ΔDPI
1	Australia	11.43	29.35	2.87	2329.68	2.87
2	Austria	12.07	23.32	4.41	1507.99	3.93
3	Belgium	13.17	23.80	4.43	2108.47	3.82
4	Bolivia	5.75	41.89	1.67	189.13	0.22
5	Brazil	12.88	42.19	0.83	728.47	4.56
6	Canada	8.79	31.72	2.85	2982.88	2.43
7	Chile	0.60	39.74	1.34	662.86	2.67
8	China (Taiwan)	11.90	44.75	0.67	289.52	6.51
9	Colombia	4.98	46.64	1.06	276.65	3.08
10	Costa Rica	10.78	47.64	1.14	471.24	2.80
11	Denmark	16.85	24.42	3.93	2496.53	3.99
12	Ecuador	3.59	46.31	1.19	287.77	2.19
13	Finland	11.24	27.84	2.37	1681.25	4.32
14	France	12.64	25.06	4.70	2213.82	4.52
15	Germany (F.R.)	12.55	23.31	3.35	2457.12	3.44
16	Greece	10.67	25.62	3.10	870.85	6.28
17	Guatemala	3.01	46.05	0.87	289.71	1.48
18	Honduras	7.70	47.32	0.58	232.44	3.19
19	Iceland	1.27	34.03	3.08	1900.10	1.12
20	India	9.00	41.31	0.96	88.94	1.54
21	Ireland	11.34	31.16	4.19	1139.95	2.99
22	Italy	14.28	24.52	3.48	1390.00	3.54
23	Japan	21.10	27.01	1.91	1257.28	8.21
24	Korea	3.98	41.74	0.91	207.68	5.81
25	Luxembourg	10.35	21.80	3.73	2449.39	1.57
26	Malta	15.48	32.54	2.47	601.05	8.12
27	Norway	10.25	25.95	3.67	2231.03	3.62
28	Netherlands	14.65	24.71	3.25	1740.70	7.66
29	New Zealand	10.67	32.61	3.17	1487.52	1.76
30	Nicaragua	7.30	45.04	1.21	325.54	2.48
31	Panama	4.44	43.56	1.20	568.56	3.61
32	Paraguay	2.02	41.18	1.05	220.56	1.03
33	Peru	12.70	44.19	1.28	400.06	0.67
34	Philippines	12.78	46.26	1.12	152.01	2.00
35	Portugal	12.49	28.96	2.85	579.51	7.48
36	South Africa	11.14	31.94	2.28	651.11	2.19
37	South Rhodesia	13.30	31.92	1.52	250.96	2.00
38	Spain	11.77	27.74	2.87	768.79	4.35
39	Sweden	6.86	21.44	4.54	3299.49	3.01
40	Switzerland	14.13	23.49	3.73	2630.96	2.70
41	Turkey	5.13	43.42	1.08	389.66	2.96
42	Tunisia	2.81	46.12	1.21	249.87	1.13

Exhibit 2.3 Continued

Index	Country	SR	POP15	POP75	DPI	ΔDPI
43	United Kingdom	7.81	23.27	4.46	1813.93	2.01
44	United States	7.56	29.81	3.43	4001.89	2.45
45	Venezuela	9.22	46.40	0.90	813.39	0.53
46	Zambia	18.56	45.25	0.56	138.33	5.14
47	Jamaica	7.72	41.12	1.73	380.47	10.23
48	Uruguay	9.24	28.13	2.72	766.54	1.88
49	Libya	8.89	43.69	2.07	123.58	16.71
50	Malaysia	4.71	47.20	0.66	242.69	5.08

statistically insignificant, while b_5 (income growth) is significant and, as expected, has a positive influence on the savings rate. Broadly speaking, these results are consistent with the life-cycle hypothesis:

$$SR = 28.56 - 0.4611POP15 - 1.691POP75 - 0.000337DPI + 0.4096\Delta DPI$$
$$(7.345)(0.1446) \quad (1.083) \quad (0.000931) \quad (0.1961)$$

$$R^2 = .33 \qquad SER = 3.802 \qquad\qquad (2.80)$$

Condition number[16] of scaled $X = 34$.

The remainder of this section is a discussion of some of the single- and multiple-row diagnostics discussed previously. The computations were performed using SENSSYS (acronym for Sensitivity System), a TROLL experimental subsystem for regression diagnostics.[17] Orthogonal decompositions are used in the least-squares regression computations, and this makes it possible to get all of the single-row diagnostic measures in addition to the usual least-squares results in roughly twice the computer time for the least-squares results alone. Multiple-row techniques involve greater expense. Further details on computation and associated costs can be found in Appendix 2B.

Single-Row Diagnostics.

Residuals. Exhibit 2.4 is a Gaussian (normal) probability plot of the studentized residuals. Departure from a fitted line (which represents a

---

[16]The significance of the condition number for assessing the presence of collinearity is discussed in great detail in Chapter 3.

[17] For more details and documentation on SENSSYS and TROLL, one may write Publications Office, Information Processing Services, MIT Room 39–484, Cambridge, MA 02139.

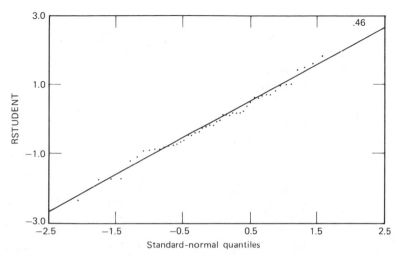

Exhibit 2.4 Normal probability plot for studentized residual: intercountry life-cycle savings data.

robust estimate of a particular Gaussian distribution with mean equal to the intercept and standard deviation equal to the slope–see Velleman and Hoaglin (1980)) is not substantial in the main body of the data for these studentized residuals, but Zambia (46) has an extreme residual. Magnitudes of the e_i^* appear in Exhibit 2.5 which reveals not only Zambia, but possibly Chile (7), as outliers; each exceeds 2.0 times its standard error. The largest values here and subsequently are starred.

Leverage and Hat-Matrix Diagonals. Exhibit 2.5 also shows the tabulated values of h_i which, as diagonals of the hat matrix, are indicative of leverage points. Most of the h_i are small, but two stand out sharply:

Exhibit 2.5 RSTUDENT and hat-matrix diagonals: intercountry life-cycle savings data

Index	Country	RSTUDENT	h_i
1	Australia	0.2327	0.0677
2	Austria	0.1709	0.1203
3	Belgium	0.6065	0.0874
4	Bolivia	−0.1903	0.0894
5	Brazil	0.9679	0.0695
6	Canada	−0.0898	0.1584
7	Chile	−2.3134*	0.0372
8	China (Taiwan)	0.6904	0.0779

Exhibit 2.5 Continued

Index	Country	RSTUDENT	h_i
9	Colombia	−0.3894	0.0573
10	Costa Rica	1.4173	0.0754
11	Denmark	1.4864	0.0627
12	Ecuador	−0.6495	0.0637
13	Finland	−0.4598	0.0920
14	France	0.6964	0.1362
15	Germany (F.R.)	−0.0491	0.0873
16	Greece	−0.8596	0.0966
17	Guatemala	−0.9085	0.0604
18	Honduras	0.1905	0.0600
19	Iceland	−1.7312	0.0704
20	India	0.1373	0.0714
21	Ireland	1.0048	0.2122*
22	Italy	0.5201	0.0665
23	Japan	1.6032	0.2233*
24	Korea	−1.6910	0.0607
25	Luxembourg	−0.4556	0.0863
26	Malta	0.8122	0.0794
27	Norway	−0.2324	0.0479
28	Netherlands	0.1160	0.0906
29	New Zealand	0.6137	0.0542
30	Nicaragua	0.1725	0.0503
31	Panama	−0.8814	0.0389
32	Paraguay	−1.7048	0.0693
33	Peru	1.8239	0.0650
34	Philippines	1.8638	0.0642
35	Portugal	−0.2104	0.0971
36	South Africa	0.1299	0.0651
37	South Rhodesia	0.3671	0.1608
38	Spain	−0.1817	0.0773
39	Sweden	−1.2029	0.1239
40	Switzerland	0.6753	0.0735
41	Turkey	−0.7113	0.0396
42	Tunisia	−0.7667	0.0745
43	United Kingdom	−0.7495	0.1165
44	United States	−0.3546	0.3336*
45	Venezuela	0.9993	0.0862
46	Zambia	2.8535*	0.0643
47	Jamaica	−0.8537	0.1407
48	Uruguay	−0.6225	0.0979
49	Libya	−1.0893	0.5314*
50	Malaysia	−0.8048	0.0652

*Exceeds cutoff values: RSTUDENT = 2.0; h_i = 0.20.

44

Libya (49) and the United States (44). Two others, Japan (23) and Ireland (21), exceed the $2p/n = 0.20$ criterion, but just barely. Deciding whether leverage is potentially detrimental will depend on what happens elsewhere in the diagnostic analysis.

Coefficient Sensitivity. We note from Exhibit 2.6, which reports DFBETAS, that seven countries—Costa Rica (10), Ireland (21), Japan (23), Peru (33), Zambia (46), Jamaica (47), and Libya (49)—show up as possibly representing influential data. Three of these (Ireland, Japan, and Libya) appear also as high-leverage candidates and in addition have DFBETAS for either three or four of the five coefficients that exceed the size-adjusted cutoff of $2/\sqrt{n} = 0.28$. One DFBETAS for Libya (49) is large in an absolute sense as well. Note that the deletion of only one data point out of 50 is causing more than one standard error of change in an estimated coefficient.

Exhibit 2.6   DFBETAS: intercountry life-cycle savings data

		DFBETAS				
		b_1	b_2	b_3	b_4	b_5
Index	Country	Intercept	POP15	POP75	DPI	ΔDPI
1	Australia	0.0123	−0.0104	−0.0265	0.0453	−0.0001
2	Austria	−0.0100	0.0059	0.0408	−0.0367	−0.0081
3	Belgium	−0.0641	0.0514	0.1207	−0.0347	−0.0072
4	Bolivia	0.0057	−0.0127	−0.0225	0.0318	0.0406
5	Brazil	0.0897	−0.0616	−0.1790	0.1199	0.0684
6	Canada	0.0054	−0.0067	0.0102	−0.0353	−0.0026
7	Chile	−0.1994	0.1326	0.2197	−0.0199	0.1200
8	China (Taiwan)	0.0211	−0.0057	−0.0831	0.0518	0.1106
9	Colombia	0.0390	−0.0522	−0.0246	0.0016	0.0090
10	Costa Rica	−0.2336	0.2842*	0.1424	0.0563	−0.0328
11	Denmark	−0.0405	0.0209	0.0465	0.1521	0.0488
12	Ecuador	0.0717	−0.0952	−0.0606	0.0195	0.0477
13	Finland	−0.1134	0.1113	0.1169	−0.0436	−0.0171
14	France	−0.1660	0.1470	0.2189	−0.0294	0.0239
15	Germany (F.R.)	−0.0080	0.0082	0.0083	−0.0069	−0.0002
16	Greece	−0.1481	0.1639	0.0286	0.1571	−0.0595
17	Guatemala	0.0155	−0.0548	0.0061	0.0058	0.0972
18	Honduras	−0.0922	0.0098	−0.0102	0.0081	−0.0018
19	Iceland	0.2478	−0.2735	−0.2326	−0.1255	0.1846
20	India	0.0210	−0.0157	−0.0143	−0.0137	−0.0189
21	Ireland	−0.3100*	0.2962*	0.4815*	−0.2573	−0.0933
22	Italy	0.0661	−0.0709	0.0030	−0.0699	−0.0286

Exhibit 2.6 Continued   DFBETAS: intercountry life-cycle savings data

		\multicolumn{5}{c}{DFBETAS}			
Index	Country	b_1 Intercept	b_2 POP15	b_3 POP75	b_4 DPI	b_5 ΔDPI
23	Japan	0.6398*	−0.6561*	−0.6739*	0.1461	0.3886*
24	Korea	−0.1689	0.1350	0.2189	0.0051	−0.1694
25	Luxembourg	−0.0682	0.0688	0.0438	−0.0279	0.0491
26	Malta	0.0365	−0.0487	0.0079	−0.0865	0.1530
27	Norway	0.0022	−0.0003	−0.0061	−0.0159	−0.0014
28	Netherlands	0.0139	−0.0167	−0.0118	0.0043	0.0225
29	New Zealand	−0.0600	0.0651	0.0941	−0.0263	−0.0647
30	Nicaragua	−0.0120	0.0179	0.0097	−0.0047	−0.0104
31	Panama	0.0282	−0.0533	0.0144	−0.0346	−0.0078
32	Paraguay	−0.2322	0.1641	0.1582	0.1436	0.2704
33	Peru	−0.0718	0.1466	0.0914	−0.0858	−0.2871*
34	Philippines	−0.1570	0.2268	0.1574	−0.1113	−0.1706
35	Portugal	−0.0213	0.0255	−0.0037	0.0399	−0.0280
36	South Africa	0.0221	−0.0202	−0.0067	−0.0204	−0.0163
37	South Rhodesia	0.1439	−0.1347	−0.0924	−0.0695	−0.0579
38	Spain	−0.0303	0.0313	0.0039	0.0351	0.0053
39	Sweden	0.1009	−0.0816	−0.0616	−0.2552	−0.0133
40	Switzerland	0.0432	−0.0464	−0.0436	0.0909	−0.0188
41	Turkey	−0.0109	−0.0119	0.0264	0.0016	0.0251
42	Tunisia	0.0737	−0.1049	−0.0772	0.0443	0.1030
43	United Kingdom	0.0467	−0.0358	−0.1712	0.1255	0.1003
44	United States	0.0690	−0.0728	0.0374	−0.2331	−0.0327
45	Venezuela	−0.0508	0.1008	−0.0336	0.1136	−0.1244
46	Zambia	0.1636	−0.0791	−0.3389*	0.0940	0.2282
47	Jamaica	0.1095	−0.1002	−0.0572	−0.0070	−0.2954*
48	Uruguay	−0.1340	0.1288	0.0295	0.1313	0.0995
49	Libya	0.5507*	−0.4832*	−0.3797*	−0.0193	−1.0244*
50	Malaysia	0.0368	−0.0611	0.0323	−0.0495	−0.0722

*Exceeds cutoff value: DFBETAS=0.28.

Covariance Matrix Sensitivity. Exhibit 2.7 presents the COVRATIO's for the intercountry life-cycle savings data. We recall from (2.36) that COVRATIO is a ratio of the determinant of the estimated coefficient covariance matrix with the ith observation deleted to that of the estimated covariance matrix based on the full data set. This magnitude is, of course, a ratio of the estimated generalized variances[18] of the regression

---

[18] See, for example, Wilks (1962) or Theil (1971, p. 124).

Exhibit 2.7 COVRATIO and DFFITS: intercountry life-cycle savings data

Index	Country	COVRATIO	DFFITS
1	Australia	1.1928	0.0627
2	Austria	1.2678	0.0632
3	Belgium	1.1762	0.1878
4	Bolivia	1.2238	−0.0596
5	Brazil	1.0823	0.2646
6	Canada	1.3283*	−0.0389
7	Chile	0.6547*	−0.4553
8	China (Taiwan)	1.1499	0.2007
9	Colombia	1.1667	−0.0960
10	Costa Rica	0.9681	0.4049
11	Denmark	0.9344	0.3845
12	Ecuador	1.1394	−0.1694
13	Finland	1.2032	−0.1464
14	France	1.2262	0.2765
15	Germany (F.R.)	1.2257	−0.0152
16	Greece	1.1396	−0.2811
17	Guatemala	1.0853	−0.2305
18	Honduras	1.1855	0.0481
19	Iceland	0.8659	−0.4767
20	India	1.2024	0.0380
21	Ireland	1.2680	0.5215
22	Italy	1.1624	0.1388
23	Japan	1.0846	0.8596*
24	Korea	0.8696	−0.4302
25	Luxembourg	1.1962	−0.1400
26	Malta	1.1282	0.2385
27	Norway	1.1680	−0.0521
28	Netherlands	1.2285	0.0366
29	New Zealand	1.1337	0.1469
30	Nicaragua	1.1743	0.0397
31	Panama	1.0667	−0.1775
32	Paraguay	0.8732	−0.4654
33	Peru	0.8313	0.4810
34	Philippines	0.8178	0.4884
35	Portugal	1.2331	−0.0690
36	South Africa	1.1945	0.0342
37	South Rhodesia	1.3131*	0.1607
38	Spain	1.2082	−0.0526
39	Sweden	1.0865	−0.4525
40	Switzerland	1.1471	0.1903
41	Turkey	1.1004	−0.1445
42	Tunisia	1.1314	−0.2176

Exhibit 2.7 Continued

Index	Country	COVRATIO	DFFITS
43	United Kingdom	1.1886	−0.2722
44	United States	1.6555*	−0.2509
45	Venezuela	1.0946	0.3070
46	Zambia	0.5116*	0.7482*
47	Jamaica	1.1995	−0.3455
48	Uruguay	1.1872	−0.2051
49	Libya	2.0906*	−1.1601*
50	Malaysia	1.1126	−0.2126

*Exceeds cutoff values: COVRATIO = 1 ± 0.30; DFFITS = 0.63.

coefficients with and without the ith observation deleted from the data, and, as such, it can be interpreted as a measure of the effect of the ith observation on the efficiency of coefficient estimation. A value of COVRATIO greater than one indicates that the absence of the associated observation impairs efficiency, while a value of less than one indicates the reverse. As was noted in the discussion surrounding (2.36), values of COVRATIO that lie outside the range defined by $1 \pm 3(p/n)$ can be considered extreme, and hence they wave a warning flag. In this instance $3p/n = 0.30$; thus we shall be interested in values that lie below 0.70 and above 1.30. Six countries produce COVRATIO's that lie outside these bounds: Canada (6), Chile (7), Southern Rhodesia (37), United States (44), Zambia (46), and Libya (49).

Of these six countries, four have been pinpointed by previous diagnostics, the two new candidates being Canada (6) and Southern Rhodesia (37). We recall that extreme values can occur for COVRATIO if the deletion of the ith observation produces a large change in the estimated regression variance s^2 or in the elements of $(X^TX)^{-1}$ or in both. Thus Libya (49) had been detected earlier because it has high leverage and large coefficient changes; Chile (7) and Zambia (46) because of large residuals; and the United States (44) because of high leverage alone. The two new countries, Canada (6) and Southern Rhodesia (37), however, possess neither unusually high leverage nor large residuals in isolation. Thus the COVRATIO diagnostic measure, combining the residual and leverage information as it does, is able to pinpoint these new countries as well as the others, indicating that COVRATIO may possess value as a comprehensive diagnostic measure.

Change in Fit.   The scaled row-deleted change in fit, DFFITS, is displayed in Exhibit 2.7. The three countries whose DFFITS exceed the size-adjusted cutoff of $2\sqrt{p/n} = 0.63$ are three that have previously occurred, namely, Japan (23), Zambia (46), and Libya (49).

Internal Scaling.   Internal scaling according to Exhibit 2.8 provides general confirmation about the most severely affected data points. The $(7/2)\bar{s}$ criterion is so stringent that no outliers were recorded for e_i. Remaining information on hat-matrix diagonals, DFFIT and DFBETA, is presented in Exhibit 2.8. The four large hat-matrix diagonals happen to include those which exceeded the size-adjusted cutoff: Ireland (21), Japan (23), United States (44), and Libya (49). DFBETA, according to the row counts in column 8, are most extreme for Japan (23) and Libya (49), which is consistent with DFFIT and the use of size-adjusted cutoffs. Thus, the internal scaling used here appears to be conservative relative to size-adjusted cutoffs, for each of the countries indicated by internal scaling as being potentially influential had already been exposed by size-adjusted diagnostics.

Additional information can be obtained from DFBETA column counts. Coefficients b_3 and b_5, for example, show noticeably more extreme behavior than do the others, while coefficient b_4 shows the least. Another point that emerges is that the extremes are skewed. The fact that most extreme magnitudes appear at either the lower or upper end of these rank-ordered diagnostics suggests that the interaction between model and data raises questions about the model specification itself, a proposition which has not appeared from earlier analysis.

A Provisional Summary.   It is now desirable to bring together the information that has been obtained thus far. The first point to note is that Japan (23) and Libya (49) have both high leverage *and* a significant influence on the estimated parameters. This is reason enough to view these observations as presenting potentially serious problems. (After the analysis had reached this point, we were informed by Arlie Sterling that a data error had been discovered for Japan (23). Once corrected, this observation became more nearly in accord with the majority of countries.)

Second, the United States (44) has high leverage combined with only meager differential effects on the estimated coefficients. Thus leverage in this instance can be viewed as neutral or beneficial. It is important to observe that not all leverage points cause large changes in the estimated coefficients or fit.

Exhibit 2.8 Extreme values from internal scaling: intercountry life-cycle savings data (L = lower tail, U = upper tail)

| | | (1) | (2) | DFBETA | | | | | (8) |
| | | Hat-Matrix | | (3) | (4) | (5) | (6) | (7) | Row |
Index	Country	Diagonals	DFFIT	b_1 Intercept	b_2 POP15	b_3 POP75	b_4 DPI	b_5 ΔDPI	Count
21	Ireland	U		U		U			3
23	Japan	U	U		L	L		U	5
33	Peru							L	1
44	United States	U							1
46	Zambia					L			1
47	Jamaica							L	1
49	Libya	U	L	U	L	L		L	6
Column Count: L		0	1	0	2	3	0	3	
Column Count: U		4	1	2	0	1	0	1	
Total		4	2	2	2	4	0	4	

We also have seen that large residuals do not necessarily coincide with large changes in coefficients; all of the large changes in coefficients are associated with studentized residuals less than two. Thus residual analysis alone is not a sufficient diagnostic tool.

Multiple-Row Diagnostics.

Partial-Regression Leverage Plots: A Preliminary Analysis. The single-row analysis of the savings data has led us to consider observation 49, and perhaps 23, as sufficiently influential to be worthy of further attention. As a preliminary to applying the multiple-row diagnostics, we turn briefly to the five partial-regression leverage plots that display how 49 and 23 relate to each other as well as to any of the other more marginal leverage points that may be indicated.

The plots for the intercept and POP15, Exhibits 2.9 and 2.10, respectively, show how observations 49 and 23 may be working together to increase the intercept and decrease the coefficient of POP15. It is also suggested that the role of 37 may be masked to some extent by 23. The effect of the large residual at point 46 is ambiguous. The plot for POP75 (Exhibit 2.11) indicates that observations 21 and 23 could be offsetting each other's leverage on this coefficient.

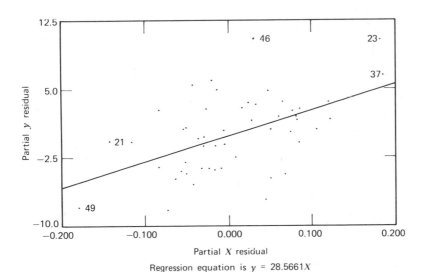

Regression equation is $y = 28.5661X$

Exhibit 2.9 Partial-regression leverage plot for b_1 (intercept), S.E. $= 7.3545$.

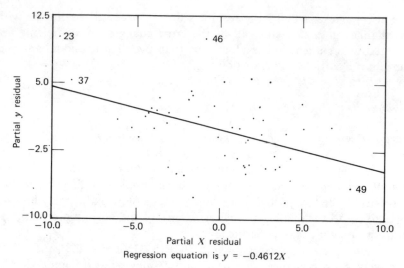

Exhibit 2.10   Partial-regression leverage plot for b_2 (POP15), S.E. $=0.14464$.

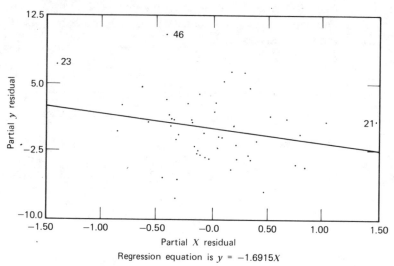

Exhibit 2.11   Partial-regression leverage plot for b_3 (POP75), S.E. $=1.0836$.

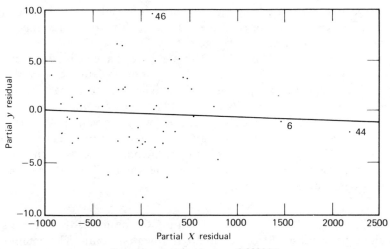

Regression equation is $y = -0.00034X$

Exhibit 2.12 Partial-regression leverage plot for b_4 (DPI), S.E. = 0.0009.

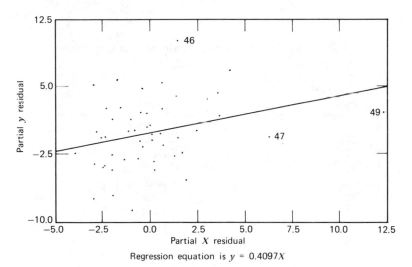

Regression equation is $y = 0.4097X$

Exhibit 2.13 Partial-regression leverage plot for b_5 (ΔDPI), S.E. = 0.1961.

The partial-regression leverage plot for DPI (Exhibit 2.12) is much more clear-cut, since observation 44 has markedly high marginal leverage, as the single-row analysis has already indicated. The role of observation 6 has possibly been masked by 44. Although we have already seen that observation 44 is not influential, this fact cannot be deduced from this plot alone. Furthermore, this partial-regression leverage plot cannot indicate what happens when observations 6 and 44 are set aside together.

The plot for ΔDPI (Exhibit 2.13) shows observation 49 with high marginal leverage, perhaps masking 47. We need nongraphical methods, to assess the difference in influence between the pair 6 and 44 in the previous partial-regression leverage plot and the pair 47 and 49 in this partial-regression leverage plot.

Using Multiple-Row Methods. To apply the multiple-row methods, we require a basic subset B^* of potentially influential points. The complement of this set defines a group of "typical" points for purposes of comparison. The set B^* is chosen by using the single-row results with relaxed cutoffs ($1.5 \, p/n$ for h_i, 1.68 for e_i^*, 0.24 for DFBETAS, 0.53 for DFFITS, and

Exhibit 2.14 Subset of potentially influential points: intercountry life-cycle savings data

Observation Index	Country
3	Belgium
6	Canada
7	Chile
10	Costa Rica
14	France
19	Iceland
21	Ireland
23	Japan
24	Korea
32	Paraguay
33	Peru
34	Philippines
37	South Rhodesia
39	Sweden
44	United States
46	Zambia
47	Jamaica
49	Libya

$1 \pm 2.5(p/n)$ for COVRATIO) and by examining the partial-regression leverage plots. The 18 points that are so chosen are given in Exhibit 2.14. Austria (2) is slightly above the relaxed COVRATIO cutoff but is not included, as it was not otherwise indicated and since we had limited our original computer codes to accept a B^* with a maximum of 18 points.

Deletion. A brief table of the results for MDFFIT is given in Exhibit 2.15. The row values have been converted to percentages by dividing all values for a given deletion set of size m by the largest MDFFIT for that size. These percentages provide a ready means for indicating gaps that separate specific groups of points from the remaining groups. For example, when $m = 1$, observations 49, 23, and 46 are quite separated from the rest. When $m = 2$, the subsets 47 and 49, and 23 and 46 attract our attention. When $m = 3$, the set 24, 47, and 49 also appears to be interesting, in part because 24 has not been given special attention before. This is a subjective analysis, designed to provide clues rather than confirmation. We note that while observations 47 and 49 do appear to cause problems, observations 6 and 44 do not form an influential group.

Exhibit 2.15   MDFFIT: intercountry life-cycle savings data

m	Subset	MDFFIT	Index Relative to Max
1	49	9.08	1.00
	23	8.02	0.88
	46	6.54	0.72
	21	3.10	0.34
	34	3.06	0.34
2	47 49	29.94	1.00
	23 46	23.74	0.79
	24 49	17.52	0.59
	19 23	17.02	0.57
	33 49	16.51	0.55
3	24 47 49	48.07	1.00
	33 47 49	41.39	0.86
	37 47 49	38.23	0.80
	19 23 46	34.94	0 73
	39 47 49	31.90	0.66
4	24 33 47 49	59.27	1.00
	24 37 47 49	55.54	0.94
	33 37 47 49	52.19	0.88
	7 24 47 49	50.49	0.85
	33 34 47 49	48.79	0.82

For completeness we give further statistics in Exhibit 2.16 with observation 49 deleted. As we might suspect, 47 is quite prominent; its leverage substantially exceeds the cutoff while its DFFITS is now noticeably large. Thus 49 has substantially masked the impact of 47. In this instance we would have discovered the masked point either by deleting 49 and running a single-row analysis or by using multiple-row methods.

Exhibit 2.16   Studentized residuals, hat-matrix diagonals and DFFITS: intercountry life-cycle savings data with Libya (49) removed

Index	Country	RSTUDENT	h_i	DFFITS
1	Australia	0.2872	0.0700	0.0788
2	Austria	0.1128	0.1229	0.0422
3	Belgium	0.5421	0.0908	0.1714
4	Bolivia	−0.0854	0.0982	−0.0282
5	Brazil	0.9235	0.0712	0.2557
6	Canada	−0.0564	0.1592	−0.0245
7	Chile	−2.2693*	0.0390	−0.4573
8	China (Taiwan)	0.5163	0.1038	0.1757
9	Colombia	−0.4639	0.0615	−0.1188
10	Costa Rica	1.3329	0.0814	0.3968
11	Denmark	1.4559	0.0635	0.3793
12	Ecuador	−0.6843	0.0645	−0.1798
13	Finland	−0.3980	0.0951	−0.1290
14	France	0.5386	0.1560	0.2316
15	Germany (F.R.)	0.0343	0.0928	0.0109
16	Greece	−0.9468	0.1016	−0.3185
17	Guatemala	−0.8641	0.0621	−0.2225
18	Honduras	0.1530	0.0612	0.0390
19	Iceland	−1.6939	0.0717	−0.4708
20	India	0.2626	0.0836	0.0793
21	Ireland	0.8714	0.2252*	0.4699
22	Italy	0.5673	0.0681	0.1534
23	Japan	1.5177	0.2284*	0.8258*
24	Korea	−1.8290	0.0707	−0.5046
25	Luxembourg	−0.2874	0.1096	−0.1008
26	Malta	0.5852	0.1248	0.2210
27	Norway	−0.2439	0.0480	−0.0548
28	Netherlands	−0.0437	0.1105	−0.0154
29	New Zealand	0.6383	0.0546	0.1534
30	Nicaragua	0.1461	0.0509	0.0338
31	Panama	−0.9458	0.0418	−0.1978
32	Paraguay	−1.5680	0.0856	−0.4800
33	Peru	1.9233	0.0700	0.5278

Exhibit 2.16 Continued

Index	Country	RSTUDENT	h_i	DFFITS
34	Philippines	1.8546	0.0643	0.4865
35	Portugal	−0.3977	0.1225	−0.1486
36	South Africa	0.2467	0.0757	0.0706
37	South Rhodesia	0.6063	0.1969	0.3002
38	Spain	−0.1706	0.0774	−0.0494
39	Sweden	−0.2000	0.1240	−0.4515
40	Switzerland	0.7565	0.0782	0.2204
47	Turkey	−0.7211	0.0396	−0.1466
42	Tunisia	−0.7417	0.0751	−0.2113
43	United Kingdom	−0.7088	0.1178	−0.2590
44	United States	−0.3582	0.3336*	−0.2535
45	Venezuela	1.0941	0.0920	0.3484
46	Zambia	2.7495*	0.0726	0.7697*
47	Jamaica	−1.4611	0.2897*	−0.9332*
48	Uruguay	−0.4702	0.1171	−0.1712
50	Malaysia	−0.9664	0.0825	−0.2899

*Exceeds cutoff values: RSTUDENT≈2.0; h_i=0.20; DFFITS=0.63.

The impact of the group consisting of observations 23 and 46 is more difficult to assess, but since each point has come to our attention separately, there is no problem of masking. Indeed, the data alone provide us with no reason to group these points (as they do for 6 (Canada) and 44 (U.S.)), but a glance at the partial-regression leverage plots shows how observation 46 reinforces the impact of observation 23 in every case. A full assessment would require checking the model with 23 and 46 omitted.

Since the computational costs for MDFFIT are of the order 2^{m^*}, where m^* is the size of B^*, it is usually necessary to keep m^* below 20. Of course, any such limitation increases the danger that some masked points may be inadvertently omitted from B^*, and MDFFIT would never find them. It is therefore of use to complement MDFFIT with the stepwise fit procedure SMDFFIT [see text surrounding (2.62)] which does not require the formation of B^* in order to remain computationally feasible.

Exhibit 2.17 lists the value of SMDFFIT for two different stepwise procedures. The first method uses the largest $m+1$ values of (2.62) for SMDFFIT at the end of the mth step (deletion sets of size m) in order to start the $m+1$ step. This method immediately highlights observations 47 and 49, for when observation 49 (the largest for $m=1$) is deleted, 47 becomes the next largest point. Method 2 uses the m largest points from

Exhibit 2.17   SMDFFIT: intercountry life-cycle savings data

	Method 1			Method 2		
m	Observation	SMDFFIT	Index Relative to Max (1)	Observation	SMDFFIT	Index Relative to Max (2)
1	49	3.21	1.00	49	3.21	1.00
	47	1.60	0.50	47	1.60	0.50
	26	0.88	0.27	26	0.88	0.27
	37	0.78	0.24	37	0.78	0.24
	8	0.66	0.21	8	0.66	0.21
2	49	6.87	1.00	23	6.83	1.00
	47	3.48	0.51	37	4.43	0.65
	26	1.96	0.29	21	4.36	0.64
	8	1.56	0.23	13	4.24	0.62
	37	1.51	0.22	28	3.22	0.47
3	49	6.33	1.00	49	5.27	1.00
	47	3.19	0.50	37	2.52	0.48
	26	1.76	0.28	47	2.39	0.45
	37	1.46	0.23	48	1.94	0.37
	8	1.42	0.22	25	1.74	0.33
4	49	7.61	1.00	49	4.24	1.00
	47	3.72	0.49	23	3.02	0.71
	37	2.20	0.29	47	2.71	0.64
	26	1.93	0.25	8	2.33	0.55
	48	1.62	0.21	24	2.02	0.48

|DFFITS| to start the mth step without further reference to the previous stage. In this case we see that observation 49 is chosen for $m = 1$ and observation 47 shows up as before. For $m = 2$, however, the process converges with observation 23 as the largest and observation 37 as the second largest. The pair 23 and 37 has already been brought to our attention by the partial-regression leverage plots and needs further examination.

This example highlights a possible weakness of stepwise procedures: they can "converge" to different sets of points depending on the starting set. In practice we have often found this useful, because it indicates different subgroups. For large data sets where even stringent single-row cutoffs might produce an m^* greater than 20, the stepwise procedures are a feasible means for uncovering masked points that would not have shown up in the stringent single-row analysis. The lack of global convergence for the stepwise procedures aids in this process.

Exhibit 2.18 Multiple COVRATIO statistics: intercountry life-cycle savings data

m	Observations	Large COVRATIO	Index Relative to Max
1	49	2.09	1.00
	44	1.66	0.79
	6	1.33	0.64
2	44 49	3.47	1.00
	37 49	2.80	0.81
	6 49	2.79	0.80
3	6 44 49	5.09	1.00
	37 44 49	4.71	0.93
	21 44 49	4.61	0.91

m	Observations	Small COVRATIO	Index Relative to Min
1	46	0.51	1.00
	7	0.65	1.27
	34	0.82	1.61
2	7 46	0.32	1.00
	34 46	0.37	1.16
	33 46	0.38	1.19
3	7 34 46	0.23	1.00
	7 33 46	0.24	1.04
	7 19 46	0.24	1.04

The results of using multiple-deletion statistics for COVRATIO are given in Exhibit 2.18. The large values of COVRATIO point to observations 44 and 49 as having a major impact on the covariance matrix. This confirms our earlier analysis using single-row methods. We should not, however, attach much significance to the fact that they are indicated together here since they are also the largest COVRATIO values determined by the single-row analyses and appear in different contexts in the partial-regression leverage plots. The smallest values of COVRATIO provide no new information since observations 46 and 7 had the smallest values in the single-row case.

Residuals. In Exhibit 2.19 we record the values of the ratio (2.65) for residual sum-of-squares, RESRATIO. No attempt is made here to compare

Exhibit 2.19   RESRATIO: intercountry life-cycle savings data

m	Subset	RESRATIO	Index Relative to Max
1	46	8.14	1.00
	7	5.35	0.66
	34	3.47	0.43
	33	3.32	0.41
	19	3.00	0.37
2	7 46	7.34	1.00
	34 46	6.68	0.91
	33 46	6.51	0.89
	23 46	6.26	0.85
3	7 34 46	6.88	1.00
	33 34 46	6.73	0.98
	7 33 46	6.73	0.98
	7 19 46	6.64	0.97
4	7 33 34 46	7.13	1.00
	10 33 34 46	6.68	0.94
	7 19 34 46	6.67	0.94
	7 19 33 46	6.61	0.93

these ratios to the F-statistic since we are looking at extreme values. Bonferroni bounds [Miller (1966)] could conceivably be employed, but the necessary tables are not readily available. The significance methods developed by Andrews and Pregibon (1978) could also be used, but do not provide much help when there are more than 30 observations.

The multiple-row residual statistic points to observations 7 and 46 and other observations with large residuals. As we have indicated earlier, this approach does not appear to provide adequate information about observations that influence the coefficients and fit. However, an examination of the larger residual correlations [cf. (2.67)] is useful for

Exhibit 2.20   Squared residual correlations: intercountry life-cycle savings data

Observations	Squared Residual Correlations
47 49	.173
6 44	.091
26 49	.049
39 44	.045
37 49	.043

qualitative assessment. Exhibit 2.20 contains the five largest squared residual correlations. The two largest, that between observations 47 and 49 and that between observations 6 and 44, confirm our earlier analysis. Of course, these calculations do not tell us the crucial information that observations 47 and 49 are influential and that observations 6 and 44 are consistent with the rest of the data and appear to be helpful in reducing variance.

Differentiation. The results for the derivative procedure, MEWDFFIT, given in Exhibit 2.21 point again to observations 47 and 49 as being influential. They do not indicate any other subgroup. We have included these results so that they may be compared to MDFFIT, which is considerably more expensive to compute. While the pair 23 and 46 is not as prominent with MEWDFFIT, we feel that MEWDFFIT provides a great deal of information when MDFFIT proves to be too expensive.

Exhibit 2.21   MEWDFFIT: intercountry life-cycle savings data

m	Subset	MEWDFFIT	Index Relative to Max
1	49	19.3	1.00
	23	10.3	0.53
	46	6.9	0.36
	21	3.2	0.20
	34	3.1	0.16
2	47 49	32.3	1.00
	24 49	27.3	0.85
	33 49	26.7	0.83
	23 46	25.7	0.80
3	24 47 49	42.8	1.00
	33 47 49	40.2	0.94
	23 47 49	36.1	0.84
	19 23 46	35.4	0.83
4	24 33 47 49	49.4	1.00
	33 34 47 49	48.2	0.98
	7 24 47 49	46.7	0.95
	19 23 39 46	45.3	0.92

Geometry. The geometric approaches are aimed more at "outliers" than at influential subsets because they adjoin **y** to **X** and ignore the regression context. Therefore we expect our results to be somewhat different.

The Wilks' Λ statistic still points to observations 47 and 49, as is seen in Exhibit 2.22. The second group of size 2 contains observations 6 and 44, and it is separated from 47 and 49. Although more testing is needed, there is reason to believe that this statistic is good for finding subsets of influential and outlying data points.

Exhibit 2.22 Wilks' Λ: intercountry life-cycle savings data

m	Subset	Λ	Index Relative to Min
1	49	0.46	1.00
	44	0.67	1.46
	23	0.74	1.61
	21	0.78	1.70
	46	0.80	1.74
2	47 49	0.39	1.00
	6 44	0.54	1.38
	21 49	0.57	1.46
	44 49	0.57	1.46
	23 49	0.59	1.51
3	24 47 49	0.45	1.00
	6 39 44	0.47	1.04
	23 47 49	0.49	1.09
	14 47 49	0.49	1.09

The Andrews-Pregibon statistic [see (2.77)] listed in Exhibit 2.23 points clearly to observation 49 but gives no evidence that there are any outlying subgroups. Of course, observations 44 and 49 and observations 47 and 49 are the top-ranked subsets of size 2, and a careful analysis would examine these more closely. This is especially indicated since the group for $m = 3$ contains observations 44, 47, and 49. Unfortunately, the role of observation 44 is confused in this statistic. It is indeed an outlier, but quite different from observations 47 and 49. Further analysis is needed to distinguish geometric outliers from influential observations.

We have not included any results from the stepwise distance measures because our focus is on influential observations rather than outliers. The stepwise deletion and derivative procedures seem to be better suited to identifying influential observations.

Exhibit 2.23 Andrews-Pregibon statistic Q: intercountry life-cycle savings data

m	Subset	Q	Index Relative to Min
1	49	0.45	1.00
	44	0.66	1.47
	23	0.73	1.62
	21	0.77	1.71
	46	0.78	1.73
2	44 49	0.30	1.00
	47 49	0.31	1.03
	23 49	0.33	1.10
	21 49	0.34	1.13
	46 49	0.36	1.20
3	44 47 49	0.20	1.00
	23 44 49	0.22	1.07
	23 47 49	0.22	1.10
	21 44 49	0.23	1.12

Final Comments

The question arises whether the approach taken here in detecting outliers is more effective than a simple examination of each data column for detached observations. The answer is clearly "yes". While extreme outliers (see p. 29) appear, for example, in the ΔDPI column of Exhibit 2.3 for Libya (49) and Jamaica (47), there is nevertheless no indication given here of their influence on other coefficients. Futhermore, additional influential observations (21 and 23) are revealed through the single-row deletion diagnostics. Jamaica, moreover, turns out to be a potentially troublesome data point even in the absence of Libya. Thus, if one restricts the analysis to single-row deletion procedures, it is prudent to reanalyze with the suspect points removed, to ascertain whether one or more observations have obscured the impact of others.

Partial-regression leverage plots provide qualitative information that pairs of observations, like (6, 44) and (47, 49), might be influential. The multiple-row diagnostics show, however, that between these two anomalous pairs, the naturally related pair, 6 (Canada) and 44 (U.S.), is not influential, while the not obviously related pair, 47 and 49, is.

APPENDIX 2A: ADDITIONAL THEORETICAL BACKGROUND

Deletion Formulas

The fundamental deletion formulas are known as the Sherman-Morrison-Woodbury Theorem [Rao (1973), p. 33]. Let $\mathbf{A}$ be a nonsingular matrix and $\mathbf{u}$ and $\mathbf{v}$ be two column vectors. Then

$$(\mathbf{A} - \mathbf{u}\mathbf{v}^T)^{-1} = \mathbf{A}^{-1} + \frac{\mathbf{A}^{-1}\mathbf{u}\mathbf{v}^T\mathbf{A}^{-1}}{1 - \mathbf{v}^T\mathbf{A}^{-1}\mathbf{u}}, \tag{2A.1}$$

or more specifically, for $\mathbf{A} = \mathbf{X}^T\mathbf{X}$, and $\mathbf{u} = \mathbf{v} = \mathbf{x}_i^T$,

$$(\mathbf{X}^T(i)\mathbf{X}(i))^{-1} = (\mathbf{X}^T\mathbf{X})^{-1} + \frac{(\mathbf{X}^T\mathbf{X})^{-1}\mathbf{x}_i^T\mathbf{x}_i(\mathbf{X}^T\mathbf{X})^{-1}}{1 - h_i}. \tag{2A.2}$$

From this comes [Miller (1964)]

$$\mathbf{b} - \mathbf{b}(i) = \frac{(\mathbf{X}^T\mathbf{X})^{-1}\mathbf{x}_i^T e_i}{1 - h_i}. \tag{2A.3}$$

Since

$$(n - p - 1)s^2(i) = \sum_{j \neq i}(y_j - \mathbf{x}_j\mathbf{b}(i))^2, \tag{2A.4}$$

we get, using (2A.3),

$$(n - p - 1)s^2(i) = \sum_{j=1}^{n}\left(e_j + \frac{h_{ij}e_i}{1 - h_i}\right)^2 - \frac{e_i^2}{(1 - h_i)^2}$$

$$= (n - p)s^2 + \frac{2e_i}{1 - h_i}\sum_{j=1}^{n}e_j h_{ij} + \frac{e_i^2}{(1 - h_i)^2}\sum_{j=1}^{n}h_{ij}^2 - \frac{e_i^2}{(1 - h_i)^2}$$

$$= (n - p)s^2 - \frac{e_i^2}{1 - h_i}, \tag{2A.5}$$

where we have used the fact that $\mathbf{H}$ annihilates the vector of residuals. A different proof of (2A.5) is given by Beckman and Trussell (1974).

To prove (2.20), repeated here for convenience,

$$\det\left(\mathbf{X}^T(i)\mathbf{X}(i)\right) = (1 - h_i)\det\left(\mathbf{X}^T\mathbf{X}\right), \qquad (2\text{A}.6)$$

we need first to show that

$$\det(\mathbf{I} - \mathbf{u}\mathbf{v}^T) = 1 - \mathbf{v}^T\mathbf{u}, \qquad (2\text{A}.7)$$

where $\mathbf{u}$ and $\mathbf{v}$ are column vectors. Let $\mathbf{Q}$ be an orthonormal matrix such that

$$\mathbf{Q}\mathbf{u} = \|\mathbf{u}\|\,\boldsymbol{\xi}_1, \qquad (2\text{A}.8)$$

where $\boldsymbol{\xi}_1$ is the first standard basis vector. Then

$$\det(\mathbf{I} - \mathbf{u}\mathbf{v}^T) = \det \mathbf{Q}[\mathbf{I} - \mathbf{u}\mathbf{v}^T]\mathbf{Q}^T \qquad (2\text{A}.9)$$

$$= \det\left[\mathbf{I} - \|\mathbf{u}\|\,\boldsymbol{\xi}_1\mathbf{v}^T\mathbf{Q}^T\right] = 1 - \mathbf{v}^T\mathbf{Q}^T\boldsymbol{\xi}_1\|\mathbf{u}\|, \qquad (2\text{A}.10)$$

which is just $1 - \mathbf{v}^T\mathbf{u}$ because of (2A.8). Now,

$$\det\left(\mathbf{X}^T(i)\mathbf{X}(i)\right) = \det\left[\left(\mathbf{I} - \mathbf{x}_i^T\mathbf{x}_i(\mathbf{X}^T\mathbf{X})^{-1}\right)\mathbf{X}^T\mathbf{X}\right]. \qquad (2\text{A}.11)$$

Hence, letting $\mathbf{u} = \mathbf{x}_i^T$ and $\mathbf{v}^T = \mathbf{x}_i(\mathbf{X}^T\mathbf{X})^{-1}$ completes the proof,[19] since $\mathbf{x}_i(\mathbf{X}^T\mathbf{X})^{-1}\mathbf{x}_i^T = h_i$. Another approach to this proof is outlined by Rao (p. 33).

Differentiation Formulas

Let

$$\mathbf{W} = \begin{bmatrix} 1 & & & & & & \\ & \ddots & & & & & \\ & & 1 & & & & \\ & & & w_i & & & \\ & & & & 1 & & \\ & & & & & \ddots & \\ & & & & & & 1 \end{bmatrix} \qquad (2\text{A}.12)$$

---

[19] We are indebted to David Gay for simplifying our original proof.

and

$$\mathbf{b}(w_i) = (\mathbf{X}^T\mathbf{W}\mathbf{X})^{-1}\mathbf{X}^T\mathbf{W}\mathbf{y}. \qquad (2A.13)$$

From (2A.1) we obtain

$$(\mathbf{X}^T\mathbf{W}\mathbf{X})^{-1} = (\mathbf{X}^T\mathbf{X})^{-1} + \frac{(1-w_i)(\mathbf{X}^T\mathbf{X})^{-1}\mathbf{x}_i^T\mathbf{x}_i(\mathbf{X}^T\mathbf{X})^{-1}}{1-(1-w_i)h_i}, \qquad (2A.14)$$

and hence

$$\frac{\partial}{\partial w_i}(\mathbf{X}^T\mathbf{W}\mathbf{X})^{-1} = \frac{-(\mathbf{X}^T\mathbf{X})^{-1}\mathbf{x}_i^T\mathbf{x}_i(\mathbf{X}^T\mathbf{X})^{-1}}{(1-(1-w_i)h_i)^2}. \qquad (2A.15)$$

Some algebraic manipulation using (2A.13) and (2A.14) gives

$$\mathbf{b}(w_i) = \mathbf{b} - (\mathbf{X}^T\mathbf{X})^{-1}\mathbf{x}_i^T e_i \frac{(1-w_i)}{1-(1-w_i)h_i}, \qquad (2A.16)$$

where $\mathbf{b}$ and e_i are the least-squares estimates obtained when $w_i = 1$. Thus

$$\frac{\partial\mathbf{b}(w_i)}{\partial w_i} = (\mathbf{X}^T\mathbf{X})^{-1}\frac{\mathbf{x}_i^T e_i}{(1-(1-w_i)h_i)^2}. \qquad (2A.17)$$

Theorems Related to the Hat Matrix

Size of the Diagonal Elements. Since $\mathbf{H}$ is a projection matrix, it is symmetric and idempotent ($\mathbf{H}^2 = \mathbf{H}$). Thus we can write

$$h_i \equiv h_{ii} = \sum_{j=1}^{n} h_{ij}^2 = h_{ii}^2 + \sum_{j \neq i} h_{ij}^2, \qquad (2A.18)$$

and it is clear that $0 \leqslant h_i \leqslant 1$. It is possible to go a little further. Let $\tilde{\mathbf{X}}$ denote the $n \times (p-1)$ matrix obtained by centering the explanatory variables. Then

$$\hat{\mathbf{y}} - \bar{\mathbf{y}} = \mathbf{H}\mathbf{y} - \bar{\mathbf{y}} = \tilde{\mathbf{H}}\mathbf{y}, \qquad (2A.19)$$

and therefore the elements of the centered hat matrix are

$$\tilde{h}_{ij} = h_{ij} - \frac{1}{n}. \qquad (2A.20)$$

This implies that $1/n \leqslant h_i \leqslant 1$. Sometimes $\tilde{h}_i$ is called the distance to the center of the data, since

$$\tilde{h}_i = (\mathbf{x}_i - \bar{\mathbf{x}})(\tilde{\mathbf{X}}^T\tilde{\mathbf{X}})^{-1}(\mathbf{x}_i - \bar{\mathbf{x}})^T. \qquad (2A.21)$$

It is easy to show that the eigenvalues of a projection matrix are either 0 or 1 and that the number of nonzero eigenvalues is equal to the rank of the matrix. In this case $\mathrm{rank}(\mathbf{H}) = \mathrm{rank}(\mathbf{X}) = p$ and hence trace $\mathbf{H} = p$ or

$$\sum_{i=1}^{n} h_i = p. \qquad (2A.22)$$

Distribution Theory. Wilks' Λ statistic [Rao (1973), p. 570] for two groups where one group consists of a single point is

$$\Lambda(\tilde{\mathbf{x}}_i) = \frac{\det\left(\tilde{\mathbf{X}}^T\tilde{\mathbf{X}} - (n-1)\bar{\tilde{\mathbf{x}}}^T(i)\bar{\tilde{\mathbf{x}}}(i) - \tilde{\mathbf{x}}_i^T\tilde{\mathbf{x}}_i\right)}{\det(\tilde{\mathbf{X}}^T\tilde{\mathbf{X}})}. \qquad (2A.23)$$

The numerator may be rewritten as

$$\det\left(\tilde{\mathbf{X}}^T\tilde{\mathbf{X}} - \frac{n^2}{(n-1)}\left(\bar{\tilde{\mathbf{x}}} - \frac{\tilde{\mathbf{x}}_i}{n}\right)^T\left(\bar{\tilde{\mathbf{x}}} - \frac{\tilde{\mathbf{x}}_i}{n}\right) - \tilde{\mathbf{x}}_i^T\tilde{\mathbf{x}}_i\right), \qquad (2A.24)$$

and using the fact that $\tilde{X}$ is centered, this reduces to

$$\det\left(\tilde{\mathbf{X}}^T\tilde{\mathbf{X}} - \frac{n}{n-1}\tilde{\mathbf{x}}_i^T\tilde{\mathbf{x}}_i\right). \qquad (2A.25)$$

Now we apply (2A.6) to show that (2A.25) is equal to

$$\left(1 - \frac{n}{n-1}\tilde{\mathbf{x}}_i(\tilde{\mathbf{X}}^T\tilde{\mathbf{X}})^{-1}\tilde{\mathbf{x}}_i^T\right)\det(\tilde{\mathbf{X}}^T\tilde{\mathbf{X}}), \qquad (2A.26)$$

and thus

$$\Lambda(\tilde{\mathbf{x}}_i) = 1 - \frac{n}{n-1}\tilde{h}_i = \frac{n}{n-1}(1 - h_i). \qquad (2A.27)$$

If the rows of $\tilde{\mathbf{X}}$ are assumed to be i.i.d. from a $(p-1)$-dimensional Gaussian distribution, then [Rao (1973), p. 570]

$$\frac{n-p}{p-1}\left[\frac{1 - \Lambda(\tilde{\mathbf{x}}_i)}{\Lambda(\tilde{\mathbf{x}}_i)}\right] \sim F_{p-1, n-p}, \qquad (2A.28)$$

and it follows that

$$\frac{n-p}{p-1} \frac{[h_i - (1/n)]}{(1-h_i)} \sim F_{p-1, n-p}. \tag{2A.29}$$

Dummy Variables and Singular Matrices. When a dummy variable consisting of all zeros except for a one at the ith observation is added to an **X** matrix of full rank to form a new matrix **N**, then the ith diagonal element of **H(N)**, the hat matrix for **N**, is 1. This follows from (2A.6) because deleting the ith row of **N** makes **N**(i) singular. If $h_i(\mathbf{N}) = 1$, then $e_i(\mathbf{N})$ is zero because $y_i(\mathbf{N}) = \hat{y}_i(\mathbf{N})$.

The least-squares coefficients using **N** are just those found by using **X**(i) plus a coefficient for the dummy column of **N** which is chosen to make $e_i(\mathbf{N})$ equal to zero. This shows that SSR(new model) $= (n-p-1)s^2(i)$, which is the result needed in (2.28).

To conclude this section we formally show that for any **X** matrix with $h_1 = 1$ (we can take $i = 1$ without loss of generality), there exists a nonsingular transformation **T**, such that $a_1 = (\mathbf{T}^{-1}\mathbf{b})_1 = y_1$ and $a_2, \ldots, a_p$ do not depend on y_1. This implies that, in the transformed coordinate system, the parameter estimate a_1 depends only on observation 1.

When $h_1 = 1$, we have for the coordinate vector $\boldsymbol{\xi}_1 = (1, 0, \ldots, 0)^T$,

$$\mathbf{H}\boldsymbol{\xi}_1 = \boldsymbol{\xi}_1, \tag{2A.30}$$

since $h_{1j} = 0$, $j \neq 1$. Let **P** be any $p \times p$ nonsingular matrix whose first column is $(\mathbf{X}^T\mathbf{X})^{-1}\mathbf{X}^T\boldsymbol{\xi}_1$. Then

$$\mathbf{XP} = \begin{bmatrix} 1 & \mathbf{c} \\ \mathbf{0} & \mathbf{A} \end{bmatrix}, \tag{2A.31}$$

where **c** is $1 \times (p-1)$ and **0** is $(p-1) \times 1$. Now let

$$\mathbf{C} = \begin{bmatrix} 1 & -\mathbf{c} \\ \mathbf{0} & \mathbf{I} \end{bmatrix}, \tag{2A.32}$$

with **I** denoting the $(p-1) \times (p-1)$ identity matrix. The transformation we seek is given by $\mathbf{T} = \mathbf{PC}$, which is nonsingular because both **P** and **C** have inverses. Now

$$\mathbf{XT} = \begin{bmatrix} 1 & \mathbf{0}^T \\ \mathbf{0} & \mathbf{A} \end{bmatrix}, \tag{2A.33}$$

and the least-squares estimate, **a**, of the parameters $\boldsymbol{\alpha} = \mathbf{T}^{-1}\boldsymbol{\beta}$ will have the first residual, $y_1 - a_1$, equal to zero because $h_1 = 1$. (The transformations do not affect the hat matrix.) Clearly $a_2, \ldots, a_p$ cannot affect this residual and thus $a_2, \ldots, a_p$ will not depend on y_1.

APPENDIX 2B: COMPUTATIONAL ELEMENTS[20]

In this computational appendix we describe algorithms that can be used to compute the diagnostic measures developed in Section 2.1. Three concerns shape our choice of algorithms for the single-row measures considered there: numerical stability, as described below; computational efficiency; and availability of crucial subroutines in widely distributed libraries. The first two of these concerns are best met by numerical analytic methods for solving linear least-squares problems, and we begin with a summary of the relevant results. For more detailed discussion of these topics the reader is directed to Golub (1969), Lawson and Hanson (1974), and Forsythe and Moler (1967). Algorithms for computing the diagonal elements of the hat matrix [see (2.15)] and the DFBETA [see (2.1)] are then presented. Summary descriptions of algorithms for computing the remaining diagnostic measures complete the treatment of the single-row techniques. Greatest detail is given to the computation of the quantities $\mathbf{b}$, $\mathbf{e}$, h_i, and $\mathbf{b} - \mathbf{b}(i)$.

The latter sections of this appendix describe the algorithms we have developed for calculating the multiple-row diagnostic measures. These procedures are the product of our ongoing research and have not yet been made part of a widely distributed software library. Rather than reproducing the FORTRAN codes here, we describe the algorithms in sufficient detail to guide the interested researcher to a straightforward implementation. Computational efficiency is the key concern in construction of these algorithms for reasons we consider in detail below. Fortunately, these efficient algorithms have stable numerical procedures for their kernels.

Computational Elements for Single-Row Diagnostics

Orthogonal Decompositions, the Least-Squares Solution, and Related Statistics. The linear least-squares problem (ordinary least-squares regression) may be cast in the following notation:

$$\min_{\mathbf{b}} \|\mathbf{y} - \mathbf{Xb}\|_2^2, \qquad (2\text{B}.1)$$

[20] This appendix was written by Stephen C. Peters.

where $\|\cdot\|_2$ is the Euclidean (square root of sum-of-squares) norm,
  $\mathbf{X}$ is the $n \times p$ matrix of explanatory variables,
  $\mathbf{y}$ is an n-vector of observations on the response variable, and
  $\mathbf{b}$ is a p-vector of coefficient estimates.

Since the Euclidean norm is invariant under an orthogonal transformation, a problem equivalent to (2B.1) is

$$\min_{\mathbf{b}} \|\mathbf{Q}\mathbf{y} - \mathbf{Q}\mathbf{X}\mathbf{b}\|_2^2, \tag{2B.2}$$

where $\mathbf{Q}$ is an $n \times n$ orthogonal matrix.

It is always possible to choose $\mathbf{Q}$ so that

$$\mathbf{Q}\mathbf{X} = \begin{bmatrix} \mathbf{R} \\ \mathbf{0} \end{bmatrix}, \tag{2B.3}$$

where $\mathbf{R}$ is $p \times p$ and upper triangular. Hence, if we define

$$\mathbf{Q}\mathbf{y} \equiv \begin{bmatrix} \mathbf{c} \\ \mathbf{d} \end{bmatrix}, \tag{2B.4}$$

where $\mathbf{c}$ is a p-vector, then a solution, $\mathbf{b}$, to (2B.1) necessarily satisfies

$$\mathbf{c} = \mathbf{R}\mathbf{b}. \tag{2B.5}$$

Further, when $\mathbf{X}$ has full rank p, this solution is unique and the triangular system is stably and efficiently solved by back-substitution. The algorithms suggested by the numerical analysts for determining $\mathbf{Q}$, $\mathbf{R}$, and $\mathbf{b}$ are stable in the sense that the computed solution, $\mathbf{b}$, is the exact solution to a nearby problem:

$$\min_{\mathbf{b}} \|(\mathbf{y} + \mathbf{f}) - (\mathbf{X} + \mathbf{E})\mathbf{b}\|_2^2, \tag{2B.6}$$

where $\mathbf{f}$ and $\mathbf{E}$ are, respectively, a vector and a matrix of very small perturbations.[21]

This method is preferred to solving the normal equations in the form

$$(\mathbf{X}^T\mathbf{X})\mathbf{b} = \mathbf{X}^T\mathbf{y}, \tag{2B.7}$$

since it avoids both the computation of the cross-products matrix, with the

---

[21] Relative to $\mathbf{X}$, the sizes of the elements of $\mathbf{E}$ are bounded by a small multiple of the relative precision of the (finite) machine arithmetic. Similarly, $\mathbf{f}$ is a vector of small perturbations relative to $\mathbf{y}$.

resulting rounding-error problems, and the need to solve a linear system whose conditioning may be worse by as much as the square of the condition of (2B.5).

The most reliable computation of $\mathbf{Q}$ and $\mathbf{R}$ is by the so-called QR decomposition algorithm employing Householder transformations [see Golub (1969) or Lawson and Hanson (1974)]. A frequent alternative, the modified Gram-Schmidt algorithm, is less useful for our needs since, without additional reorthogonalization steps, the $\mathbf{Q}$ delivered by modified Gram-Schmidt can differ greatly from orthogonality. Since we require an orthogonal $\mathbf{Q}$ matrix for subsequent computations, we prefer algorithms based on Householder transformations which do not suffer this defect [see Wilkinson (1965)]. The cost of determining the least-squares solution by Householder transformations is approximately $np^2 - (p^3/3)$ operations, where we adopt the convention that an operation is one floating-point multiplication and one floating-point addition. FORTRAN subroutines which reliably compute $\mathbf{Q}$ and $\mathbf{R}$ are available for a variety of machine types in the LINPACK (1979) and ROSEPACK (1980) subroutine libraries.

Directly related to the least-squares solution are the residuals e and the estimated coefficient covariance matrix $s^2(\mathbf{X}^T\mathbf{X})^{-1}$. The residuals are formed directly by

$$e = y - \mathbf{X}b \tag{2B.8}$$

and s^2 is calculated as

$$s^2 = e^T e / (n - p). \tag{2B.9}$$

We compute $(\mathbf{X}^T\mathbf{X})^{-1}$ (when $\mathbf{X}$ has full rank) by exploiting the QR decomposition, for if $\mathbf{Q}\mathbf{X} = \begin{bmatrix} \mathbf{R} \\ \mathbf{0} \end{bmatrix}$, then

$$\mathbf{X} = \mathbf{Q}^T \begin{bmatrix} \mathbf{R} \\ \mathbf{0} \end{bmatrix} \equiv \tilde{\mathbf{Q}}\mathbf{R}, \tag{2B.10}$$

where $\tilde{\mathbf{Q}}$ is the first p columns of $\mathbf{Q}^T$ and $\tilde{\mathbf{Q}}^T\tilde{\mathbf{Q}} = \mathbf{I}$. Then

$$(\mathbf{X}^T\mathbf{X})^{-1} = (\mathbf{R}^T\tilde{\mathbf{Q}}^T\tilde{\mathbf{Q}}\mathbf{R})^{-1} = (\mathbf{R}^T\mathbf{R})^{-1} = \mathbf{R}^{-1}\mathbf{R}^{-T}. \tag{2B.11}$$

The upper triangular matrix $\mathbf{R}$ is stably inverted by back-substitution against the identity matrix in $p^3/6$ operations. Multiplication by the transpose requires $p^3/6$ additional operations.

In Chapter 3 the Singular-Value Decomposition (SVD) is introduced in the development of diagnostics for collinearity. This orthogonal decomposition can also be advantageously used to solve the least-squares problem and to reliably identify rank deficiency. For a full treatment of the latter capability see Golub, Klema, and Peters (1980). The SVD determines orthogonal matrices $\mathbf{U}(n \times n)$ and $\mathbf{V}(p \times p)$ such that

$$\mathbf{U}^T \mathbf{X} \mathbf{V} = \begin{bmatrix} \mathbf{D} \\ \mathbf{0} \end{bmatrix}, \tag{2B.12}$$

where $\mathbf{D}$ is a $p \times p$ diagonal matrix whose nonnegative diagonal elements, μ_i are called the *singular values* of $\mathbf{X}$ (which in turn are equivalent to the square roots of the eigenvalues of $\mathbf{X}^T \mathbf{X}$). The least-squares solution is

$$\mathbf{b} = \mathbf{V} \mathbf{D}^{-1} \tilde{\mathbf{U}}^T \mathbf{y}, \tag{2B.13}$$

where $\tilde{\mathbf{U}}$ is the first p columns of $\mathbf{U}$. While algorithms that compute the SVD are computationally just as stable as those that compute the QR decomposition, their cost is significantly greater; $2np^2 + 4p^3$ operations are needed for the algorithm described in Golub and Reinsch (1970). FORTRAN subroutines which perform the SVD are available for a variety of machine types in EISPACK II (1976), LINPACK (1979), and ROSEPACK (1980).

For $n > 2p$, the cost of the SVD can be reduced by first using the QR decomposition. As is readily shown, the singular values of $\mathbf{X}$ are the same as those of $\mathbf{R}$. Hence, first apply the QR decomposition to $\mathbf{X}$ at a cost of $np^2 - (p^3/3)$ operations. Then compute the SVD of $\mathbf{R}$, at an additional cost of $6p^3$ operations, and determine the least-squares solution as

$$\mathbf{b} = \mathbf{V}_R \mathbf{D}_R^{-1} \mathbf{U}_R^T \tilde{\mathbf{Q}}^T \mathbf{y}, \tag{2B.14}$$

where $\tilde{\mathbf{Q}}$ is the first p columns of $\mathbf{Q}^T$ and the subscript R denotes the matrices that result when the SVD is applied to the matrix $\mathbf{R}$. Finally,

$$(\mathbf{X}^T \mathbf{X})^{-1} = \mathbf{V}_R \mathbf{D}_R^{-2} \mathbf{V}_R^T \tag{2B.15}$$

is formed with $p^3/2$ operations.

The Diagonal Elements of the Hat Matrix. Once the decomposition $\mathbf{QX} = \begin{bmatrix} \mathbf{R} \\ \mathbf{0} \end{bmatrix}$ has been performed the hat matrix is readily seen to be

$$\mathbf{X}(\mathbf{X}^T \mathbf{X})^{-1} \mathbf{X}^T = \tilde{\mathbf{Q}} \mathbf{R}(\mathbf{R}^{-1} \mathbf{R}^{-T}) \mathbf{R}^T \tilde{\mathbf{Q}}^T = \tilde{\mathbf{Q}} \tilde{\mathbf{Q}}^T, \tag{2B.16}$$

where $\tilde{Q}$ is the first p columns of Q^T. The diagonal elements of $\tilde{Q}\tilde{Q}^T$, that is, the h_i, are just the row sum-of-squares of $\tilde{Q}$:

$$h_i = \sum_{j=1}^{p} \tilde{Q}_{ij}^2, \qquad i = 1, \ldots, n. \qquad (2B.17)$$

The h_i may be simply calcuated by forming, in turn, each column of $\tilde{Q}$, squaring its n elements, and accumulating these new squares, so that $\tilde{Q}$ need never be stored in its entirety.

When $QX = \begin{bmatrix} R \\ 0 \end{bmatrix}$ has been determined by Householder transformations, the matrix Q is formed as a product of p elementary symmetric orthogonal transformations, M_k, where $k = 1 \ldots p$; that is,

$$Q = M_p M_{p-1} \cdots M_2 M_1. \qquad (2B.18)$$

It is a condensed form of the M's that are stored and returned by the algorithms cited. To find $\tilde{Q}$ we note

$$Q^T = (M_p M_{p-1} \cdots M_2 M_1)^T = M_1 M_2 \cdots M_{p-1} M_p. \qquad (2B.19)$$

The first p columns of Q^T (i.e., $\tilde{Q}$) are found by applying the elementary transformations $M_1 M_2 \cdots M_{p-1} M_p$ to the first p columns of the identity matrix. The cost of forming the h_i in this way is about $2np^2$ operations. FORTRAN subroutines which compute the diagonal elements of the hat matrix after the QR decomposition are available in the LINPACK and ROSEPACK subroutine libraries.

When the singular-value decomposition of R has been formed to deal with ill-conditioning and collinearity, the same procedure may be followed to compute the h_i, since

$$X(X^T X)^{-1} X^T = \tilde{Q} U_R D_R V_R^T (V_R D_R^{-2} V_R^T) V_R D_R U_R^T \tilde{Q}^T = \tilde{Q}\tilde{Q}^T. \qquad (2B.20)$$

If the entire hat matrix is desired, $\tilde{Q}$ should be formed explicitly and multiplied by its transpose in $n^2 p / 2$ operations.

Computing the **DFBETA.** The QR decomposition can also be used to advantage in the computation of the DFBETA as given in (2.1). The $p \times n$ matrix **B** whose columns contain the $\mathbf{b} - \mathbf{b}(i)$ is

$$B = (X^T X)^{-1} X^T G = (R^{-1} R^{-T}) R^T \tilde{Q}^T G = R^{-1} \tilde{Q}^T G, \qquad (2B.21)$$

where $\mathbf{G}$ is a diagonal matrix with $G_{ii} = e_i/(1-h_i)$ and $\tilde{\mathbf{Q}}$ is the first p columns of $\mathbf{Q}^T$. The columns of $\mathbf{B}$ may be found by backsolving the system

$$\mathbf{RB} = \tilde{\mathbf{Q}}^T\mathbf{G}. \tag{2B.22}$$

Having computed the h_i, $\tilde{\mathbf{Q}}$ is already available and $\mathbf{B}$ is formed with approximately $np^2/2$ additional operations. Note that care is needed in computing $e_i/(1-h_i)$ when h_i is very near unity, but this occurrence is seldom found in practice.

When the SVD has been applied to $\mathbf{R}$, $\mathbf{B}$ is obtained from

$$\mathbf{B} = \mathbf{V}_R\mathbf{D}_R^{-1}\mathbf{U}_R^T\tilde{\mathbf{Q}}^T\mathbf{G} \tag{2B.23}$$

using straightforward matrix multiplication in about np^2 operations.

One other strategy for computing $\mathbf{b} - \mathbf{b}(i)$ is noteworthy. Given the QR factorization of $\mathbf{X}$, one can apply updating and downdating procedures to $\mathbf{R}$ and $\mathbf{y}$ effectively to remove row i. The resulting triangular system can then be solved for $\mathbf{b}(i)$. Updating and downdating techniques require about $(9/2)p^2$ operations for each removal [see Chambers (1971)].

Exhibit 2B.1 summarizes the costs for computing the basic regression and diagnostic elements just described. The remaining measures developed in Chapter 2 are all easily computed by combining these basic elements.

Exhibit 2B.1   Summary of computational costs

Element	Algorithm	Approximate Cost
Basic Regression Elements		
$\mathbf{b}$	$\left\{\begin{array}{l}QR\text{ decomposition}\\ \text{SVD}\end{array}\right.$	$np^2 - (p^3/3)$ $2np^2 + 4p^3$
$\mathbf{e}$	matrix multiplication	np
s^2	inner product $\mathbf{e}^T\mathbf{e}/(n-p)$	n
$(\mathbf{X}^T\mathbf{X})^{-1}$	$\left\{\begin{array}{l}\text{back substitution using }\mathbf{R}\\ \mathbf{VD}^{-2}\mathbf{V}^T\text{ from SVD}\end{array}\right.$	$p^3/3$ $p^3/2$
Basic Diagnostics Elements		
h		$2np^2$
$\mathbf{b} - \mathbf{b}(i)$	$\left\{\begin{array}{l}\text{After }QR\\ \text{After SVD}\end{array}\right.$	$np^2/2$ np^2

Exhibit 2B.1 Continued

Derived Diagnostic Elements

Element	Formula	Approximate Cost
$s^2(i)$	$s^2(i) = \dfrac{n-p}{n-p-1}s^2 - \dfrac{e_i^2}{(n-p-1)(1-h_i)}$	$\mathcal{O}(n)$
DFBETAS	$\dfrac{\mathbf{b} - \mathbf{b}(i)}{s(i)\sqrt{(\mathbf{X}^T\mathbf{X})_{jj}^{-1}}}$	$\mathcal{O}(np)$
DFFIT	$\dfrac{h_i e_i}{1 - h_i}$	$\mathcal{O}(n)$
DFFITS	$\left(\dfrac{h_i}{1-h_i}\right)^{1/2}\dfrac{e_i}{s(i)\sqrt{1-h_i}}$	$\mathcal{O}(n)$
e_i^*	$\dfrac{e_i}{s(i)\sqrt{1-h_i}}$	$\mathcal{O}(n)$
COVRATIO	$\left[\left(\dfrac{n-p-1}{n-p} + \dfrac{e_i^{*2}}{n-p}\right)^p (1-h_i)\right]^{-1}$	$\mathcal{O}(n)$

The formulas for these measures are repeated and approximate costs are summarized. As indicated by the exhibit, the computational expense (in operation counts) for computing all diagnostic measures is roughly two and one-half times the cost of determining the basic regression elements $\mathbf{b}$, $\mathbf{e}$, s^2, and $(\mathbf{X}^T\mathbf{X})^{-1}$.

Computational Elements for Multiple-Row Diagnostics

In the second part of Section 2.1 it is noted that the influence of one point could be masked by another and that the true impact and nature of a group of influential observations might not be fully diagnosed by single-row diagnostics. This led to a consideration of techniques that examine the influence of subsets of more than one row. The computational burden of these multiple-row methods increases dramatically over that of the single-row measures. Whereas single-row diagnostic computations deal with only n cases and are usually bounded in complexity by simple, low-order polynomials in n (e.g., $\mathcal{O}(np^2)$), multiple-row techniques must consider vastly more cases. If influential groups of largest size m are suspected, there are $\sum_{k=1}^{m}\binom{n}{k}$ such cases for which calculations are required. Clearly, computing budgets (and patience!) can be rapidly depleted by problems having only moderate values of n and m.

Our computational approach to this problem is two-fold. First we identify a base set, B^*, of rows which we suspect participate in group influences. The size of B^* is substantially smaller than n, and our computations will involve only this smaller set of rows. The criteria whereby B^* is constructed are discussed in Section 2.1. Second, in our algorithms we trade increased storage for improved speed, and we employ algorithms which, overall, require a fixed number of arithmetic operations for any subset of B^* considered. This is the best that can be achieved when an exhaustive search is carried out over these subsets.

In Section 2.1 we described three different approaches for detecting influential groups: multiple-deletion, derivative measures, and geometric measures. The geometric approach to Wilks' Λ statistic is both computationally and conceptually the simplest, and it extends immediately to an algorithm for a derivative measure, so we consider its implementation in some detail below. With the structure of that algorithm in hand, a straightforward decomposition technique will then be introduced to provide the core of the algorithms for the remaining measures. There is a strong correspondence between these row-selection techniques and procedures for computing all-possible-(column)-subsets regression. The paper by Furnival and Wilson (1974) gives thorough consideration to this related area.

Notation and the Subset Tree. As noted above, we restrict our consideration of influential-observation groups to a base set of rows, B^*, extracted from the entire sample. The size of B^*, say m^*, is usually substantially smaller than n. All possible subsets of m elements taken from B^* are then examined for $m = 1, 2, \ldots, m_{max} \leqslant m^*$ ($m_{max} \approx m^*/2$ has generally been chosen in our quantative work). The organization of subset selections utilized in our algorithms is illustrated in Exhibit 2B.2 for a base set B^* which contains four elements (rows) and where $m_{max} = 3$. Each node in the tree corresponds to a selected subset. Following Furnival and

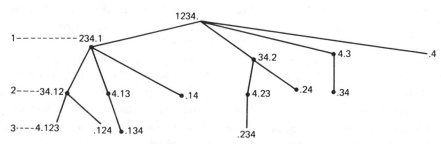

Exhibit 2.B.2   The subset tree.

Wilson (1974), a dot notation similar to that employed for partial correlation coefficients is used to label the nodes in the tree. The indexes following the dot specify the particular subset taken from B^*. Preceding the dot are the indexes of the elements of B* that are available for selection, but have not yet appeared as part of a subset in the path from the root (the root, 1234, is just the null subset). Hence, 34.2 represents a subset containing a single element (element 2) and further specifies that elements 3 and 4 are available for selection in subsets descendant from this node. Missing indexes ("1" in this case) indicate that the corresponding element has been *removed from consideration* as a participant in subsets on the paths to the terminal nodes.

It is not difficult to see that, at horizontal level 1, the tree depicts the $\binom{4}{1} = 4$ possible subsets of 1 element chosen from B^*; at horizontal level 2, the $\binom{4}{2} = 6$ possible subsets of 2 elements; and so on. The number of nodes in a tree of this kind is

$$\sum_{k=1}^{m_{max}} \binom{m^*}{k}.$$

Our algorithms examine subsets by traversing this tree in what is commonly called a lexicographic or preordering: visit the parent node, then its eldest (leftmost) child. This is applied recursively, i.e., parent, eldest child, eldest grandchild. When a terminal node is reached, the next younger sibling is visited or, if there is none, the parent's next younger sibling, and so on. The search is depth first, and in the case of Exhibit 2B.2, yields the subset sequence: 1234, 234.1, 34.12, 4.123, .124, 4.13, .134,..., .4. The traversal algorithm is simply implemented with the use of a push-down stack. [See Knuth (1973).]

It is crucial to recognize that from any subset, the constituents of descendant subsets visited by this traversal are known (their indexes precede the dot) and the number of these available elements *decreases* with the depth of the search. Our algorithms at a given subset will "work ahead" on these known constituents allowing for considerable economies in computation at the (generally more numerous) descendant subsets.

An Algorithm for the Geometric Measure, Wilks' Λ. In Section 2.1, (2.75) was derived for computing Wilks' Λ statistic from a subset of observations D taken from B^*. Recall that $\tilde{Z}$ is the matrix [y X] centered by column means and $\tilde{P}$ the projection matrix $\tilde{Z}(\tilde{Z}^T\tilde{Z})^{-1}\tilde{Z}^T$. Let $\tilde{P}^*$ be the $m^* \times m^*$ submatrix of $\tilde{P}$ chosen according to the elements of B^*.

Following the procedures specified earlier in this appendix, $\tilde{\mathbf{P}}^*$ can be formed once and for all in $\mathcal{O}(np^2) + \mathcal{O}(m^{*2}p)$ operations. Let l_1 be an m-vector consisting of ones where elements of B^* appear in D and with zeros elsewhere. Then $\Lambda(D)$ depends on D through the quadratic form $l_1^T \tilde{\mathbf{P}}^* l_1$, and we now consider the efficient computation of such quantities over all subsets D chosen from B^* with size $m \leqslant m_{\max}$.

The quadratic form $l_1^T \tilde{\mathbf{P}}^* l_1$ is itself easy to compute since it involves only the (unweighted) sum of elements of $\tilde{\mathbf{P}}^*$. The summands are, of course, the elements of the $m \times m$ submatrix of $\tilde{\mathbf{P}}^*$ selected by D. Since this matrix is symmetric, there are $m(m+1)/2$ distinct elements to be summed. A naive algorithm that computes the sum for each subset independently of the others requires about

$$\sum_{k=1}^{m_{\max}} \frac{k^2 + k}{2} \binom{m^*}{k}$$

floating-point additions.

A somewhat more sophisticated approach takes advantage of the fact that, having computed $l_1^T \tilde{\mathbf{P}}^* l_1$ for a particular D, when a subset, D^+, with one additional element is selected (e.g., any child of D in Exhibit 2B.2), only a "stripe" of $\tilde{\mathbf{P}}^*$ need be summed with the previous $l_1^T \tilde{\mathbf{P}}^* l_1$ to obtain the new statistic. Exhibit 2B.3 illustrates the previously summed submatrix selected by D and the additional summation required for D^+.

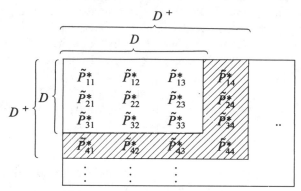

Exhibit 2B.3 The transition from $\Lambda(D)$ to $\Lambda(D^+)$

The stripe summed with the previous $l_1^T \tilde{\mathbf{P}}^* l_1$ is shaded.

As we traverse the tree, we save, for use by all its children, the value of $l_1^T \tilde{\mathbf{P}}^* l_1$ computed for each parent on the path back to the root. This requires, at most, m_{max} additional storage locations. The improved algorithm will require about

$$\sum_{k=1}^{m_{max}} k \binom{m^*}{k}$$

floating-point additions.

A further economy can be achieved for a given subset D by taking into account the elements of B^* that might appear in descendant subsets (their indexes precede the dot in Exhibit 2B.2). For each parent in the path back to the root we store, in addition to the value of $l_1^T \tilde{P}^* l_1$ computed, the partial sums $\sum_i \tilde{P}_{ij}^*$, where i ranges over the indexes of elements already in D (i.e., those following the dot) and j ranges over the indexes which are yet to be selected (i.e., those preceding the dot). Updating the statistic as one goes from parent to a child requires the addition of only three terms. The summands are the value of $l_1^T \tilde{P}^* l_1$ for the parent, the diagonal of $\tilde{P}^*$ selected by the index of the new element appearing in the offspring, and twice the partial-sum corresponding to the index i of the new element. Referring to Exhibit 2B.3, we have effectively collapsed the column standing above $\tilde{P}_{44}^*$ by forming the partial-sum $\tilde{P}_{14}^* + \tilde{P}_{24}^* + \tilde{P}_{34}^*$ in the previous stages. Each partial-sum can be updated by a single addition to give the new partial-sum stored for the child. The partial-sum corresponding to the new offspring element is excluded from the update so that the number of sums so maintained decreases as the tree is descended. Fewer than $m^* \cdot m_{max}$ additional storage locations are required to store the previous statistics and partial-sums. It can be shown that this algorithm requires about

$$4 \sum_{k=1}^{m_{max}} \binom{m^*}{k}$$

floating-point additions. This is proportional to the number of subsets considered and we have found no way to reduce further the number of operations. We look in further research to tree-pruning or branch-and-bound techniques that might allow us to exclude subtrees based on characteristics of their parents. An algorithm for computing the derivative measure in (2.73) can be constructed in a completely analogous way. The matrix whose elements are $h_{ij} e_i e_j / (1 - h_i)(1 - h_j)$ for $i, j \in B^*$ simply takes the place of $\tilde{P}^*$ in the above procedure.

Dummy Variables, Sequential Choleski Decomposition, and the Andrews-Pregibon Statistic. The Andrews-Pregibon statistic for detecting influential-observation groups is [see (2.77)]

$$Q(D_m) = \frac{\text{SSR}(D) \det(\mathbf{X}^T(D)\mathbf{X}(D))}{\text{SSR} \det(\mathbf{X}^T\mathbf{X})},$$

where D is again an m-element subset of rows taken from B^*. Two results from linear algebra point toward a computational strategy. The first concerns the determinant

$$\det(\mathbf{X}^T(D)\mathbf{X}(D)) = \det\begin{pmatrix} \mathbf{X}^T\mathbf{X} & \mathbf{X}_D^T \\ \mathbf{X}_D & \mathbf{I} \end{pmatrix} = \det\left([\mathbf{X}\ \mathbf{E}_D]^T[\mathbf{X}\ \mathbf{E}_D]\right),$$

$$(2B.24)$$

where $\mathbf{X}_D$ is the $m \times p$ matrix of the influential rows specified by D, and $\mathbf{E}_D$ is an $n \times m$ matrix consisting of columns of the identity matrix where ones appear in the rows specified by D. These columns are the dummy variables discussed in Section 2.1. We observe that the determinant of the row-reduced cross-products matrix is identically that of a suitably column-augmented cross-products matrix.

The second result also depends on a duality relating row-reduced and column-augmented cross-products matrices. Consider the (symmetric) cross-products matrix $[\mathbf{X}\ \mathbf{E}_D\ \mathbf{y}]^T[\mathbf{X}\ \mathbf{E}_D\ \mathbf{y}]$; a triangular decomposition for it can be effected for which the following identity holds:

$$\begin{bmatrix} [\mathbf{X}\ \mathbf{E}_D]^T[\mathbf{X}\ \mathbf{E}_D] & [\mathbf{X}\ \mathbf{E}_D]^T\mathbf{y} \\ \mathbf{y}^T[\mathbf{X}\ \mathbf{E}_D] & \mathbf{y}^T\mathbf{y} \end{bmatrix} = \begin{bmatrix} \mathbf{L}_D & \mathbf{0} \\ \mathbf{l}_D^T & \lambda_D \end{bmatrix}\begin{bmatrix} \mathbf{L}_D^T & \mathbf{l}_D \\ \mathbf{0}^T & 1 \end{bmatrix},$$

$$(2B.25)$$

where $\mathbf{L}_D$ is $(p+m) \times (p+m)$ upper triangular and $\mathbf{l}_D$ is $(p+m) \times 1$. It can be shown that λ_D is the residual sum-of-squares for the least-squares regression of $\mathbf{y}$ on $[\mathbf{X}\ \mathbf{E}_D]$, and that *this is identically SSR(D)*. Furthermore, it follows that

$$\det\left([\mathbf{X}\ \mathbf{E}_D]^T[\mathbf{X}\ \mathbf{E}_D]\right) = \det(\mathbf{L}_D\mathbf{L}_D^T), \qquad (2B.26)$$

and using (2B.24) the determinant of $\mathbf{X}^T(D)\mathbf{X}(D)$ is just the square of the product of the diagonals of the Choleski factor, $\mathbf{L}_D$.

It is possible to organize the computation of the decomposition and the formation of the Andrews-Pregibon statistic so that economies similar to those detailed in the previous section can be obtained when all subsets D taken from B^* of size $m \leqslant m_{max}$ are examined. Specifically, in the tree traversal, the sequential Choleski method is employed which operates on the cross-products matrix $[\mathbf{X}\ \mathbf{E}_{B^*}\ \mathbf{y}]^T[\mathbf{X}\ \mathbf{E}_{B^*}\ \mathbf{y}]$ and forms, at each node, a column of $\mathbf{L}_D$ corresponding to the new element appearing in the subsets. It then adjusts a submatrix (specified by the constituents of possible *descendent* subsets) for the effect of the column just considered. The adjustments are simply elementary row operations just like Gaussian elimination. Initially, p stages serve to decompose $\mathbf{X}^T\mathbf{X}$ (the leading principal submatrix). Further stages result in the augmented regressions as the tree is traversed. The adjustments to submatrices are analogous to the formation of partial-sums described previously. That is, adjustments are performed only over those elements which are members of descendent subsets; and the number of these elements, hence the work for each adjustment, decreases with depth in the tree.

As in the previous algorithm, results computed at parent nodes in the path back to the root must be stored. Here the storage cost is somewhat greater since adjusted submatrices, rather than vectors, must be maintained. Roughly $(m^{*2}/2)m_{max}$ storage locations are required. It can be shown that an algorithm constructed in this way will require approximately $14\sum_{k=1}^{m_{max}}\binom{m^*}{k}$ operations, where an operation consists of a floating-point multiplication followed by a floating-point addition (we have taken the square-root operation to be equivalent to six floating-point operations). The Choleski algorithm is numerically stable in the sense that the computed triangular factors are exact for a matrix very near to the cross-products matrix being decomposed, differing only by a small multiple of the machine rounding error. It is critically important, however, that the cross-products matrix itself be computed and stored with extra precision, since only in this way can we be assured that it accurately represents the configuration of $[\mathbf{X}\ \mathbf{E}_{B^*}\ \mathbf{y}]$.

Further Elements Computed from the Triangular Factors. When the triangular factorization, (2B.25), has been determined, a number of other group-influence measures are readily available at only moderate additional cost. The generalized fit statistic, (2.60), is

$$\text{MDFFIT} \equiv [\mathbf{b}-\mathbf{b}(D)]^T\mathbf{X}^T(D)\mathbf{X}(D)[\mathbf{b}-\mathbf{b}(D)].$$

Bingham (1977) provides several useful identities, one of which yields

$$\text{MDFFIT} = e_D^T U_D e_D,$$

where e_D is the $m \times 1$ subvector of the ordinary least-squares residuals selected by D, and $U_D \equiv X_D[X^T(D)X(D)]^{-1}X_D^T$, is an $m \times m$ matrix. Now,

$$\left([X \ E_D]^T[X \ E_D]\right)^{-1} = \left(L_D L_D^T\right)^{-1}$$

$$= \begin{bmatrix} [X^T(D)X(D)]^{-1} & -[X^T(D)X(D)]^{-1}X_D^T \\ -X_D[X^T(D)X(D)]^{-1} & I + X_D[X^T(D)X(D)]^{-1}X_D^T \end{bmatrix}.$$

Notice that $I + U_D$ appears as the $k \times k$ trailing submatrix. Exploiting the triangular factorization, we then have

$$\text{MDFFIT} = \begin{bmatrix} 0^T & e_D^T \end{bmatrix} L_D^{-T} L_D^{-1} \begin{bmatrix} 0 \\ e_D \end{bmatrix} - e_D^T e_D,$$

where 0 is a p-vector of zeros. The product $\begin{bmatrix} 0 \\ c_D \end{bmatrix} = L_D^{-1} \begin{bmatrix} 0 \\ e_D \end{bmatrix}$ is most stably and economically determined by back-substitution in $L_D \begin{bmatrix} 0 \\ c_D \end{bmatrix} = \begin{bmatrix} 0 \\ e_D \end{bmatrix}$, whence, $\text{MDFFIT} = c_D^T c_D - e_D^T e_D$. The back-substitution involves just the trailing $k \times k$ lower triangle of L_D. These computations can again be arranged so that much of the back-substitution work for descendant subsets can be performed at the parent nodes. This work actually proceeds concurrently with the formation of L_D. Fewer than $m^* \cdot m_{max}$ additional storage locations are required for storing intermediate results (essentially the c_D). The operation count is $16 \sum_{k=1}^{m_{max}} \binom{m^*}{k}$ operations.

***Inequalities Related to* MDFFIT..** Since $X^T(D)X(D) = X^TX - X_D^TX_D$, we obtain [Rao (1973), p. 33],

$$[X^T(D)X(D)]^{-1} = (X^TX)^{-1} - (X^TX)^{-1}X_D^T[X_D(X^TX)^{-1}X_D - I]^{-1}X_D(X^TX)^{-1},$$

and therefore,

$$U_D = X_D[X^T(D)X(D)]^{-1}X_D^T = H_D + H_D(I - H_D)^{-1}H_D.$$

If we decompose $\mathbf{H}_D$ as

$$\mathbf{H}_D = \mathbf{V}^T \mathbf{S}_1 \mathbf{V},$$

where $\mathbf{S}_1$ is $\mathrm{diag}\{s_i\}$ and $\mathbf{V}$ is orthonormal, then

$$\mathbf{U}_D = \mathbf{V}^T \mathbf{S}_2 \mathbf{V}$$

with $\mathbf{S}_2 = \mathrm{diag}\left\{ \dfrac{s_i}{1-s_i} \right\}$.

When all $s_i \neq 1$,

$$\text{MDFFIT} = \sum_{i \in D} \frac{s_i}{1-s_i} (\mathbf{V}\mathbf{e}_D)_i^2.$$

Several useful inequalities result from this. One example occurs if trace $\mathbf{H}_D \equiv \sum_{i \in D} h_i < 1$. Then

$$\text{MDFFIT} \leqslant \frac{\displaystyle\sum_{i \in D} h_i}{1 - \displaystyle\sum_{i \in D} h_i} \sum_{i \in D} e_i^2,$$

since $\max_{i \in D} s_i \leqslant$ trace $\mathbf{H}_D$. This is related to an inequality for (2.59) derived by Cook and Weisberg (1979).

Similar arguments can be used to show that

$$\sum_{i,j \in D_m} h_{ij} e_i e_j \leqslant \sum_{i \in D_m} h_i \sum_{i \in D_m} e_i^2,$$

and

$$\Lambda(D_m) = 1 - \frac{n}{m(n-m)} \mathbf{l}^T \tilde{\mathbf{P}} \mathbf{l}$$

$$\geqslant 1 - \frac{n}{n-m} \sum_{i \in D_m} \left[\tilde{h}_i + \frac{e_i^2}{SSR} \right]$$

$$= \frac{n}{n-m} \left[1 - \sum_{i \in D_m} \left(h_i + \frac{e_i^2}{SSR} \right) \right],$$

where $SSR = \displaystyle\sum_{k=1}^{n} e_k^2$.

Although still computationally complex, inequalities of this form can reduce the overall computational load, since they are based only on the h_i and the e_i and can remove many subsets from consideration. The reader is reminded, however, that once these subsets have been excluded, the algorithms developed above for dealing with all subsets cannot be used efficiently. Trade-offs among these approaches have not been established, but branch-and-bound techniques hold promise.

CHAPTER 3

Detecting and Assessing Collinearity

In this chapter we develop means (1) for detecting one or more collinear relations among a set of explanatory variables $\mathbf{X}$ to be used in a linear regression, (2) for identifying the subsets of explanatory variables involved in each collinear relation, and (3) for assessing the extent to which each of the least-squares estimates $\mathbf{b} = (\mathbf{X}^T\mathbf{X})^{-1}\mathbf{X}^T\mathbf{y}$ is potentially harmed by the presence of such collinear relations.

3.1 INTRODUCTION AND HISTORICAL PERSPECTIVE

No precise definition of collinearity[1] has been firmly established in the literature. Literally, two variates are collinear if the data vectors representing them lie on the same line (i.e., subspace of dimension one). More generally, k variates are collinear if the vectors that represent them lie in a subspace of dimension less than k, that is, if one of the vectors is a linear combination of the others. In practice, such "exact collinearity" rarely occurs, and it is certainly not necessary in order that a collinearity problem exist. A broader notion of collinearity is therefore needed to deal with the problem as it affects statistical estimation. More loosely, then, two variates are collinear if they lie almost on the same line, that is, if the angle

---

[1]The terms *collinearity, multicollinearity,* and *ill conditioning* are all used to denote this situation. The well-defined term of the numerical analysts, *ill conditioning,* would seem to be the term of choice, but *collinearity* is so widely employed that it is difficult to ignore it altogether and we use it frequently. The term *ill conditioning* will also be employed once it is defined. The term *multicollinearity* is also popular, particularly in econometrics, but, being inherently redundant, this term has been avoided in this work.

between them is small. In the event that one of the variates is not constant, this is equivalent to saying that they have a high correlation between them. We can readily generalize this notion to more than two variates by saying that collinearity exists if there is a high multiple correlation when one of the variates is regressed on the others.[2] This intuitive view of collinearity is adequate for the immediate discussion, but we eventually define collinearity in terms of the *conditioning* of the data matrix $\mathbf{X}$.

It is clear from the preceding discussion that collinearity has to do with specific characteristics of the data matrix $\mathbf{X}$ and not the statistical aspects of the linear regression model $\mathbf{y} = \mathbf{X}\boldsymbol{\beta} + \boldsymbol{\varepsilon}$. That is, collinearity is a data problem, not a statistical problem. This, together with the fact that, in some applications of regression models, experimental data can be obtained that are relatively free from collinearity, has allowed the topic of collinearity to receive very uneven textbook treatment. In many applications of linear regression, however, such as in econometrics, oceanography, geophysics, and other fields that rely on nonexperimental data, collinear data frequently arise and cause problems. Collinearity, then, is a nonstatistical problem that is nevertheless of great importance to the efficacy of least-squares estimation.

Intuitively we can understand the potential harm that results from collinear data by realizing that the collinear variates do not provide information that is very different from that already inherent in the others. It becomes difficult, therefore, to infer the separate influence of such explanatory variates on the response variate.

---

[2]An ambiguity immediately arises in defining the coefficient of multiple correlation. Let $\mathbf{e}$ be the residuals that result when any n-vector $\mathbf{z}$ is regressed on any set of p n-vectors $\mathbf{W}_i$ comprising the columns of the $n \times p$ matrix $\mathbf{W}$, that is, $\mathbf{e} \equiv \mathbf{z} - \mathbf{W}\mathbf{c}$, where $\mathbf{c} = (\mathbf{W}^T\mathbf{W})^{-1}\mathbf{W}^T\mathbf{z}$. Consider the two measures

$$(a) \quad R^2 \equiv 1 - \frac{\mathbf{e}^T\mathbf{e}}{\mathbf{z}^T\mathbf{z}}$$

and

$$(b) \quad \tilde{R}^2 \equiv 1 - \frac{\mathbf{e}^T\mathbf{e}}{\mathbf{z}^T\mathbf{A}\mathbf{z}},$$

where $\mathbf{A} \equiv \mathbf{I} - (\boldsymbol{\iota}\boldsymbol{\iota}^T/n)$, $\boldsymbol{\iota}$ a vector of ones. In the latter measure, the matrix $\mathbf{A}$ causes the $\mathbf{z}$-data to be centered about their mean. Each of these measures has variously been defined to be the square of the coefficient of multiple correlation. The latter, however, lacks generality, applying only when there is a constant column in $\mathbf{W}$. Since the diagnostics developed later in this chapter require no such limitation on the data matrix, we, along with Theil (1971, p. 164), adopt definition (a) as our measure of the squared coefficient of multiple correlation. Should the data already be centered about their means, of course, the two measures will coincide.

We can view the nature of collinearity geometrically with reference to Exhibit 3.1. Here we have pictured several situations relevant to the model

$$y_i = \beta_0 + \beta_1 x_{i1} + \beta_2 x_{i2} + \varepsilon_i,$$

that is, the $p = 3$ case. In these exhibits we show scatters of the n data points. In the x_1, x_2 "floor" are the (x_1, x_2) scatters (points denoted by •), while above we show the data "cloud" that results when the y dimension is included (points denoted by ∘). Exhibit 3.1a depicts the well-behaved case where the x_1 and x_2 data are not collinear. The data cloud above provides a well-defined least-squares plane, that is, the plane that minimizes the sum of squared errors in the y direction between the actual y_i and the plane. The y-intercept of this plane estimates β_0, and the partial slopes in the x_1 and x_2 directions, respectively, estimate β_1 and β_2. Since the plane is well defined, the various parameters are estimated with precision. Exhibit 3.1b depicts a case of perfect collinearity between x_1 and x_2. The data cloud now has no width, and the resulting least-squares plane is not defined; any

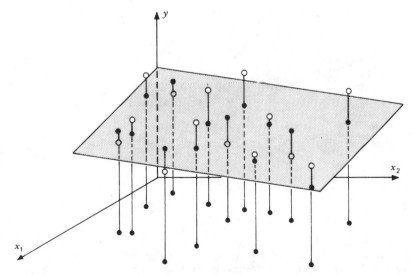

Exhibit 3.1a   No collinearity—all regression coefficients well-determined. A small change in any parameter of the regression plane will cause a relatively large change in the residual sum-of-squares.

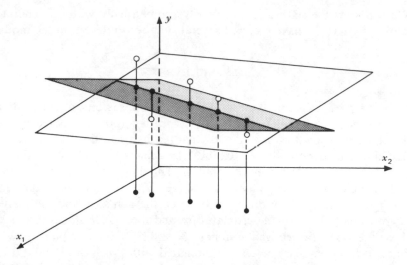

Exhibit 3.1b Exact collinearity—all regression coefficients undetermined. A simultaneous change in all the parameters of the regression plane can leave the residual sum-of-squares unchanged.

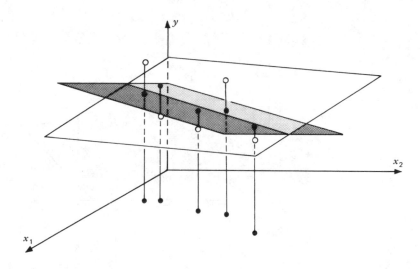

Exhibit 3.1c Strong collinearity—all regression coefficients ill-determined. A simultaneous change in all the parameters of the regression plane can cause little change in the residual sum-of-squares.

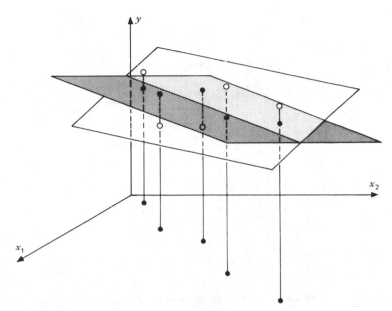

Exhibit 3.1d Strong collinearity—constant term well-determined. Only changes in the slope parameters of the regression plane can leave the residual sum-of-squares little affected.

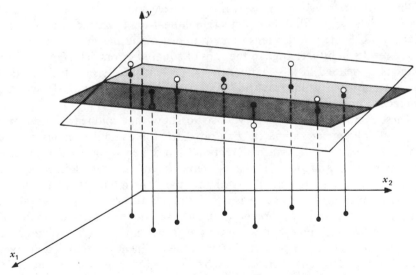

Exhibit 3.1e Strong collinearity—b_2 well-determined. Only changes in the intercept and b_1 parameters of the regression plane can leave the residual sum-of-squares little affected.

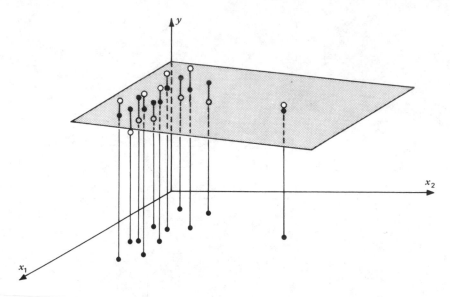

Exhibit 3.1f   Strong collinearity—except for an outlier.

plane lying along the "axis" of the cloud results in the same minimum sum of squared errors. This illustrates the well-known fact that perfect collinearity destroys the uniqueness of the least-squares estimator. Exhibit 3.1c depicts strong (but not perfect) collinearity. Here the least-squares plane is ill defined in the sense that tilting it along the major axis of the cloud results in little change in the sum of squared errors. The fact that the plane is ill defined in this manner translates statistically into the fact that the least-squares estimates (the y-intercept and the partial slopes) are imprecise; that is, they have high variance.

These simple illustrations serve also to show that collinearity need not harm all parameter estimates. Exhibit 3.1d, for example, shows a case where the partial slopes are ill defined—and hence one has imprecise estimates of β_1 and β_2—but the y-intercept remains well defined as the plane is tilted along the cloud axis. Here, then, the intercept term remains estimated with precision. Likewise, in Exhibit 3.1e, we have a case where the partial slope in the x_2 direction remains well defined. Here estimates of β_0 and β_1 will lack precision while that of β_2 will not. A little thought here will convince the reader that, in this case, the collinear relation is no longer between x_1 and x_2, but rather is now between x_1 and the constant term. In Exhibit 3.1f we have included a situation that suggests the potential interaction of the diagnostics for collinearity with those developed in Chapter 2 for influential data points. We say more about this interaction in Chapters 4 and 5.

It is clear, then, that the ability to diagnose collinearity is important to many users of least-squares regression. Such diagnosis consists of two related but separate elements: (1) detecting the presence of collinear relationships among the data series, and (2) assessing the extent to which these relationships have degraded[3] estimated parameters. Diagnostic information of this sort would aid the investigator in determining whether and where corrective action is necessary and worthwhile.[4] This chapter presents and examines a method for treating both diagnostic elements. First, it provides numerical indexes whose magnitudes signify the presence of one or more collinear or near dependencies[5] among the columns of a data matrix X; second, it provides a means for determining, within the linear regression model, the extent to which each such near dependency is potentially degrading the least-squares estimation of each regression coefficient. In most instances this latter information also enables the investigator to determine specifically which columns of the data matrix are involved in each near dependency; that is, it isolates the variates involved (and therefore also those not involved) and the specific relationships in which they are included.

Overview

The remainder of this section places the work reported here in its historical perspective. Section 3.2 provides an analytic background for the concepts employed and culminates in an empirically based procedure for diagnosing the presence of collinearity and assessing its potential harm to regression estimates. The basic building blocks employed in developing the diagnostic techniques of Section 3.2 have long been known to numerical analysts, but only recently are they becoming part of the working vocabulary of

---

[3]This term is given meaning later on.

[4]While the emphasis of this book is on diagnostics, remedial or corrective mechanisms are discussed and exemplified in Chapter 4.

[5]The relations among the data that result in collinearity or ill conditioning have variously been referred to as *multicollinear relations, collinear relations, dependencies, near dependencies, near collinearity,* and *near singularity.* The first of these terms is avoided here in line with footnote 1. The last of these terms is specious, since a matrix is either singular or not. The remaining terms are used interchangeably here with a preference for *near dependency.* This term is defined in the next section.

econometrics and statistics. Section 3.3 presents a series of experiments designed to illuminate the empirical properties of the collinearity diagnostics suggested in Section 3.2 and results in a set of interpretive tools. Finally, the process is summarized and exemplified in Section 3.4. There we learn, for example, that everything that macroeconomists thought was bad about the data used to estimate the consumption function is true.

Historical Perspective

Many procedures have been employed to detect collinearity.[6] We discuss here those most commonly used and indicate certain of their problems and weaknesses.

1. Hypothesized signs are incorrect, "important" explanatory variables have low t-*statistics, and various regression results are sensitive to deletion of a row or a column of* **X**.

Any of the above conditions is frequently cited as evidence of collinearity, and, even worse, collinearity is often cited as an explanation for these conditions. Unfortunately, none of these conditions is either necessary or sufficient for the existence of collinearity, and more refined techniques are required both to detect the presence of collinearity and to assess its potentially harmful effects.

2. Examine the correlation matrix **R**, *of the explanatory variables—or the inverse of this correlation matrix,* **R**$^{-1}$.

Examining the correlation matrix of the explanatory variables is a commonly employed procedure since this matrix is a standard output of most regression packages. If we assume the **X** data have been centered and scaled to have unit length, the correlation matrix **R** is simply $X^T X$. While a high correlation coefficient between two explanatory variates can indeed point to a possible collinearity problem, the absence of high correlations cannot be viewed as evidence of no problem. It is clearly possible for three or more variates to be collinear while no two of the variates taken alone are highly correlated. The correlation matrix is wholly incapable of diagnosing such a situation. We encounter a phenomenon of this sort in

---

[6]The reader who is interested in a more extensive survey of the collinearity problem is referred to Kumar (1975a).

Section 3.4, where we examine the consumption-function data. The correlation matrix is also unable to reveal the presence of several coexisting near dependencies among the explanatory variates. High pairwise correlations between, say, X_1 and X_2 and between X_3 and X_4 could be due to a single near dependency involving all four variates, or to two separate near dependencies, one between X_1 and X_2 and one between X_3 and X_4. The procedures we develop in the following section allow separate near dependencies to be uncovered as well as provide information on the involvement of each variate in each such near dependency.

Clearly the shortcomings just mentioned in regard to the use of R as a diagnostic measure for collinearity would seem also to limit the usefulness of R^{-1}, and this is the case. The prevalence of this measure, however, justifies its separate treatment. Recalling that we are currently assuming the X data to be centered and scaled for unit length, we are considering $R^{-1} = (X^T X)^{-1}$. The diagonal elements of R^{-1}, the r^{ii}, are often called the variance inflation factors, VIF_i, [Chatterjee and Price (1977)], and their diagnostic value follows from the relation

$$VIF_i = \frac{1}{1 - R_i^2},$$

where R_i^2 is the multiple correlation coefficient of X_i regressed on the remaining explanatory variables. Clearly a high VIF indicates an R_i^2 near unity, and hence points to collinearity.[7] This measure is therefore of some use as an overall indication of collinearity. Its weaknesses, like those of R, lie in its inability to distinguish among several coexisting near dependencies and in the lack of a meaningful boundary to distinguish between values of VIF that can be considered high and those that can be considered low.

3. The technique of Farrar and Glauber.

A diagnostic technique employing information from R and R^{-1} that has had great impact in the econometric literature and a growing impact in the

---

[7] The term "variance inflation factor" derives from the fact that the variance of the ith regression coefficient $\sigma_{\beta_i}^2$ obeys the relation

$$\sigma_{\beta_i}^2 = \frac{\sigma^2}{X_i^T X_i} VIF_i,$$

where σ^2 is the variance of the regression disturbance term. This is shown in Theil (1971, p. 166).

statistical literature is that of Farrar and Glauber (1967). We treat it separately here. Farrar and Glauber devised a measure of collinearity based on an assumption that the $n \times p$ data matrix $\mathbf{X}$ is a sample of size n from a p-variate Gaussian (normal) distribution. Under the further assumption that $\mathbf{X}$ has orthogonal columns, they contend that a transformation of $\det(\mathbf{R})$ is approximately χ^2 distributed, and hence provides a measure of the deviation from orthogonality or the presence of collinearity. In addition, they make use of the VIF_i (though not by that name) as indicators of the variates involved. They further propose the use of the measure $r_{ij.} \equiv -r^{ij}/(\sqrt{r^{ii}} \sqrt{r^{jj}})$, that is, the partial correlation between $\mathbf{X}_i$ and $\mathbf{X}_j$, adjusted for all other $\mathbf{X}$-variates, to investigate the patterns of interdependence in greater detail. The suggested technique of Farrar and Glauber falls prey to several interesting criticisms.

First, the use of $\mathbf{R}$ and the VIF_i and their weaknesses have already been discussed. The use of $\det(\mathbf{R})$ is, however, new to our discussion. But, like $\mathbf{R}$ itself, $\det(\mathbf{R})$ cannot diagnose the presence of several coexisting near dependencies, for the existence of any one near dependency will make $\det(\mathbf{R})$ near zero.

Second, Haitovsky (1969) has criticized a measure based on a deviation from orthogonality of the columns of $\mathbf{X}$, and has proposed a widely accepted change in emphasis to a measure of deviation from perfect singularity in $\mathbf{X}$. This change reflects the fact that the Farrar and Glauber measure frequently indicates collinearity when, as a practical matter, no problem exists. While the Haitovsky modification seemingly strengthens the Farrar and Glauber process, more recent criticism has proved more troublesome.

Third, Kumar (1975b) highlights (1) the obvious fact that the Farrar and Glauber technique, in assuming the data matrix $\mathbf{X}$ to be stochastic, has no relevance to the standard regression model in which $\mathbf{X}$ is assumed fixed, and (2) the less obvious fact that even when $\mathbf{X}$ is stochastic, a truly rigorous derivation of the statistic employed by Farrar and Glauber would depend on an assumption that the rows of the $\mathbf{X}$ matrix are independently distributed.

Fourth, the use of the elements $r_{ij.} \equiv -r^{ij}/(\sqrt{r^{ii}} \sqrt{r^{jj}})$ by Farrar and Glauber to display detail of variate involvement lacks discrimination. As we show in Appendix 3C, these correlations all approach unity (± 1) as collinearity becomes more troublesome. Thus $r_{ij.}$ can be near unity even if variates $\mathbf{X}_i$ and $\mathbf{X}_j$ are not involved in any collinear relation.

Fifth, in quite another vein, O'Hagan and McCabe (1975) question the validity of Farrar and Glauber's "statistical" interpretation of a measure of collinearity, concluding that their procedure misinterprets the use of a *t*-statistic as providing a cardinal measure of the severity of collinearity.

Indeed this latter criticism is correctly placed, for it is not proper, as some are wont, to interpret the Farrar and Glauber procedure as a statistical test for collinearity. Indeed Farrar and Glauber never make any such claim for their procedure. A statistical test, of course, must be based on a *testable hypothesis*; that is, the probability for the outcome of a relevant test statistic, calculated with actual sample data, is assessed in light of the distribution implied for it *by the model* under a specific null hypothesis. The Farrar and Glauber technique differs critically from this classical procedure exactly in the fact that the linear regression model makes no testable assumptions on the data matrix $\mathbf{X}$.[8] There are no distributional implications from the linear regression model for specific null hypotheses (such as orthogonality) on the nature of the data matrix $\mathbf{X}$ against which tests can be made.

The preceding highlights the fact (made clear in Appendix 3D) that collinearity can cause computational problems and reduce the precision of statistical estimates, but, in the context of the linear regression model, collinearity is not itself a statistical phenomenon subject to statistical test. A solution to the problem of diagnosing collinearity, then, must be sought elsewhere, in methods that deal directly with the numerical properties of the data matrix that can cause calculations based on it to be unstable or sensitive (in ways to be discussed later).

4. Examine Bunch Maps.

Historical completeness certainly requires inclusion of Ragnar Frisch's (1934) bunch-map analysis. Frisch's technique of graphically investigating the possible relationships among a set of data series was among the first major attempts to uncover the sources of near linear dependencies in economic data series. Frisch's work addresses itself to the first of collinearity's diagnostic problems—the location of dependencies—but makes no attempt to determine the degree to which regression results are degraded by their presence. Bunch-map analysis has not become a major tool in regression analysis because its extension to dependencies among more than two variates is time consuming and quite subjective.

---

[8]The data matrix $\mathbf{X}$ is, of course, assumed to have full rank, but this is not testable, for its absence, the null hypothesis, renders the regression model invalid.

5. Examine the eigenvalues and eigenvectors (or principal components) of the correlation matrix, $\mathbf{R}$.

For many years the eigensystem (eigenvalues and eigenvectors) of the cross-products matrix $\mathbf{X}^T\mathbf{X}$ or its related correlation matrix $\mathbf{R}$ has been employed in dealing with collinearity. Kloeck and Mennes (1960), for example, depict several ways of using the principal components of $\mathbf{X}$ or related matrices to reduce some ill effects of collinearity. In a direction more useful for diagnostic purposes, Kendall (1957) and Silvey (1969) have suggested using the eigenvalues of $\mathbf{X}^T\mathbf{X}$ as a key to the presence of collinearity: collinearity is indicated by the presence of a "small" eigenvalue. Unfortunately we are not informed what "small" is, and there is a natural tendency to compare small to the wrong standard, namely, zero. At least in some cases [e.g., Chatterjee and Price (1977)], it is indicated, but without justification, that collinearity may exist if "one eigenvalue is small in relation to the others." Here, small is interpreted in relation to larger eigenvalues rather than in relation to zero. This fundamental distinction lies at the heart of the methods discussed later in this chapter and requires much additional theory to provide its justification. This additional material comes from the extremely rich literature in numerical analysis showing the relevance of the *condition number* of a matrix (a measure related to the ratio of the maximal to minimal eigenvalues) to problems akin to collinearity.

A Basis for a Diagnostic

None of the above approaches has been fully successful at diagnosing the presence of collinearity or assessing its potential harm. The basis for a successful diagnostic is, however, close at hand, for various concepts developed in the field of numerical analysis are capable of putting real meaning to the measures discussed above based on eigenvalues.

Recent efforts of numerical analysts have provided a very useful set of tools for a rigorous examination of collinearity. While their attention has not been directly focused on collinearity, it has been directed at related topics. Numerical analysts are, for example, interested in the properties (conditioning) of a matrix $\mathbf{A}$ of a linear system of equations $\mathbf{A}\mathbf{z}=\mathbf{c}$ that allow a solution for $\mathbf{z}$ to be obtained with numerical stability. The relevance of this to collinearity in least squares is readily apparent, for the least-squares estimator is a solution to the linear system $(\mathbf{X}^T\mathbf{X})\mathbf{b}=\mathbf{X}^T\mathbf{y}$ with variance-covariance matrix $\sigma^2(\mathbf{X}^T\mathbf{X})^{-1}$. To the extent, then, that collinearity among the data series of $\mathbf{X}$ results in a matrix $\mathbf{A}=\mathbf{X}^T\mathbf{X}$ whose ill conditioning causes both the solution for $\mathbf{b}$ and its variance-covariance matrix to be numerically unstable, the techniques of the numerical analysts have direct bearing on understanding the econometrician's and statistician's problems with collinearity. The important efforts of the numerical analysts relevant to this study are contained in Businger and Golub (1965), Golub and Reinsch (1970), Lawson and Hanson (1974), Stewart (1973), and Wilkinson (1965).

Few of the techniques of the numerical analysts have been directly absorbed in applied statistical fields such as econometrics. This is strange since one of the principal tools of numerical analysis, the singular-value decomposition (SVD) of the data matrix X, has an intimate connection to the eigensystem of $X^T X$, a connection that we discuss in detail below. This lack of communication among the disciplines is explained in part by awkward differences in notation, and in part by seemingly different interests. Numerical analysts, for example, tend to work with nonstochastic equation systems. Furthermore, they have placed much of their emphasis on determining which columns of a data matrix can be discarded with least sacrifice to subsequent analysis,[9] a solution that is rarely open to the econometrician or applied statistician whose theory has already determined those variates that must always be present—or those that may be deleted, but not on grounds of numerical stability. Nevertheless, as we see, the numerical analysts' techniques, fused with the Kendall-Silvey line of research employing eigenvalues, have much to offer the user of least squares in diagnosing collinearity.

We recall from above that Silvey concludes that collinearity is discernable by the presence of a "small" eigenvalue of $X^T X$, a fact first noted by Kendall. This conclusion is correct, but falls short of the mark, since Silvey fails to inform us when an eigenvalue is small. We show that the numerical-analytic notion of the condition number can be applied to solve this shortcoming. Furthermore, Silvey provides the basis for diagnosing the involvement of the individual explanatory variates; namely, he examines a decomposition of the estimated variance of each regression coefficient in a manner that can illuminate the degradation of each coefficient caused by collinear relationships. Silvey, however, fails to exploit this use of the decomposition.

This chapter, then, (1) applies the relevant techniques of numerical analysis to Silvey's suggestion in order to provide a set of indexes (condition indexes) that signal the presence of one or more near dependencies among the columns of X and (2) adapts the Silvey regression-variance decomposition in a manner that can be combined with the above indexes to uncover those variates that are involved in particular

---

[9]See, for example, Hawkins (1973) or Golub, Klema, and Stewart (1976), or Webster, Gunst, and Mason (1974), (1976).

near dependencies and to assess the degree to which the estimated coefficients are being degraded by the presence of the near dependencies.

3.2   TECHNICAL BACKGROUND

In this section we develop the two principal tools of analysis employed in this chapter, the singular-value decompostion (SVD) of a matrix **X** (and its associated notion of the *conditioning* of **X**), and the decomposition of the estimated regression variance in a manner corresponding to the SVD. As noted, none of these concepts is new; the innovation is their combination in a manner that helps the user of linear regression solve the two diagnostic problems of collinearity stated at the outset: detection and assessment of damage.

The Singular-Value Decomposition

Any $n \times p$ matrix **X**, considered here to be a matrix of n observations on p variates, may be decomposed[10] as

$$\mathbf{X} = \mathbf{U}\mathbf{D}\mathbf{V}^T, \tag{3.1}$$

where $\mathbf{U}^T\mathbf{U} = \mathbf{V}^T\mathbf{V} = \mathbf{I}_p$ and **D** is diagonal with nonnegative diagonal elements μ_k, $k = 1, \ldots, p$,[11, 12] called the *singular values* of **X**. The preceding holds whether or not **X** has been scaled or centered. For the purposes of the collinearity diagnostics that follow, however, we shall discover (below and in Appendix 3B) that it is always desirable to scale **X** to have equal (unit) column lengths. Furthermore, if the data are relevant to a model *with a constant term*, **X** should contain uncentered data along with a column of ones; indeed, the use of the centered data matrix $\tilde{\mathbf{X}}$ in this situation is to be avoided, since centering can mask the role of the constant in any underlying near dependencies and produce misleading diagnostic

---

[10]See, for example, Golub (1969), Golub and Reinsch (1970), Hanson and Lawson (1969), and Becker et al. (1974).

[11]This decomposition is efficiently and stably effected by a program called MINFIT, a part of EISPACK II (1976).

[12]In (3.1) **U** is $n \times p$, **D** is $p \times p$, and **V** is $p \times p$. Alternative formulations are also possible and may prove more suitable to other applications. Hence one may have

$$\begin{array}{cccccccc} & n\times p & n\times n & n\times p & p\times p & & n\times p & n\times r & r\times r & r\times p \\ \mathbf{X} & = & \mathbf{U} & \mathbf{D} & \mathbf{V}^T & \text{or} & \mathbf{X} & = & \mathbf{U} & \mathbf{D} & \mathbf{V}^T \end{array}$$

where $r = \text{rank } \mathbf{X}$. In this latter formulation **D** is always of full rank, even if **X** is not.

results. Throughout this analysis X may include lagged values of the response variable y; that is, X is a matrix of predetermined variables.

The singular-value decomposition is closely related to the familiar concepts of eigenvalues and eigenvectors, but it has useful differences. Noting that $X^T X = V D^2 V^T$, we see that V is an orthogonal matrix that diagonalizes $X^T X$ and hence the diagonal elements of D^2, the squares of the singular values, must be the eigenvalues of the real symmetric matrix $X^T X$.[13] Further, the orthogonal columns of V must be the eigenvectors of $X^T X$ (and, similarly demonstrated, the columns of U are the p eigenvectors of XX^T associated with its p nonzero eigenvalues).

The singular-value decomposition of the matrix X, therefore, provides information that encompasses that given by the eigensystem of $X^T X$. As a practical matter, however, there are reasons for preferring the use of the singular-value decomposition. First, it applies directly to the data matrix X that is the focus of our concern, and not to the cross-product matrix $X^T X$. Second, as we shall see, the notion of a condition number of X is properly defined in terms of the singular values of X (spectral norm) and not the square roots of the eigenvalues of $X^T X$. Third, the concept of the singular-value decomposition has such great practical, as well as analytical, use in matrix algebra that we take this opportunity to extend knowledge of it to a wider audience than the numerical analysts. And fourth, whereas the eigensystem and the SVD of a given matrix are mathematically equivalent, computationally they are not. Algorithms exist that allow the singular-value decomposition of X to be computed with much greater numerical stability than is possible in computing the eigensystem of $X^T X$, particularly in the case focused upon here where X is ill conditioned.[14] As a practical matter, then, the collinearity diagnostics we discuss should always be carried out using the stable algorithm for the singular-value decomposition of X rather than an algorithm for determining the eigenvalues and eigenvectors of $X^T X$. The use of the SVD throughout this chapter reminds us of this fact.

Exact Linear Dependencies: Rank Deficiency. In the first instance let us assume X (which may contain a constant column of ones) has exact linear dependencies among its columns, a case rarely encountered in actual practice, so that rank $X = r < p$. Since, in the SVD of X, U and V are each orthogonal (and hence are necessarily of full rank), we must have rank X = rank D. There will therefore be exactly as many zero elements along

---

[13]See, for example, Theil (1971, p. 27).
[14]Furthermore, in operating directly on the $n \times p$ matrix X, the SVD avoids the additional computational burden of forming $X^T X$, an np^2 operation.

the diagonal of $\mathbf{D}$ as the nullity of $\mathbf{X}$, and the SVD in (3.1) may be partitioned as

$$\mathbf{X} = \mathbf{U}\mathbf{D}\mathbf{V}^T = \mathbf{U}\begin{bmatrix} \mathbf{D}_{11} & \mathbf{0} \\ \mathbf{0} & \mathbf{0} \end{bmatrix}\mathbf{V}^T, \tag{3.2}$$

where $\mathbf{D}_{11}$ is $r \times r$ and nonsingular. Postmultiplying by $\mathbf{V}$ and further partitioning, we obtain

$$\mathbf{X}\begin{bmatrix} \mathbf{V}_1 & \mathbf{V}_2 \end{bmatrix} = \begin{bmatrix} \mathbf{U}_1 & \mathbf{U}_2 \end{bmatrix}\begin{bmatrix} \mathbf{D}_{11} & \mathbf{0} \\ \mathbf{0} & \mathbf{0} \end{bmatrix}, \tag{3.3}$$

where $\mathbf{V}_1$ is $p \times r$, $\mathbf{U}_1$ is $n \times r$, $\mathbf{V}_2$ is $p \times (p-r)$, and $\mathbf{U}_2$ is $n \times (p-r)$. Equation (3.3) results in the two matrix equations

$$\mathbf{X}\mathbf{V}_1 = \mathbf{U}_1\mathbf{D}_{11} \tag{3.4}$$

$$\mathbf{X}\mathbf{V}_2 = \mathbf{0}. \tag{3.5}$$

Interest centers on (3.5), for it displays all of the linear dependencies of $\mathbf{X}$. The $p \times (p-r)$ matrix $\mathbf{V}_2$ provides an orthonormal basis for the null space associated with the columns of $\mathbf{X}$.

If , then, $\mathbf{X}$ possessed $p - r$ exact linear relations among its columns (and computers possessed exact arithmetic), there would also be exactly $p - r$ zero singular values in $\mathbf{D}$, and the variates involved in each of these dependencies would be determined by the nonzero elements of $\mathbf{V}_2$ in (3.5).

Needless to say, in most statistical applications, the interrelations among the columns of $\mathbf{X}$ are not exact dependencies, and computers deal in finite, not exact, arithmetic. Exact zeros for the singular values or for the elements of $\mathbf{V}_2$ will therefore rarely, if ever, occur. In general, then, it will be difficult to determine the nullity of $\mathbf{X}$ (as determined by zero μ's) or those columns of $\mathbf{X}$ which do not enter into specific linear relationships (as determined by the zeros of $\mathbf{V}_2$). Nevertheless, it is suggested in the foregoing that each near linear dependence among the columns of $\mathbf{X}$ will manifest itself in a small singular value, a small μ. This corresponds to Silvey's notion that the presence of collinearity is revealed by the existence of a small eigenvalue. The question now is to determine what is small. Although what ultimately is to be judged as large or small must remain an empirical question, we are greatly aided in answering this question by the notion of a *condition number* of a matrix $\mathbf{X}$.

The Condition Number.   Intuition is pressed to define a notion of an ill-conditioned matrix. One is tempted to say a matrix is ill conditioned if it "almost is not of full rank," or "if its inverse almost does not exist"—two obviously absurd statements. Yet in effect this is what is meant when it is said that an ill-conditioned square matrix is one with a small determinant (or an ill-conditioned rectangular matrix is one with a small $\det(X^TX)$). A small determinant, of course, has nothing to do with the invertibility of a matrix, for the matrix αI_n has as its determinant the number α^n, which can be made arbitrarily small; and yet it is clear that A^{-1} always exists for $\alpha \neq 0$ and is readily calculated as $\alpha^{-1}I_n$.[15]

A means for defining the conditioning of a matrix that accords somewhat with intuition and avoids the pitfalls of the above techniques is afforded by the singular-value decomposition. The motivation behind this technique derives from a more meaningful method for determining when an inverse of a given matrix "blows up." As we see, it becomes reasonable to consider a matrix A to be ill conditioned if the product of its spectral norm (defined below) with that of A^{-1} is large. This measure, called the *condition number* of A, provides summary information on the potential difficulties to be encountered in various calculations based on A; the larger the condition number, the more ill conditioned the given matrix. In particular, as we show below, the condition number provides a measure of the potential sensitivity of the solution vector z of the linear system $Az = c$ to small changes in the elements of c and A. Furthermore, as we show in Appendix 3A, the condition number of a data matrix X provides an upper bound on the elasticity (a measure of sensitivity frequently employed in economics) of the diagonal elements of the matrix $(X^TX)^{-1}$ with respect to any element of the data matrix X.

As an aid to understanding the relevance of the condition number, we summarize several of its important properties here in the context of solutions to linear systems of equations. We then provide two illuminating examples. For a more detailed treatment of what follows, along with the

---

[15]It is also sometimes thought that the ill conditioning of a given matrix can be discovered by the presence of small diagonal elements in a triangular factorization of the matrix. This, too, is not true. Two examples from Golub and Reinsch (1970) and Wilkinson (1965) illustrate this point. Consider

$$
\begin{bmatrix}
.501 & -1 & & & \\
 & .502 & -1 & & \Large 0 \\
 & & \ddots & & \\
\Large 0 & & & .599 & -1 \\
 & & & & .600
\end{bmatrix}
\quad \text{and} \quad
\begin{bmatrix}
1 & -1 & -1 & \cdots & -1 \\
 & & -1 & \cdots & -1 \\
 & & \ddots & & \\
 & & & & 1
\end{bmatrix}
$$

Each of these matrices can be shown by the singular-value decomposition, in a way described later, to be quite ill conditioned even though neither possesses a small diagonal element.

proofs and derivations that are omitted here, the reader is referred to Forsythe and Moler (1967).[16]

The familiar Euclidean norm of an n-vector $\mathbf{z}$, denoted $\|\mathbf{z}\|$, is defined as

$$\|\mathbf{z}\| \equiv (\mathbf{z}^T\mathbf{z})^{1/2}.$$

An important generalization of the Euclidean norm to an $n \times n$ matrix $\mathbf{A}$ is the *spectral norm*, denoted $\|\mathbf{A}\|$ and defined as

$$\|\mathbf{A}\| \equiv \sup_{\|\mathbf{z}\|=1} \|\mathbf{A}\mathbf{z}\|.$$

It is readily shown that $\|\mathbf{A}\| = \mu_{max}$, that is, the maximal singular value of $\mathbf{A}$. Similarly, if $\mathbf{A}$ is square, $\|\mathbf{A}^{-1}\| = 1/\mu_{min}$. Further, like the Euclidean norm, the spectral norm can be shown to be a true norm; that is, it possesses the following properties:

1. $\|\lambda\mathbf{A}\| = |\lambda| \cdot \|\mathbf{A}\|$ for all real λ and all $\mathbf{A}$.
2. $\|\mathbf{A}\| = 0$ if and only if $\mathbf{A} = \mathbf{0}$, the matrix of zeros.
3. $\|\mathbf{A} + \mathbf{B}\| \leqslant \|\mathbf{A}\| + \|\mathbf{B}\|$ for all $m \times n$ matrices $\mathbf{A}$ and $\mathbf{B}$.

And, in addition, the spectral norm obeys the following relations:

4. $\|\mathbf{A}\mathbf{z}\| \leqslant \|\mathbf{A}\| \cdot \|\mathbf{z}\|$ (which follows directly from the definition).
5. $\|\mathbf{A}\mathbf{B}\| \leqslant \|\mathbf{A}\| \cdot \|\mathbf{B}\|$ for all commensurate $\mathbf{A}$ and $\mathbf{B}$.

We shall now see that the spectral norm is directly relevant to an analysis of the conditioning of a linear system of equations $\mathbf{A}\mathbf{z} = \mathbf{c}$, $\mathbf{A}$ $n \times n$ and nonsingular with solution $\mathbf{z} = \mathbf{A}^{-1}\mathbf{c}$. We can ask how much the solution vector $\mathbf{z}$ would change ($\delta\mathbf{z}$) if there were small changes or perturbations in the elements of $\mathbf{c}$ or $\mathbf{A}$, denoted $\delta\mathbf{c}$ and $\delta\mathbf{A}$. In the event that $\mathbf{A}$ is fixed but $\mathbf{c}$ changes by $\delta\mathbf{c}$, we have $\delta\mathbf{z} = \mathbf{A}^{-1}\delta\mathbf{c}$, or

$$\|\delta\mathbf{z}\| \leqslant \|\mathbf{A}^{-1}\| \cdot \|\delta\mathbf{c}\|.$$

Further, employing property 4 above to the equation system, we have

$$\|\mathbf{c}\| \leqslant \|\mathbf{A}\| \cdot \|\mathbf{z}\|;$$

and from multiplying these last two expressions we obtain

$$\frac{\|\delta\mathbf{z}\|}{\|\mathbf{z}\|} \leqslant \|\mathbf{A}\| \cdot \|\mathbf{A}^{-1}\| \cdot \frac{\|\delta\mathbf{c}\|}{\|\mathbf{c}\|}.$$

---

[16]For even more detailed treatments, see Faddeeva (1959) and Wilkinson (1965).

That is, the magnitude $\|A\| \cdot \|A^{-1}\|$ provides a bound for the relative change in the length of the solution vector z that can result from a given relative change in the length of c. A similar result holds for perturbations in the elements of the matrix A. Here it can be shown that

$$\frac{\|\delta z\|}{\|z + \delta z\|} \leqslant \|A\| \cdot \|A^{-1}\| \cdot \frac{\|\delta A\|}{\|A\|}.$$

Because of its usefulness in this context, the magnitude $\|A\| \cdot \|A^{-1}\|$ is defined to be the *condition number* of the nonsingular matrix A and is denoted as $\kappa(A)$. The preceding expressions show that $\kappa(A)$ provides a measure of the potential sensitivity of the solution of a linear system of equations (which includes the least-squares normal equations) to changes in the elements of c and A of the linear system.

Further understanding of the condition number is provided by the following two examples. Consider first the matrix $A = \begin{bmatrix} 1 & \alpha \\ \alpha & 1 \end{bmatrix}$. Clearly, as $\alpha \to 1$, this matrix tends toward perfect singularity. The singular values of A are readily shown[17] to be $(1 \pm \alpha)$ and those of A^{-1} to be $(1 \pm \alpha)^{-1}$. Hence, as $\alpha \to 1$ the product $\|A\|\|A^{-1}\| = (1 + \alpha)(1 - \alpha)^{-1}$ explodes; A is ill conditioned for α near unity.

By way of contrast consider the admittedly well-conditioned matrix introduced above, $B = \begin{bmatrix} \alpha & 0 \\ 0 & \alpha \end{bmatrix}$. As we have seen, the often-held intuitive feeling that B becomes ill conditioned as $\alpha \to 0$ is incorrect, and this is correctly reflected in the condition number, for $\|B\| = \alpha$ and $\|B^{-1}\| = \alpha^{-1}$ and the product $\|B\|\|B^{-1}\| = \alpha\alpha^{-1} = 1$ as $\alpha \to 0$. In this case, then, the product of the spectral norm of B with that of B^{-1} does not blow up, and B is well conditioned for all $\alpha \neq 0$.

The conditioning of any square matrix A can be summarized, then, by a condition number $\kappa(A)$ defined as the product of the maximal singular value of A (its spectral norm) and the maximal singular value of A^{-1}. This concept is readily extended to a rectangular matrix and can be calculated without recourse to an inverse. From the SVD, $X = UDV^T$, it is easily shown that the generalized inverse X^+ of X is VD^+U^T, where D^+ is the generalized inverse of D and is simply D with its nonzero diagonal elements inverted.[18] Hence the singular values of X^+ are merely the reciprocals of those of X, and the maximal singular value of X^+ is the reciprocal of the minimum (nonzero) singular value of X. Thus for any

[17]It is readily apparent from the application of the SVD (3.1) to a real symmetric matrix A that the singular values of A are also its eigenvalues.
[18]See Golub and Reinsch (1970) or Becker et al. (1974).

$n \times p$ matrix $\mathbf{X}$ we may define its condition number to be

$$\kappa(\mathbf{X}) = \frac{\mu_{\max}}{\mu_{\min}} \geqslant 1. \tag{3.6}$$

It is readily shown that the condition number of any matrix with orthonormal columns is unity, and hence $\kappa(\mathbf{X})$ reaches its lower bound in this cleanest of all possible cases. Furthermore, it is clear that $\kappa(\mathbf{X}) = \kappa(\mathbf{X}^+)$, and hence the condition number has the highly desirable property of telling us the same story whether we are dealing with $\mathbf{X}$ or its (generalized) inverse.

Near Linear Dependencies: How Small is Small?   We have seen that for each exact linear dependency among the columns of $\mathbf{X}$ there is one zero singular value. Extending this property to near dependencies leads one to suggest, as did Kendall (1957) and Silvey (1969), that the presence of near dependencies will result in "small" singular values (or eigenvalues). This suggestion does not include a means for determining what small is. The preceding discussion of condition numbers, however, does provide a basis for assessing smallness. The degree of ill conditioning depends on how small the minimum singular value is relative to the maximum singular value; that is, $\mu_{\max}$ provides a yardstick against which smallness can be measured. In this connection, it proves useful to define

$$\eta_k \equiv \frac{\mu_{\max}}{\mu_k} \qquad k = 1, \ldots, p \tag{3.7}$$

to be the kth *condition index* of the $n \times p$ data matrix $\mathbf{X}$. Of course $\eta_k \geqslant 1$ for all k, with the lower bound necessarily occurring for some k. The largest value for η_k is also the condition number of the given matrix. A singular value that is small relative to its yardstick $\mu_{\max}$, then, has a high condition index.

We may therefore extend the Kendall-Silvey suggestion as follows: there are as many near dependencies among the columns of a data matrix $\mathbf{X}$ as there are high condition indexes (singular values small relative to $\mu_{\max}$). Two points regarding this extension must be emphasized.

First, we have not merely redirected the problem from one of determining when small is small to one of determining when large is large. As we saw above, taken alone, the singular values (or eigenvalues) shed no light on the conditioning of a data matrix: equally well-conditioned problems can have arbitrarily low singular values.[19] Determining when a singular value is small, then, has no relevance to determining the presence of a near dependency causing a data matrix to be ill conditioned. We did see, however, in our discussion of the condition number, that determining when a singular value is small relative to μ_{max} (or, equivalently, determining when a condition index is high) is directly related to this problem. The meaningfulness of the condition index in this context is verified in the empirical studies of Section 3.3.[20]

Second, even if there is measurable meaning to the term "large" in connection with condition indexes, there is no a priori basis for determining how large a condition index must be before there is evidence of collinear data or, even more importantly, evidence of data so collinear that its presence is degrading or harming regression estimates. Just what is to be considered a large condition index is a matter to be empirically determined, and the experiments of Section 3.3 are aimed at aiding such an understanding. There we learn that weak dependencies are associated with condition indexes around 5 or 10, whereas moderate to strong relations are associated with condition indexes of 30 to 100.

The use of the condition index, then, extends the Kendall–Silvey suggestion in two ways. First, practical experience will allow an answer to the question of when small is small (or large is large), and second, the simultaneous occurrence of several large η's keys the simultaneous presence of more than one near dependency.

The Regression-Coefficient Variance Decomposition

As we have seen, when any one singular value of a data matrix is small relative to μ_{max}, we interpret it as indicative of a near dependency (among the columns of **X**) associated with that singular value. In this section, reinterpreting and extending the work of Silvey (1969), we show how the estimated variance of each regression coefficient may be decomposed into a sum of terms each of which is associated with a singular value, thereby providing means for determining the extent to which near dependencies (having high condition indexes) degrade (become a dominant part of) each variance. This decomposition provides the link between the numerical analysis of a data matrix **X**, as embodied in its singular-value decomposition, and the quality of the subsequent regression analysis using

---

[19]The matrix $\mathbf{B} = \alpha\mathbf{I}$ employed above provides an excellent example here.

[20]The problem of determining when, as a practical matter, a matrix may be considered rank deficient (i.e., when μ_{min} may be considered *zero* relative to μ_{max}) is treated in Golub, Klema, and Stewart (1976).

$\mathbf{X}$ as a data matrix, as embodied in the variance-covariance matrix of $\mathbf{b}$.[21] We recall from our discussion in the introduction to this chapter that the first of these is an important extension to the use of eigenvalues and the second is an important extension to the use of the correlation matrix $\mathbf{R}$ and the VIFs.

The variance-covariance matrix of the least-squares estimator $\mathbf{b} = (\mathbf{X}^T\mathbf{X})^{-1}\mathbf{X}^T\mathbf{y}$ is, of course, $\sigma^2(\mathbf{X}^T\mathbf{X})^{-1}$, where σ^2 is the common variance of the components of ε in the linear model $\mathbf{y} = \mathbf{X}\boldsymbol{\beta} + \varepsilon$. Using the SVD, $\mathbf{X} = \mathbf{UDV}^T$, the variance-covariance matrix of $\mathbf{b}, \mathbf{V}(\mathbf{b})$, may be written as

$$\mathbf{V}(\mathbf{b}) = \sigma^2(\mathbf{X}^T\mathbf{X})^{-1} = \sigma^2\mathbf{VD}^{-2}\mathbf{V}^T, \tag{3.8}$$

or, for the kth component of $\mathbf{b}$,

$$\operatorname{var}(b_k) = \sigma^2 \sum_j \frac{v_{kj}^2}{\mu_j^2}, \tag{3.9}$$

where the μ_j's are the singular values and $\mathbf{V} \equiv (v_{ij})$.

Note that (3.9) decomposes $\operatorname{var}(b_k)$ into a sum of components, each associated with one and only one of the p singular values μ_j (or eigenvalues μ_j^2). Since these μ_j^2 appear in the denominator, other things being equal, those components associated with near dependencies—that is, with small μ_j —will be large relative to the other components. This suggests, then, that an unusually high *proportion* of the variance of two or more coefficients[22] concentrated in components associated with the same small singular value provides evidence that the corresponding near dependency is causing problems. Let us pursue this suggestion.

Define the k, jth *variance-decomposition proportion* as the proportion of the variance of the kth regression coefficient associated with the jth component of its decomposition in (3.9). These proportions are readily calculated as follows.

First let

$$\phi_{kj} \equiv \frac{v_{kj}^2}{\mu_j^2} \quad \text{and} \quad \phi_k \equiv \sum_{j=1}^{p} \phi_{kj} \qquad k = 1,\ldots,p. \tag{3.10}$$

---

[21]This link is obviously of the utmost importance, for ill conditioning is a numeric property of a data matrix having, in itself, nothing directly to do with least-squares estimation. To have meaning in a regression context, then, there must be some means by which the numeric information on ill conditioning can be shown to affect directly the quality (variances) of the regression estimates. It is this link, for example, that is lacking in the Farrar and Glauber techniques described in the introduction to this chapter.

[22]"Two or more," since there must be at least two columns of $\mathbf{X}$ involved in any dependency.

Exhibit 3.2 Variance-decomposition proportions

Associated Singular Value	Proportions of			
	$\mathrm{var}(b_1)$	$\mathrm{var}(b_2)$	$\cdots$	$\mathrm{var}(b_p)$
μ_1	π_{11}	π_{12}	$\cdots$	π_{1p}
μ_2	π_{21}	π_{22}	$\cdots$	π_{2p}
$\vdots$	$\vdots$	$\vdots$		$\vdots$
μ_p	π_{p1}	π_{p2}	$\cdots$	π_{pp}

Then, the variance-decomposition proportions are

$$\pi_{jk} \equiv \frac{\phi_{kj}}{\phi_k}, \qquad k,j = 1,\ldots,p. \tag{3.11}$$

The investigator seeking patterns of high variance-decomposition proportions will be aided by a summary table (a Π matrix) in the form of Exhibit 3.2.[23]

Notice that the π_{jk} make use of the SVD information on near dependencies in a way that is directly applicable to examining their effects on regression estimates.

Two Interpretive Considerations

Section 3.3 reports detailed experiments using the two tools developed here, the SVD and its associated Π matrix of variance-decomposition proportions. These experiments are designed to provide experience in the behavior of these tools when employed for analyzing collinearity, for detecting it, and for assessing the damage it has caused to regression estimates. Before proceeding to these experiments, however, it is necessary to develop two important interpretive properties of the Π matrix of variance-decomposition proportions. An example of these two properties completes this section.

Near Collinearity Nullified by Near Orthogonality. In the variance decomposition given by (3.9), small μ_j, other things equal, lead to large components of $\mathrm{var}(b_k)$. However, not all $\mathrm{var}(b_k)$ need be adversely affected by a small μ_j, for the v_{kj}^2 in the numerator may be even smaller. In

---

[23]For convenience in displaying the tables in this book, we have adopted the order of subscripts j and k as shown in (3.11). In an efficient computer code effecting these diagnostics, it may prove desirable to reverse their order, particularly when p is large.

the extreme case where $v_{kj} = 0$, var(b_k) would be unaffected by any near dependency among the columns of $\mathbf{X}$ that would cause μ_j to become even very small. As is shown by Belsley and Klema (1974), the $v_{kj} = 0$ when columns k and j of $\mathbf{X}$ belong to mutually orthogonal partitions. The proof to this intuitively plausible statement is lengthy and need not be repeated here, for it reflects a fact well known to users of linear regression; namely, that the introduction into regression analysis of a variate orthogonal to all preceding variates will not change the regression estimates or the true standard errors of the coefficients of the preceding variates.[24] Thus, if two very nearly collinear variates (near multiples of one another) that are also mutually orthogonal to all prior variates are added to a regression equation, the estimates of the prior coefficients and their variances must be unaffected. In terms of the variance decomposition in (3.9), this situation results in at least one μ_j (corresponding to the two closely collinear variates) which is very small, and which has no weight in determining any of the var(b_k), for k corresponding to the initially included variates. Clearly, the only way this can occur is for the v_{kj} between the prior variates and the additional variates to be zero. Hence we have the result that, in the SVD of $\mathbf{X} = [\mathbf{X}_1 \, \mathbf{X}_2]$ with $\mathbf{X}_1^T \mathbf{X}_2 = 0$, it is always possible to find[25] a $\mathbf{V}$ matrix with the form

$$\mathbf{V} = \begin{bmatrix} \mathbf{V}_{11} & \mathbf{0} \\ \mathbf{0} & \mathbf{V}_{22} \end{bmatrix}.$$

Thus we see that the bad effects of collinearity, resulting in relatively small μ's, may be mitigated for some coefficients by near orthogonality, resulting in small v_{kj}'s.

At Least Two Variates Must Be Involved.   At first it would seem that the concentration of the variance of any one regression coefficient in any one of its components could signal that collinearity may be causing problems. However, since two or more variates are required to create a near dependency, it must be that two or more variances are adversely

[24]This and the following statement refer to the true standard errors or variances taken from the diagonal elements of $\sigma^2(\mathbf{X}^T\mathbf{X})^{-1}$ and not to the estimated standard errors based on s^2, which will, of course, not necessarily remain unaltered. Of course, all that is needed here is the invariance of the diagonal elements of $(\mathbf{X}^T\mathbf{X})^{-1}$ that correspond to the preceding variates.
[25]The careful wording "it is always possible to find" is required here. As is shown in Belsley and Klema (1974), if there are multiple eigenvalues of $\mathbf{X}$, there is a class of $\mathbf{V}$'s in the SVD of $\mathbf{X}$, one, but not all, of which takes the partitioned form shown. Such multiplicities are therefore of theoretical importance but of little practical consequence since they will occur with probability zero for a "real life" economic data matrix.

affected by high variance-decomposition proportions associated with a single singular value (i.e., a single near dependency).

To illuminate this, consider a data matrix X with mutually orthogonal columns—the best possible experimental data. Our previous result immediately implies that the V matrix of the singular-value decomposition of X is diagonal, since all $v_{ij} = 0$ for $i \neq j$. Hence the associated Π matrix of variance-decomposition proportions must take the following form:

Associated Singular Value	Proportions of			
	$\mathrm{var}(b_1)$	$\mathrm{var}(b_2)$	$\cdots$	$\mathrm{var}(b_p)$
μ_1	1			
μ_2		1	0	
$\vdots$			$\ddots$	
μ_p	0			1

It is clear that a high proportion of any variance associated with a *single* singular value is hardly indicative of collinearity, for the variance-decomposition proportions here are those for an ideally conditioned, orthogonal data matrix. Reflecting the fact that two or more columns of X must be involved in any near dependency, the degradation of a regression estimate due to collinearity can be observed only when a singular value μ_j is associated with a large proportion of the variance of *two or more* coefficients. If, for example, in a case for $p = 5$, columns 4 and 5 of X are highly collinear and all columns are otherwise mutually orthogonal, we would expect a variance-decomposition Π matrix that has the following form:

Associated Singular Value	Proportions of				
	$\mathrm{var}(b_1)$	$\mathrm{var}(b_2)$	$\mathrm{var}(b_3)$	$\mathrm{var}(b_4)$	$\mathrm{var}(b_5)$
μ_1	1.0	0	0	0	0
μ_2	0	1.0	0	0	0
μ_3	0	0	1.0	0	0
μ_4	0	0	0	1.0	0.9
μ_5	0	0	0	0	0.1

Here μ_4 plays a large role in both $\mathrm{var}(b_4)$ and $\mathrm{var}(b_5)$.

An Example. An example of the preceding two interpretive considerations is useful at this point. Consider the 6×5 data matrix given in Exhibit 3.3.

Exhibit 3.3   The modified Bauer matrix

$$
\mathbf{X} = [\mathbf{X}_1\,\mathbf{X}_2] =
\begin{bmatrix}
-74 & 80 & 18 & -56 & -112 \\
14 & -69 & 21 & 52 & 104 \\
66 & -72 & -5 & 764 & 1528 \\
-12 & 66 & -30 & 4096 & 8192 \\
3 & 8 & -7 & -13276 & -26552 \\
4 & -12 & 4 & 8421 & 16842
\end{bmatrix}.
$$

This matrix, essentially due to Bauer (1971), has the property that its fifth column is exactly twice its fourth, and both of these are in turn orthogonal to the first three columns (which are not, however, orthogonal to each other). That is, $\mathbf{X}_2$ is of rank 1 and $\mathbf{X}_1^T\mathbf{X}_2 = \mathbf{0}$. We therefore know from the foregoing that, in the SVD of $\mathbf{X}$, (1) one of the singular values associated with $\mathbf{X}_2$ will be zero (i.e., within the machine tolerance of zero[26]), and (2) in $\mathbf{V} = \begin{bmatrix} \mathbf{V}_{11} & \mathbf{V}_{12} \\ \mathbf{V}_{21} & \mathbf{V}_{22} \end{bmatrix}$, $\mathbf{V}_{12} = \mathbf{V}_{21}^T = \mathbf{0}$.

Indeed application of the program MINFIT[27] to obtain the singular-value decomposition of $\mathbf{X}$ gives for $\mathbf{V}$ (partitioned)

$$
\begin{bmatrix}
0.548 & -0.625 & 0.556 & 0.15 \times 10^{-18} & -0.54 \times 10^{-14} \\
-0.836 & -0.383 & 0.393 & 0.22 \times 10^{-19} & -0.47 \times 10^{-14} \\
0.033 & 0.680 & 0.733 & 0.16 \times 10^{-18} & -0.73 \times 10^{-14} \\
 & & & & \\
-0.64 \times 10^{-15} & -0.22 \times 10^{-15} & 0.91 \times 10^{-14} & -0.477 & 0.894 \\
0.32 \times 10^{-15} & 0.10 \times 10^{-15} & -0.46 \times 10^{-14} & -0.894 & -0.447
\end{bmatrix}
$$

with the following singular values:[28]

$$\mu_1 = \quad 170.7$$

$$\mu_2 = \quad\ 60.5$$

$$\mu_3 = \quad\ \ 7.6$$

$$\mu_4 = 36368.4$$

$$\mu_5 = \quad 1.3 \times 10^{-12}.$$

---

[26]10^{-14} on the IBM 370/168 in double precision.

[27]Golub and Reinsch (1970), Becker et al. (1974), and EISPACK II, (1976).

[28]The reader is warned against interpreting the condition indexes from these singular values at this point. For reasons explained in Section 3.3 the data should first be scaled to have equal column lengths, and the resulting singular values subjected to analysis. For the analysis of this section, however, scaling is unnecessary.

Exhibit 3.4   Variance-decomposition proportions: modified Bauer matrix

Associated Singular Value	Proportions of				
	var(b_1)	var(b_2)	var(b_3)	var(b_4)	var(b_5)
μ_1	.002	.009	.000	.000	.000
μ_2	.019	.015	.013	.000	.000
μ_3	.976	.972	.983	.000	.000
μ_4	.000	.000	.000	.000	.000
μ_5	.003	.005	.003	1.000	1.000

A glance at V verfies that the off-diagonal blocks are small—all are of the order of 10^{-14} or smaller—and well within the effective zero tolerance of the computational precision. Only somewhat less obvious is that one of the μ_j is zero. Since μ_5 is of the order of 10^{-12}, it would seem to be nonzero relative to the machine tolerance, but, as we have seen, the size of each μ_j has meaning only relative to μ_{max}, and in this case $\mu_5 / \mu_{max} = \eta_5^{-1} < 10^{-16}$, well within the machine tolerance of zero.

The Π matrix of variance-decomposition proportions for this data matrix is given in Exhibit 3.4.

Several of its properties are noteworthy. First, we would expect that the small singular value μ_5 associated with the exact linear dependency between columns 4 and 5 (C4 $=0.5$ C5) would dominate several variances —at least those of the two variates involved—and this is seen to be the case; the component associated with μ_5 accounts for virtually all the variance of both b_4 and b_5.

Second, we would expect that the orthogonality of the first three columns of X to the two involved in the linear dependency would isolate their estimated coefficients from the deleterious effects of collinearity. Indeed, the components of these three variances associated with μ_5 are very small, .003, .005, and .003, respectively.[29] This point serves also to exemplify that the analysis suggested here aids the user in determining not only which regression estimates are degraded by the presence of collinearity, but also which are not adversely affected and may therefore be salvaged.

Third, a somewhat unexpected result is apparent. The singular value μ_3 accounts for 97% or more of var(b_1), var(b_2), and var(b_3). This suggests that a second near dependency is present in X, one associated with μ_3, that involves the first three columns. This, in fact, turns out to be the case. We

---

[29]That these components are nonzero at all is due only to the finite arithmetic of the machine. In theory these components are an undefined ratio of zeros that would be defined to be zero for this case.

reexamine this example in Section 3.4, once we have gained further experience in interpreting the magnitudes of condition indexes and variance-decomposition proportions.

Fourth, to the extent that there are two separate near dependencies in X (one among the first three columns, one between the last two), the Π matrix provides a means for determining which variates are involved in which near dependency. This property of the analytic framework presented here is important, because it is not true of alternative means of analyzing near dependencies among the columns of X. One could hope, for example, to investigate such near dependencies by regressing selected columns of X on other columns or to employ partial correlations. But to do this in anything other than a shotgun manner would require prior knowledge of which columns of X would be best preselected to regress on the others, and to do so when there are several coexisting near dependencies would prove a terrible burden. Typically, the user of linear regression, when presented with a specific data matrix, will have no rational means for preselecting offending variates. Fortunately the problem can be avoided entirely through the use of the Π matrix, which displays all such near dependencies, treating all columns of X symmetrically and requiring no prior information on the numbers of near dependencies or their composition.

A Suggested Diagnostic Procedure

The foregoing discussion suggests a practical procedure for (1) testing for the presence of one or more near dependencies among the columns of a data matrix, and (2) assessing the degree to which each such dependency potentially degrades the regression estimates based on that data matrix.

The Diagnostic Procedure. It is suggested that an appropriate means for diagnosing degrading collinearity is the following double condition:

1* A singular value judged to have a high condition index, and which is associated with

2* High variance-decomposition proportions for *two or more* estimated regression coefficient variances.

The number of condition indexes deemed large (say, greater than 30) in 1* identifies the number of near dependencies among the columns of the data matrix X, and the magnitudes of these high condition indexes provide a measure of their relative "tightness." Furthermore, the determination in 2* of large variance-decomposition proportions (say, greater than .5) associated with each high condition index identifies those variates that are involved in the corresponding near dependency, and the magnitude of these proportions in conjunction with the high condition index provides a measure of the degree to which the corresponding regression estimate has been degraded by the presence of collinearity.[30]

Examining the Near Dependencies.   Once the variates involved in each near dependency have been identified by their high variance-decomposition proportions, the near dependency itself can be examined—for example, by regressing one of the variates involved on the others. Another procedure is suggested by (3.5). Since V_2 in (3.5) has rank $(p - r)$, we may partition X and V_2 to obtain

$$\begin{bmatrix} X_1 & X_2 \end{bmatrix} \begin{bmatrix} V_{21} \\ V_{22} \end{bmatrix} = X_1 V_{21} + X_2 V_{22} = 0, \qquad (3.12)$$

where V_{21} is chosen nonsingular and square. Hence the dependencies among the columns of X are displayed as

$$X_1 = -X_2 V_{22} V_{21}^{-1} \equiv X_2 G \qquad \text{where } G \equiv -V_{22} V_{21}^{-1}. \qquad (3.13)$$

The elements of G, calculated directly from those of V, provide alternative estimates of the linear relation between those variates in X_1 and those in X_2. Of course, (3.12) holds exactly only in the event that the linear dependencies of X are exact. It is also straightforward to show in this event that (3.13) provides identical estimates to those given by OLS. It seems reasonable, therefore, because of the relative simplicity involved, to employ OLS as the descriptive mechanism (*auxiliary regressions*) for displaying the linear dependencies once the variates involved are discerned in 2*.

It is important to reiterate the point made earlier that OLS applied to columns of the X matrix does not and cannot substitute for the diagnostic procedure suggested above, for OLS can be rationally applied only after it has first been determined how many dependencies there are among the columns of X and which variates are involved. The diagnostic procedure suggested here requires no prior knowledge of the numbers of near dependencies involved or of the variates involved in each; it discovers this information—treating all columns of X symmetrically and requiring that none be chosen to become the "dependent" variable, as is required by OLS applied to columns of the X matrix.

---

[30]Section 3.3 is devoted to experiments that help us put meaning to "high" and "large," two terms whose meaning in this context can only be determined empirically and which necessarily must be used loosely here.

What is "large" or "high"?   Just what constitutes a "large" condition index or a "high" variance-decomposition proportion are matters that can only be decided empirically. We turn in Section 3.3 to a set of systematic experiments designed to shed light on this matter. To provide a meaningful background against which to interpret those empirical results, it is first useful to give a more specific idea of what it means for collinearity to harm or to degrade a regression estimate.

The Ill Effects of Collinearity

The ill effects that result from regression based on collinear data are two: one computational, one statistical.

Computational Problems.   When we introduced the condition number earlier in this chapter, we showed that it provides a measure of the potential sensitivity of a solution of a linear equation to changes in the elements of the system. Computationally, this means that solutions to a set of least-squares normal equations (or, in general, a solution to a system of linear equations) contain a number of digits whose meaningfulness is limited by the conditioning of the data in a manner directly related to the condition number.[31] Indeed, the condition number gives a multiplication factor by which imprecision in the data can work its way through to imprecision in the solution to a linear system of equations. Somewhat loosely, if data are known to d significant digits, and the condition number of the matrix $\mathbf{A}$ of a linear system $\mathbf{Az} = \mathbf{c}$ is of order of magnitude 10^r, then a small change in the data in its last place can (but need not) affect the solution $\mathbf{z} = \mathbf{A}^{-1}\mathbf{c}$ in the $(d - r)$th place. Thus, if GNP data are trusted to four digits, and the condition number of $(\mathbf{X}^T\mathbf{X})$ is 10^3, then a shift in the fifth place of GNP (which, since only the first four digits count, results in what must be considered an *observationally equivalent* data matrix), could affect the least-squares solution in its second $(5 - 3)$ significant digit. Only the first digit is therefore trustworthy, the others potentially being worthless, arbitrarily alterable by modifications in $\mathbf{X}$ that do not affect the degree of accuracy to which the data are known. Needless to say, had the condition number of $\mathbf{X}^T\mathbf{X}$ been 10^4 or 10^5 in this case, one could trust none of $\mathbf{c}$'s significant digits. This computational problem in the calculation of least-squares estimates may be minimized[32] but never removed. The

---

[31]A more detailed discussion of this topic is contained in Belsley and Klema (1974).

[32]The condition number of the moment matrix $\mathbf{X}^T\mathbf{X}$ is the square of that of $\mathbf{X}$. This is seen from the SVD of $\mathbf{X} = \mathbf{UDV}^T$. Hence $\mathbf{X}^T\mathbf{X} = \mathbf{VD}^2\mathbf{V}^T$, and, by definition, this must also be the SVD of $\mathbf{X}^T\mathbf{X}$. Clearly, then, $\kappa(\mathbf{X}^T\mathbf{X}) = \mu_{max}^2 / \mu_{min}^2 = \kappa^2(\mathbf{X})$. Hence, any ill conditioning of $\mathbf{X}$ can be greatly compounded in its ill effects on a least-squares solution calculated as $\mathbf{b} = (\mathbf{X}^T\mathbf{X})^{-1}\mathbf{X}^T\mathbf{y}$. Procedures for calculating $\mathbf{b}$ that do not require forming $\mathbf{X}^T\mathbf{X}$ or its inverse exist, however. See Golub (1969) or Belsley (1974).

intuitive distrust held by users of least squares of estimates based on ill-conditioned data is therefore justified. A discussion of this topic in the context of the Longley (1967) data is to be found in Beaton, Rubin, and Barone (1978).

Statistical Problems. Statistically, as is well known, the problem introduced by the presence of collinearity in a data matrix is the decreased precision with which statistical estimates conditional[33] upon those data may be known; that is, collinearity causes the conditional variances to be high (see Appendix 3D). This problem reflects the fact that when data are ill conditioned, some data series are nearly linear combinations of others and hence add very little new, independent information from which additional statistical information may be gleaned.

Needless to say, inflated variances are quite harmful to the use of regression as a basis for hypothesis testing, estimation, and forecasting. All users of linear regression have had the suspicion that an important test of significance has been rendered inconclusive through a needlessly high error variance induced by collinear data, or that a confidence interval or forecast interval is uselessly large, reflecting the lack of properly conditioned data from which appropriately refined intervals could conceivably have been estimated.

Both of the above ill effects of collinear data are most directly removed through the introduction of new and well-conditioned data.[34] In many applications, however, new data are either unavailable, or available but unable to be acquired except at great cost in time and effort. The usefulness of having diagnostic tools that signal the presence of collinearity and even isolate the variates involved is therefore apparent, for with them the investigator can at least determine whether the effort to correct for collinearity (collect new data or apply Bayesian techniques) is potentially worthwhile, and perhaps he can learn a great deal more. But just how much can be learned? To what extent can diagnostics tell the degree to which collinearity has caused harm?

Harmful Versus Degrading Collinearity. At the outset it should be noted that not all collinearity need be harmful. We have already seen, in the example of the Bauer matrix given in Exhibit 3.3, that near orthogonality

---

[33]To avoid any possible confusion it is worth highlighting that this is the statistical use of the word *conditional*, having nothing directly to do with (and thus to be contrasted with) the numerical-analytic notion of ill-conditioned data.

[34]In addition, this statistical problem (but not necessarily the computational problem) can be alleviated by the introduction of Bayesian prior information. See Zellner (1971), Leamer (1973) and Section 4.1.

can isolate some regression estimates from the presence of even extreme collinearity. Also, it is well known [Theil (1971), pp. 152–154] that specific linear combinations of estimated regression coefficients may well be determined even if the individual coefficients are not.[35] If by chance, the investigator's interest centers only on unaffected parameter estimates or on well-determined linear combinations of the estimated coefficients, clearly no problem exists.[36] The estimate of the constant term in Exhibit 3.1d, or of β_2 in Exhibit 3.1e, provides a simple example of this. In a less extreme, and therefore a practically more useful example, we recall from (3.9) that the estimated variance of the kth regression coefficient, var(b_k), is $s^2 \sum_j v_{kj}^2 / \mu_j^2$, where s^2 is the estimated error variance. If s^2 is sufficiently small, it may be that particular var(b_k)'s are small enough for specific testing purposes in spite of large components in the v_{kj}^2 / μ_j^2 terms resulting from near dependencies. One can see from Exhibit 3.1c that, if the height of the cloud, determined by s^2, is made small relative to its "width," the regression plane will become better defined even though the conditioning of the **X** data remains unchanged. Thus, for example, if an investigator is only interested in whether a given coefficient is significantly positive, and is able, even in the presence of collinearity, to accept that hypothesis on the basis of the relevant t-test, then collinearity has caused no problem. Of course, the resulting forecasts or point estimates may have wider confidence intervals than would be needed to satisfy a more ambitious researcher, but for the limited purpose of the test of significance intitially proposed, collinearity has caused no practical harm (see Appendix 3D). These cases serve to exemplify the pleasantly pragmatic philosophy that collinearity doesn't hurt so long as it doesn't bite.

Providing evidence that collinearity has actually harmed estimation, however, is significantly more difficult. To do this, one must show, for example, that a prediction interval that is too wide for a given purpose could be appropriately narrowed if made statistically conditional on better conditioned data (or that a confidence interval could be appropriately narrowed, or the computational precision of a point estimator appropriately increased). To date no procedure provides such information. If, however, the researcher were provided with information that (1) there are strong near dependencies among the data, so that collinearity is

---

[35]The effect on the collinearity diagnostics of linear transformations of the data matrix **X** ($=$ **ZG**, for **G** nonsingular) or, equivalently, of reparameterizations of the model $y = X\beta + \varepsilon$ into $y = Z\delta + \varepsilon = (XG^{-1})G\beta + \varepsilon$ is treated in Appendix 3B.

[36]No problem exists, that is, as long as a regression algorithm is used that does not blow up in the presence of highly collinear data. Standard routines based on solving $b = (X^T X)^{-1} X^T y$ are quite sensitive to ill-conditioned **X**. This problem is greatly overcome by regression routines based on the SVD of **X**, or a QR decomposition [see Golub (1969) and Belsley (1974)].

potentially a problem, and (2) that variances of parameters (or confidence intervals based on them) that are of interest to him have a large proportion of their magnitude associated with the presence of the collinear relation(s), so that collinearity is *potentially harmful*, then he would be a long way toward deciding whether the costs of corrective action were warranted. In addition, such information would help to indicate when variances of interest were not being adversely affected and so could be relied on without further action. The above information is, of course, precisely that provided by the condition indexes and high variance-decomposition proportions used in the two-pronged diagnostic procedure suggested earlier. Thus we say that when this joint condition has been met, the affected regression coefficients have been *degraded* (but not necessarily harmed) by the presence of collinearity, degraded in the sense that the magnitude of the estimated variance is being determined primarily by the presence of a collinear relation. Therefore, there is a presumption that confidence intervals, prediction intervals, and point estimates based on this estimate could be refined, if need be, by introducing better conditioned data. A practical example of the distinction between degrading and harmful collinearity is provided by the data for an equation from the IBM econometric model analyzed in Section 3.4.

At what point do estimates become degraded? Future experience may provide a better answer to this question, but for the experiments of the next section we take as a beginning rule of thumb that estimates are degraded when two or more variances have at least one-half of their magnitude associated with a single, large singular value.

3.3 EXPERIMENTAL EXPERIENCE

The test for the presence of degrading collinearity suggested at the end of the prior section requires the joint occurrence of high variance-decomposition proportions for two or more coefficients associated with a single singular value having a "high" condition index. Knowledge of what constitutes a high condition index must be empirically determined, and it is the purpose of this section to describe a set of experiments that have been designed to provide such experience.

The Experimental Procedure

Each of the three experiments reported below examines the behavior of the singular values and variance-decomposition proportions of a series of data matrices that are made to become systematically more and more ill conditioned by the presence of one or more near dependencies constructed to become more nearly exact.

Each experiment begins with a "basic" data set $\mathbf{X}$ of n observations, on p_1 variates. The number of observations, n, which is unimportant, is around 24–27, and p_1 is 3–5, depending on the experiment. In each case the basic data series are chosen either as actual economic time series (not centered) or as constructs that are generated randomly but having similar means and variances as actual economic time series.

These basic data series are used to construct additional collinear data series displaying increasingly tighter linear dependencies with the basic series as follows. Let $\mathbf{c}$ be a p_1-vector of constants and construct

$$\mathbf{w}_i = \mathbf{Xc} + \mathbf{e}_i, \tag{3.14}$$

where the components of $\mathbf{e}_i$ are generated i.i.d. normal with mean zero and variance $\sigma_i^2 = 10^{-i}s_{Xc}^2$, $s_{Xc}^2 \equiv \mathrm{var}(\mathbf{Xc})$, $i = 0, \dots, 4$. Each $\mathbf{w}_i$ then is, by construction, a known linear combination, $\mathbf{Xc}$, of the basic data series plus a zero-mean random error term, $\mathbf{e}_i$, whose variance becomes smaller and smaller (that is, the dependency becomes tighter and tighter) with increasing i. In the $i = 0$ case the variance in $\mathbf{w}_i$ due to the error term $\mathbf{e}_i$ is seen to be equal to the variance in $\mathbf{w}_i$ due to the systematic part $\mathbf{Xc}$. In this case, the imposed linear dependency is weak. The sample correlation between $\mathbf{w}_i$ and $\mathbf{Xc}$ in these cases tends toward .4 to .6. By the time $i = 4$, however, only $1/10,000$ of $\mathbf{w}_i$'s variance is due to additive noise, and the dependency between $\mathbf{w}_i$ and $\mathbf{Xc}$ is tight, displaying correlations very close to unity. A set of data matrices that become systematically more ill conditioned may therefore be constructed by augmenting the basic data matrix $\mathbf{X}$ with each $\mathbf{w}_i$, that is, by constructing the set

$$\mathbf{X}\{i\} = [\mathbf{X} \ \mathbf{w}_i] \qquad i = 0, \dots, 4. \tag{3.15}$$

The experiments are readily extended to the analysis of matrices possessing two or more simultaneous near dependencies by the addition of more data series similarly constructed from the basic series. Thus, for a given p_1-vector $\mathbf{b}$, let

$$\mathbf{z}_j = \mathbf{Xb} + \mathbf{u}_j, \tag{3.16}$$

where the components of $\mathbf{u}_j$ are i.i.d normal with mean zero and variance $\sigma_j^2 = 10^{-j}s_{Xb}^2$, $j = 0, \dots, 4$. Experimental matrices with two linear dependencies of varying strengths are constructed as

$$\mathbf{X}\{i,j\} = [\mathbf{X} \ \mathbf{w}_i \ \mathbf{z}_j] \qquad i, j = 0, \dots, 4. \tag{3.17}$$

In the third experiment that follows, three simultaneous dependencies are examined.

The Choice of the X's. As mentioned, the data series chosen for the basic matrices **X** were either actual economic time series or variates constructed to have similar means and variances as actual economic time series. The principle of selection was to provide a basic data matrix that was reasonably well conditioned so that all significant ill conditioning could be controlled through the introduced dependencies, such as (3.14).[37]

The various series of matrices that comprise any one experiment all have the same basic data matrix and differ only in the constructed near dependencies used to augment them. Within any one such series, the augmenting near dependencies become systematically tighter with increased i or j, and it is in this sense that we can speak meaningfully of what happens to condition indexes and variance-decomposition proportions as the data matrix becomes "more ill conditioned," or "more nearly singular," or "the near dependencies get tighter," or "the degree of collinearity increases."

Experimental Shortcomings. The experiments given here, while not Monte Carlo experiments,[38] are nevertheless subject to a similar weakness; namely, the results depend on the specific experimental matrices chosen and cannot be generalized to different situations with complete assurance. It has been attempted, therefore, within the necessarily small number of experiments reported here, to choose basic data matrices using data series and combinations of data series representing as wide a variety of economic circumstances as possible. Needless to say, not all meaningful cases can be considered, and the reader will no doubt think of cases he would rather have seen analyzed. The results also depend on the particular sample realizations that occur for the various e_i and u_j series that are used to generate the dependencies, such as (3.14) or (3.16). However, the cases offered here are sufficiently varied that any systematic patterns that emerge from them are worthy of being reported and will certainly provide a good starting point for any refinements that subsequent experience may suggest. In fact we find the stability of the results across all cases to be very reassuring.

---

[37]As we see in experiment 2, this objective is only partially achieved, leading to an unexpected set of dependencies that nevertheless provides a further successful test of the usefulness of this analytical procedure.

[38]No attempt is made here to infer any statistical properties through repeated samplings.

The Need for Column Scaling. Data matrices that differ from one another only by the scale assigned the columns (matrices of the form **XB** where **B** is a nonsingular diagonal matrix) represent essentially equivalent model structures; it does not matter, for example, whether one specifies an econometric model in dollars, cents, or billions of dollars. Such scale changes do, however, affect the numerical properties of the data matrix and result in very different singular-value decompositions and condition indexes.[39] Without further adjustment, then, we have a situation in which near dependencies among structurally equivalent economic variates (differing only in the units assigned them) can result in greatly differing condition indexes. Clearly the condition indexes can provide no stable information to the user of linear regression on the degree of collinearity among the **X** variates in such a case. It is necessary, therefore, to standardize the data matrices corresponding to equivalent model structures in a way that makes comparisons of condition indexes meaningful. A natural standardization process is to scale each column to have equal length. This scaling is natural because it transforms a data matrix **X** with mutually orthogonal columns, the standard of ideal data, into a matrix whose condition indexes would be all unity, the smallest (and therefore most ideal) condition indexes possible. Any other scaling would fail to reflect this desirable property.[40] As a matter of practice, we effect equal column lengths by scaling each column of **X** to have unit length.[41] A full justification of such column equilibration is lengthy and is deferred to Appendix 3B.

In all the experiments that follow, then, the data are scaled to have unit column length before being subjected to an analysis of their condition indexes and variance-decomposition proportions. In the event that the linear relations between the variates are displayed through auxiliary regressions, the regression coefficients have been rescaled to their original units.

---

[39]Scale changes do not, however, affect the presence of exact linear dependencies among the columns of **X**, since for any nonsingular matrix **B** there exists a nonzero **c** such that $\mathbf{Xc} = \mathbf{0}$ if and only if $[\mathbf{XB}][\mathbf{B}^{-1}\mathbf{c}] \equiv \overline{\mathbf{X}}\overline{\mathbf{c}} = \mathbf{0}$, where $\overline{\mathbf{X}} = \mathbf{XB}$ and $\overline{\mathbf{c}} = \mathbf{B}^{-1}\mathbf{c}$. For a more general discussion of the effects of column scaling, see Appendix 3B.

[40]Furthermore an important converse is true with column-equilibrated data; namely, when all condition indexes of a data matrix are equal to unity, the columns are mutually orthogonal. This is readily proved by noting that all condition indexes equal to 1 implies $\mathbf{D} = \lambda\mathbf{I}$, for some λ. Hence, in the SVD of **X**, we have $\mathbf{X} = \mathbf{UDV}^T = \lambda\mathbf{UV}^T$, or $\mathbf{X}^T\mathbf{X} = \lambda^2\mathbf{VU}^T\mathbf{UV}^T = \lambda^2\mathbf{I}$, due to the orthogonality of **U** and **V**. This result is important, because it rules out the possibility that several high variance-decomposition proportions could be associated with a very low (near unit) condition index.

[41]This scaling is similar to that used to transform the cross products matrix $\mathbf{X}^T\mathbf{X}$ into a correlation matrix, except that the "mean zero" property is not needed, and, indeed, would be inappropriate in the event that **X** contains a constant column.

The Experimental Report. Selected tables displaying variance-decomposition proportions (Π matrices) and condition indexes are reported for each experiment in order to show how these two principal pieces of information change as the near dependencies get tighter.

Additional statistics, such as the simple correlations of the contrived dependencies and their R^2's as measured from relevant regressions, are also reported to provide a link between the magnitudes of condition indexes and these more familiar notions. It cannot be stated too strongly, however, that these additional statistics cannot substitute for information provided by the variance-decomposition proportions and the condition indexes. In the experiments that follow, we know a priori which variates are involved in which relations and what the generating constants (the **c** in (3.14)) are. It is therefore possible to compute simple correlations between $\mathbf{w}_i$ and $\mathbf{Xc}$ and run regressions of $\mathbf{w}_i$ on $\mathbf{X}$. In practice, of course, **c** is unknown and one does not know which elements in the data matrix are involved in which dependencies. These auxiliary statistics are, therefore, not available to the investigator as independent analytic or diagnostic tools. However, one can learn from the variance-decomposition proportions which variates are involved in which relationships, and regressions may then be run among these variates to display the dependency. Furthermore, the *t*-statistics that result from these regressions can be used in the standard way for providing additional *descriptive* evidence of the "significance" of each variate in the specific linear dependency. Once the analysis by condition indexes and variance-decomposition proportions has been conducted, then, it can suggest useful auxiliary regressions as a means of exhibiting the near dependencies; but regression by itself, particularly if there are two or more simultaneous near dependencies, cannot provide similar information.[42]

The Individual Experiments

Three experiments are conducted, each using a separate series of data matrices designed to represent different types of economic data and different types of collinearity. Thus "levels" data (manufacturing sales), "trended" data (GNP), "rate of change" data (inventory investment), and "rates" data (unemployment) are all represented. Similarly, the types of collinearity generated include simple relations between two variates, relations involving more than two variates, simultaneous near dependencies, and dependencies among variates with essential scaling problems. The different cases of relations involving more than two variates

---

[42]Although, a costly and time-consuming set of tests based on partial correlations or block regressions on the columns of the data matrix encompassing all possible combinations could perhaps be of some value.

have been chosen to involve different mixes of the different types of economic variables listed above. In each case the dependencies are generated from the unscaled (natural) economic data, and hence the various test data sets represent as closely as possible economic data with natural near dependencies.

Experiment 1: The X Series.   The basic data set employed here is

$$X = [MFGS, *IVM, MV]^{43}$$

where   MFGS   is manufacturers' shipments, total
        *IVM   is manufacturers' inventories, total
        MV   is manufacturers' unfilled orders, total,

and each series is in millions of dollars, annual 1947–1970 ($n = 24$). This basic data set provides the type of series that would be relevant, for example, to an econometric study of inventory investment.[44]

Two sets of additional dependency series are generated from X as follows:

$$w_i = MV + v_i, \qquad i = 0, \ldots, 4, \tag{3.18a}$$

with v_i generated normal with mean zero and variance-covariance matrix $\sigma_i^2 I = 10^{-i}s_{MV}^2 I$ [denoted $v_i \leftrightarrow N(0, 10^{-i}s_{MV}^2 I)$], s_{MV}^2 being the sample variance of the MV series, and

$$z_j = 0.8MFGS + 0.2*IVM + v_j, \tag{3.18b}$$

$$v_j \leftrightarrow N(0, 10^{-j}s_z^2 I),$$

s_z^2 being the sample variance of $0.8MFGS + 0.2*IVM$.

The w_i and z_j series were used to augment the basic data set to produce three sequences of matrices:

$$X1\{i\} \equiv [X \ w_i], \qquad i = 0, \ldots, 4,$$

$$X2\{j\} \equiv [X \ z_j], \qquad j = 0, \ldots, 4, \tag{3.19}$$

$$X3\{i, j\} \equiv [X \ w_i \ z_j], \qquad i, j = 0, \ldots, 4,$$

each of which is subjected to analysis.

---

[43]A * before a series name indicates a dummy series was used having the same mean and variance as the given series, but generated to provide a well-conditioned basic data matrix.
[44]See, for example, Belsley (1969).

The dependency (3.18a) is a commonly encountered simple relation between two variates. Unlike more complex relations, it is a dependency whose presence can be discovered through examination of the simple correlation matrix of the columns of the X1$\{i\}$ or X3$\{i, j\}$. Its inclusion, therefore, allows us to learn how condition indexes and simple correlations compare with one another.

The dependency (3.18b) involves three variates, and hence would not generally be discovered through an analysis of the simple correlation matrix. Equation (3.18b) was designed to present no difficult scaling problems; that is, the two basic data series MFGS and *IVM have roughly similar magnitudes and variations, and the coefficients (0.8 and 0.2) are of the same order of magnitude. No one variate, therefore, dominates the linear dependency, masking the effects of others. This situation should allow both the identification of the variates involved and the estimation of the relation among them to be accomplished with relative ease.

Experiment 2: The Y Series.   The basic data set employed here is

$$Y \equiv [*GNP58, *GAVM, *LHTUR, *GV58],$$

where   GNP58   is GNP in 1958 dollars,
             GAVM   is net corporate dividend payments,
             LHTUR   is unemployment rate,
             GV58   is annual change in total inventories, 1958 dollars.

Each basic series has been constructed from the above series to have similar means and variances, but has been chosen to produce a reasonably well-conditioned Y matrix. Data are annual, 1948–1974 ($n=27$). The variates included here, then, represent "levels" variates, (*GNP58), rates of change (*GAVM), and "rates" (*LHTUR).

Three additional dependency series are constructed as $(i, j, k = 0, \ldots, 4)$

$$u_i = *GNP58 + *GAVM + v_i,$$

$$v_i \leftrightarrow N(0, 10^{-i}s_u^2 I), \tag{3.20a}$$

$$s_u^2 = \mathrm{var}(*GNP58 + *GAVM),$$

$$w_j = 0.1*GNP58 + *GAVM + v_j,$$

$$v_j \leftrightarrow N(0, 10^{-j}s_v^2 I), \tag{3.20b}$$

$$s_v^2 = \mathrm{var}(0.1*GNP58 + *GAVM),$$

$$z_k = *GV58 + v_k,$$

$$v_k \leftrightarrow N(0, 10^{-k}s_{*GV58}^2 I). \tag{3.20c}$$

These data were used to augment **Y** to produce four series of test matrices:

$$
\begin{aligned}
\mathbf{Y}1\{i\} &= [\mathbf{Y}\ \mathbf{u}_i], & i &= 0,\ldots,4, \\
\mathbf{Y}2\{j\} &= [\mathbf{Y}\ \mathbf{w}_j], & j &= 0,\ldots,4, \\
\mathbf{Y}3\{k\} &= [\mathbf{Y}\ \mathbf{z}_k], & k &= 0,\ldots,4, \\
\mathbf{Y}4\{i,k\} &= [\mathbf{Y}\ \mathbf{u}_i\,\mathbf{z}_k], & i,k &= 0,\ldots,4.
\end{aligned}
\tag{3.21}
$$

Dependency (3.20a) presents a relation among three variates with an essential scaling problem; namely, in the units of the basic data, the variation introduced by GNP58 is less than 1% that introduced by GAVM. The inclusion of GNP58 is therefore dominated by GAVM, and its effects should be somewhat masked and difficult to discern. Dependency (3.20b) is of a similar nature except that the scaling problem has been made more extreme. These are "essential" scaling problems in that their effects cannot be undone through simple column scaling. Dependency (3.20c) is a simple relation between two variates, except in this case the variate is a rate of change, exhibiting frequent shifts in sign.

Experiment 3: The Z Series.   The basic data matrix here is an expanded version of that in the previous experiment:

$$
\mathbf{Z} = [\,*\text{GNP58}, *\text{GAVM}, *\text{LHTUR}, \text{DUM1}, \text{DUM2}\,],
$$

where DUM1 is generated to be similar to GV58, and DUM2 similar to GNP58, except that DUM1 and DUM2 are generated to have very low intercorrelation with the first three variates. This configuration allows examination of the case described in Section 3.2 when some variates are isolated by near orthogonality from dependencies among others.

The additional dependency series are $(i, j, k, m = 0,\ldots,4)$:

$$
\mathbf{u}_i = \text{DUM1} + \mathbf{e}_i,
$$
$$
\mathbf{e}_i \leftrightarrow N(\mathbf{0}, 10^{-i}s^2_{\text{DUM1}}\mathbf{I}),
\tag{3.22a}
$$

$$
\mathbf{v}_j = \text{DUM2} - \text{DUM1} + \mathbf{e}_j,
$$
$$
\mathbf{e}_j \leftrightarrow N(\mathbf{0}, 10^{-j}s^2_{\text{DUM2}-\text{DUM1}}\mathbf{I}),
\tag{3.22b}
$$

$$
\mathbf{w}_k = 3*\text{GNP58} + 1.5*\text{LHTUR} + \mathbf{e}_k,
$$
$$
\mathbf{e}_k \leftrightarrow N(\mathbf{0}, 10^{-k}s^2_{3*\text{GNP58}+1.5*\text{LHTUR}}\mathbf{I}),
\tag{3.22c}
$$

$$
\mathbf{z}_m = *\text{GAVM} + 0.7*\text{DUM2} + \mathbf{e}_m,
$$
$$
\mathbf{e}_m \leftrightarrow N(\mathbf{0}, 10^{-m}s^2_{*\text{GAVM}+0.7*\text{DUM2}}\mathbf{I}),
\tag{3.22d}
$$

These data are used to augment $\mathbf{Z}$ to produce seven series of test matrices:

$$
\begin{aligned}
\mathbf{Z}1\{i\} &\equiv [\mathbf{Z}\,\mathbf{u}_i], & \mathbf{Z}5\{j,k\} &\equiv [\mathbf{Z}\,\mathbf{v}_j\,\mathbf{w}_k], \\
\mathbf{Z}2\{j\} &\equiv [\mathbf{Z}\,\mathbf{v}_j], & \mathbf{Z}6\{i,m\} &\equiv [\mathbf{Z}\,\mathbf{u}_i\,\mathbf{z}_m], \\
\mathbf{Z}3\{k\} &\equiv [\mathbf{Z}\,\mathbf{w}_k], & \mathbf{Z}7\{i,k,m\} &\equiv [\mathbf{Z}\,\mathbf{u}_i\,\mathbf{w}_k\,\mathbf{z}_m]. \\
\mathbf{Z}4\{m\} &\equiv [\mathbf{Z}\,\mathbf{z}_m],
\end{aligned}
\tag{3.23}
$$

Each of the first three dependencies (3.22a–c) possesses essential scaling problems, with DUM2, 3*GNP58, and *GAVM, respectively, being the dominant terms. The problem is extreme in the relation of DUM2 and DUM1, where DUM1 introduces much less than 0.1% of the total variation, and difficult in the other cases. The relation defined by (3.22b) is isolated by near orthogonality from the one defined by (3.22c), and these relations occur separately in the $\mathbf{Z}2$ and $\mathbf{Z}3$ test series and together in the $\mathbf{Z}5$ series. Relation (3.22d) bridges the two subseries.

The Results

Space limitations obviously prevent reporting the full set of experimental results. Fortunately, after reporting experiment 1 in some detail, it is possible to select samples of output from experiments 2 and 3 that convey what generalizations are possible. Even so, the experimental report that follows is necessarily lengthy and may be difficult to digest fully on a first reading. With the exception of the concept of the dominant dependency, however, the essential behavior of the diagnostic procedure can be appreciated from the results of experiment 1, that on the $\mathbf{X}$ series, alone. The reader may wish, therefore, to skip experiments 2 and 3 on a first reading, and, after reading experiment 1, go directly to the summary in Section 3.4 beginning on p. 152. Ultimately, however, experiments 2 and 3 (and especially those surrounding data set $\mathbf{Y}3$) are necessary to introduce the important concept of the dominant dependency and other refinements in interpretation and to demonstrate the empirical stability of the procedure in a wider set of situations.

Experiment 1: The X Matrices. $X1$:   Let us begin with the simplest series of experimental matrices, the $\mathbf{X}1\{i\}, i = 0, \ldots, 4$. Here the data series of column 4, which we denote as $C4$, is related to that of column 3, $C3$, by (3.18a), that is, $C4 = C3 + \mathbf{e}_i, i = 0, \ldots, 4$; and this is the only contrived dependency among the four columns of $\mathbf{X}1$. We would therefore expect

Exhibit 3.5a Variance-decomposition proportions and condition indexes.*
X1 series. One constructed near dependency (3.18a): $C4 = C3 + e_i$

Associated Singular Value	var(b_1)	var(b_2)	var(b_3)	var(b_4)	Condition Index, η
			X1{0}		
μ_1	.005	.012	.002	.003	1
μ_2	.044	.799	.004	.032	5
μ_3	.906	.002	.041	.238	8
μ_4	.045	.187	.954	.727	14
			X1{1}		
μ_1	.005	.011	.001	.001	1
μ_2	.094	.834	.003	.002	5
μ_3	.899	.117	.048	.035	9
μ_4	.002	.038	.948	.962	27
			X1{2}		
μ_1	.005	.012	.000	.000	1
μ_2	.086	.889	.000	.000	5
μ_3	.901	.083	.003	.003	9
μ_4	.007	.016	.997	.997	95
			X1{3}		
μ_1	.005	.012	.000	.000	1
μ_2	.078	.903	.000	.000	5
μ_3	.855	.079	.000	.000	9
μ_4	.061	.006	.999	.999	461
			X1{4}		
μ_1	.005	.010	.000	.000	1
μ_2	.084	.792	.000	.000	5
μ_3	.906	.070	.000	.000	9
μ_4	.004	.127	1.000	1.000	976

*Columns may not add to unity due to rounding error.

one "high" condition index and a large proportion of var(b_3) and var(b_4) associated with it. Exhibit 3.5a presents the variance-decomposition proportions and the condition indexes for this series as i goes from 0 to 4.

A glance at these results confirms our expectations. In each case there is a highest condition index that accounts for a high proportion of variance for two or more of the coefficients, and these are var(b_3) and var(b_4). Furthermore, the pattern is observable in the weakest case X1{0}, and becomes increasingly clearer as the near dependency becomes tighter: all condition indexes save one remain virtually unchanged while the condition index corresponding to the imposed dependency increases strongly with each jump in i; the variance-decomposition proportions of the two "involved" variates C3 and C4 become larger and larger, eventually becoming unity.

To help interpret the condition indexes in Exhibit 3.5a, we present in Exhibit 3.5b the simple correlation between C3 and C4 for each of the X1{i} matrices and also the multiple regressions of C4 on C1, C2, and C3.

In addition to observing the general pattern that was expected, the following points are noteworthy:

1. The relation between C3 and C4 of X1{0},.having a simple correlation of .766 and a regression R^2 of .6229 (not very high in comparison with simple correlations present in most real-life economic data matrices),

Exhibit 3.5b   Regression of C4 on C1, C2, and C3.*

Data Matrix	r(C3,C4)	C1	C2	C3	R^2
X1{0}	.766	0.3905 [1.11]	−0.1354 [0.91]	0.9380 [4.76]	.6229
X1{1}	.931	0.1481 [0.97]	0.0925 [1.38]	0.8852 [10.10]	.8765
X1{2}	.995	−0.0076 [−0.17]	0.0142 [0.72]	0.9982 [38.80]	.9893
X1{3}	.999	0.0111 [1.22]	0.0015 [0.37]	0.9901 [188.96]	.9996
X1{4}	1.000	−0.0012 [−0.28]	0.0033 [1.76]	0.9976 [400.56]	.9999

*The figures in square brackets are t's. Since interest in these results centers on "significance," it seems proper to show t's rather than estimated standard deviations.

shows itself in a condition number of 14 and is sufficiently high to account for large proportions (.95 and .73) of the variance of the affected coefficients, $\text{var}(b_3)$ and $\text{var}(b_4)$.

2. Also at this lowest level $(i = 0)$, the diagnostic test proposed at the end of Section 3.2 does correctly indicate the existence of the one near dependency and correctly key the variates involved.

3. In light of point 2 above, the regressions of Exhibit 3.5b that were run for comparative purposes are also those that would be suggested by the results for displaying the near dependencies. Exhibit 3.5b verifies that even in the $X1\{0\}$ case with an η of 14, the proper relation among the columns of $X1\{0\}$ is being clearly observed.

4. With each increase in i (corresponding to a tenfold reduction in the variance of the noise in the near dependency), the simple correlations and R^2's increase one step, roughly adding another 9, in the series .9, .99, .999 and so on, and the condition index increases roughly along the progression 10, 30, 100, 300, 1000, a pattern we consistently observe in future examples.[45] This relation suggests a means for comparing the order of magnitude of the "tightness" of a near dependency.

5. Also with each increase in i, the variance-decomposition proportion of the affected coefficients associated with the highest η increases markedly (again roughly adding one more 9 with each step).

6. As noted in Section 3.2, it is the joint condition of high variance-decomposition proportions for two or more coefficients associated with a high condition index that signals the presence of degrading collinearity. In the case of $X1\{0\}$, the second highest condition index, 8, is not too different from the highest, but it is a dominant component in only one variance, $\text{var}(b_1)$. In this case, then, a condition index of 14 (roughly 10) is "high enough" for the presence of collinearity to begin to be observed.

$X2$: The $X2\{i\}$ series also possesses only one constructed near dependency, (3.18b), but involving three variates, columns 1, 2, and 4 in the form $C4 = 0.8C1 + 0.2C2 + e_i$. We expect, then, high variance-decomposition proportions for these three variates to be associated with a single high condition index. Exhibit 3.6a presents the Π matrix of variance-decomposition proportions and the condition indexes for the $X2\{i\}$ data series, and Exhibit 3.6b gives the corresponding simple

---

[45] This progression corresponds closely to roughly equal increments in $\log_{10} \eta_i$ of $1/2$, that is, $\log_{10} \eta_i = 1 + (i/2)$. Alternatively, and somewhat more directly, this progression results from the sequence $10^{i/2}$, $i > 0$, and corresponds to $[1/(1 - R^2)]^{i/2}$ for $R^2 = 0, .9, .99, .999$, and so on.

Exhibit 3.6 a Variance-decomposition proportions and condition indexes. *X2 series*. One constructed near dependency (3.18*b*): **C4** = 0.8C1 + 0.2C2 + **e***ᵢ*

Associated Singular Value	var(b_1)	var(b_2)	var(b_3)	var(b_4)	Condition Index, η
			X2{0}		
μ_1	.003	.012	.004	.005	1
μ_2	.027	.735	.001	.068	4
μ_3	.009	.223	.636	.526	9
μ_4	.960	.030	.359	.401	11
			X2{1}		
μ_1	.001	.004	.003	.000	1
μ_2	.021	.297	.011	.001	5
μ_3	.091	.026	.767	.006	10
μ_4	.887	.673	.219	.993	31
			X2{2}		
μ_1	.000	.001	.004	.000	1
μ_2	.001	.039	.012	.000	5
μ_3	.004	.002	.983	.002	9
μ_4	.995	.958	.001	.998	102
			X2{3}		
μ_1	.000	.000	.004	.000	1
μ_2	.000	.002	.014	.000	5
μ_3	.000	.000	.976	.000	9
μ_4	1.000	.997	.006	1.000	381
			X2{4}		
μ_1	.000	.000	.004	.000	1
μ_2	.000	.000	.013	.000	5
μ_3	.000	.000	.938	.000	9
μ_4	1.000	1.000	.046	1.000	1003

Exhibit 3.6b Regression of C4 on C1, C2 and C3

Data Matrix	$r(C4, \widehat{C4})$ $\widehat{C4} = 0.8C1 + 0.2C2$	C1	C2	C3	R^2
X2{0}	.477	0.8268 [3.84]	−0.0068 [−0.07]	0.1089 [0.88]	.2864
X2{1}	.934	0.6336 [10.14]	0.1776 [6.49]	0.1032 [2.87]	.8976
X2{2}	.995	0.8186 [40.90]	0.1879 [21.44]	0.0007 [0.06]	.9911
X2{3}	.999	0.7944 [149.03]	0.2023 [86.69]	0.0012 [0.40]	.9993
X2{4}	1.000	0.7990 [393.94]	0.1992 [224.32]	0.0012 [1.02]	.9999

correlations and regressions. In this case the correlations are between C4 in (3.18b) and $\widehat{C4} = 0.8C1 + 0.2C2$. The regressions are C4 regressed on C1, C2, and C3.

The following points are noteworthy:

1. Once again the expected results are clearly observed, at least for $i \geqslant 1$.
2. In the case of X2{0}, the constructed dependency is weak, having a simple correlation of less than .5. The resulting condition index, 11, is effectively the same as the second highest condition index, 9, and hence this dependency is no tighter than the general background conditioning of the basic data matrix. We see as we proceed that in the case where *several condition indexes are effectively the same*, the procedure can have trouble distinguishing among them, and the variance-decomposition proportions of the variates involved can be arbitrarily distributed among the nearly equal condition indexes. In X2{0} the two condition indexes 9 and 11 account for over 90% of the variance in b_1, b_3, and b_4. We can explain b_3's presence in this group by the fact that the simple correlation between C1 and C3 is .58 (greater than the constructed correlation between C4 and $\widehat{C4}$). The absence of b_2 is explained by a minor scaling problem; C2 accounts for only one half the variance of the constructed variate $C4 = 0.8C1 + 0.2C2 + e$, and, in this case, its influence is being dominated by other correlations.
3. By the time the simple correlation between C4 and $\widehat{C4}$ becomes .934, in the case of X2{1}, the above problem completely disappears. The contrived relation involving columns 1, 2, and 4 now dominates the other correlations among the columns of the basic data matrix, and the

variance-decomposition proportions of these variates associated with the largest condition index, 31, are all greater than .5.

4. We again observe that, with each increase in i, the condition index corresponding to this ever-tightening relation jumps in the same progression noted before, namely, 10, 30, 100, 300, 1000. In this regard it is of interest to observe that the contrived relation among columns 1, 2, and 4, becomes clearly distinguishable from the background in case X2{1}, when its condition index becomes one step in this progression above the "noise," that is, when it becomes 31 versus the 10 associated with the background dependencies.

5. Once again, the presence of collinearity begins to be observable with condition indexes around 10. In this instance, however, an unintended relation (the .58 correlation between C1 and C3) also shows itself, confounding clear identification of the intended dependency among C4, C1, and C2.

6. In both this and the X1 case, a condition index of 30 signals clear evidence of the presence of the linear dependency and degraded regression estimates.

X3:   This series of matrices combines the two dependencies (3.18a) and (3.18b) just examined into a single five-column matrix with $C4 = C3 + e$ and $C5 = 0.8C1 + 0.2C2 + u$. This, then, offers the first constructed example of simultaneous or coexisting dependencies.[46] We expect that there should be two high condition indexes, one associated with high variance-decomposition proportions in $var(b_3)$ and $var(b_4)$—due to dependency (3.18a)—and one associated with high variance-decomposition proportions between $var(b_1)$ and $var(b_2)$ and $var(b_5)$—due to dependency (3.18b).

In an effort to reduce the increasing number of Π matrices relevant to this case we concentrate our reported results in two ways. First, we report only representative information from among the 25 cases (X3$\{i,j\}, i, j = 0, \ldots, 4$), and second, where possible, we report only the rows of the variance-decomposition proportions that correspond to ·the condition indexes of interest. We note in the previous two series that many of the rows, those corresponding to lower condition indexes, are effectively unchanging as i varies, and convey no useful additional information for the analysis at hand.

Let us begin by holding i constant at 2 and varying $j = 0, \ldots, 4$; dependency (3.18a) is therefore moderately tight, while (3.18b) varies. Exhibit 3.7 presents the results for this case.

---

[46]Although we have already seen something like it above in the case of X2{0}.

Exhibit 3.7 Variance-decomposition proportions and condition indexes. *X3 series*. Two constructed near dependencies (3.18a) and (3.18b): $C4 = C3 + e_2$ (unchanging); $C5 = 0.8C1 + 0.2C2 + u_i (i = 0, \ldots, 4)$

Associated Singular Value	var(b_1)	var(b_2)	var(b_3)	var(b_4)	var(b_5)	Condition Index, η
			X3{2, 0}			
μ_1	.002	.007	.000	.000	.003	1
μ_2	.025	.734	.000	.000	.064	5
μ_3	.025	.240	.003	.003	.303	8
μ_4	.941	.002	.000	.001	.630	12
μ_5	.007	.016	.996	.996	.001	106
			X3{2, 1}*			
μ_3	.076	.008	.003	.003	.009	9
μ_4	.765	.537	.000	.004	.792	34
μ_5	.147	.192	.997	.992	.199	118
			X3{2, 2}*			
μ_3	.004	.001	.003	.003	.002	8
μ_4	.746	.750	.459	.464	.758	127
μ_5	.249	.211	.537	.533	.241	99
			X3{2, 3}*			
μ_3	.002	.000	.003	.003	.000	8
μ_4	.999	.997	.173	.182	.999	469
μ_5	.000	.001	.824	.816	.001	83
			X3{2, 4}*			
μ_3	.000	.000	.003	.003	.000	8
μ_4	1.000	1.000	.003	.001	1.000	1124
μ_5	.000	.000	.994	.996	.000	106

*The unchanging and inconsequential rows corresponding to μ_1 and μ_2 have not been repeated. See text.

The auxiliary correlation and regression statistics need not, of course, be repeated, for these are the same as the relevant portions of Exhibits 3.5b and 3.6b. In particular, the correlations and regressions for the unchanging X3$\{2, j\}$ relation for $i = 2$ between C4 and C3 are those for X1$\{2\}$ in Exhibit 3.5b, and the regressions for column 5 on the basic columns 1, 2, and 3 for $j = 0, \ldots, 4$ are those given in Exhibit 3.6b for X2$\{j\}, j = 0, \ldots, 4$.

The following points are noteworthy:

1. The unchanging "tight" relation between columns 3 and 4 is observable throughout, having a correlation of .995 and a large condition index in the neighborhood of 100.

2. The relation with varying intensity among C5, C1, and C2 begins weakly for the X3$\{2, 0\}$ case and, as before, is somewhat lost in the background. Still, the involvement of var(b_1) and var(b_5) with the condition index 12 is observable even here, although it is being confounded with the other condition index, 8, of roughly equal value. The unchanging tight relation between C4 and C3 with index 106 is unobscured by these other relations.

3. When the relation between columns 1, 2, and 5 becomes somewhat tighter than the background, as in the case of X3$\{2, 1\}$, its effects become separable. This case clearly demonstrates the ability of the procedure to correctly identify two simultaneous dependencies and indicate the variates involved in each: the η of 34 is associated with the high variance-decomposition proportions in var(b_1), var(b_2), and var(b_5), and the η of 118 is associated with those of var(b_3) and var(b_4).

4. When the two contrived dependencies become of roughly equal intensity, as in the case of X3$\{2, 2\}$, both having η's in the neighborhood of 100, the involvement of the variates in the two relations once again becomes confounded. However, only the information on the separate involvement of the variates is lost through this confounding. It is still possible to determine that there are two near dependencies among the columns of X, and it is still possible to determine which variates are involved; in this case all of them, for the two condition indexes together account for well over 90% of the variance in b_1, b_2, b_3, b_4, and b_5, indicating the involvement of each. The only information being lost here is which variates enter which dependency.

5. When the relation among columns 1, 2, and 5 again becomes strong relative to the unchanging relation between columns 3 and 4, as in the cases X3$\{2, 3\}$ and X3$\{2, 4\}$, their separate identities reemerge.

6. Once again, the order of magnitude of the relative tightness of a near dependency seems to increase with a progression in the condition index in

the scale of $10, 30, 100, 300, 1000$. Dependencies of roughly equal magnitude can be confounded; dependencies of differing magnitudes are able to be separately identified.

The preceding analysis examines $X3\{i, j\}$ by varying the second dependency and holding the first constant at $i = 2$. Let us reverse this order and examine $X3\{i, 1\}$ for $i = 0, \ldots, 4$, and j held constant at 1.

As i increases, we would expect there to be two high condition indexes. The one corresponding to the unchanging dependency between columns 1, 2, and 5 will not be too high, since j is held at 1. The relation between

Exhibit 3.8 Variance-decomposition proportions and condition indexes. *X3 series*. Two constructed near dependencies $(3.18a)$ and $(3.18b)$: $C4 = C3 + e_i$ $(i = 0, \ldots, 4)$; $C5 = 0.8C1 + 0.2C2 + u_1$ (unchanging)

Associated Singular Value	$var(b_1)$	$var(b_2)$	$var(b_3)$	$var(b_4)$	$var(b_5)$	Condition Index, η
			$X3\{0, 1\}$			
μ_1	.001	.002	.001	.002	.000	1
μ_2	.008	.274	.003	.032	.000	5
μ_3	.092	.004	.044	.250	.011	9
μ_4	.885	.643	.171	.008	.988	35
μ_5	.015	.076	.781	.708	.001	15
			$X3\{1, 1\}$			
μ_4	.821	.656	.226	.077	.900	35
μ_5	.073	.018	.720	.880	.089	30
			$X3\{2, 1\}$			
μ_4	.765	.537	.000	.004	.792	34
μ_5	.147	.192	.997	.992	.199	118
			$X3\{3, 1\}$			
μ_4	.871	.673	.000	.000	.985	34
μ_5	.028	.010	.999	.999	.005	519
			$X3\{4, 1\}$			
μ_4	.842	.549	.000	.000	.901	34
μ_5	.062	.191	1.000	1.000	.089	1148

columns 3 and 4 will get tighter and more highly defined as i increases from 0 to 4.

Exhibit 3.8 reports these results. Exhibit 3.5b and the second row of Exhibit 3.6b provide the relevant supplementary correlations and regressions.

The following points are noteworthy:

1. Both relations are observable from the outset.

2. A condition index of 15 (greater than 10) corresponds to a dependency that is tight enough to be observed.

3. The confounding of the two dependencies is observable when the condition indexes are close in magnitude, the case of X3{1, 1}, but it is not as pronounced here as in the previous examples.

4. The rough progression of the condition indexes in the order of 10, 30, 100, 300, 1000 is observed again.

To complete the picture on the behavior of X3{i, j}, we report in Exhibit 3.9 the variance-decomposition proportions for selected values of i and j increasing together.

Exhibit 3.9 Variance-decomposition proportions and condition indexes. $X3$ series. Two constructed near dependencies (3.18a) and (3.18b): C4 = C3 + e$_i$ (selected values); C5 = 0.8C1 + 0.2C2 + u$_j$ (selected values)

Associated Singular Value	var(b_1)	var(b_2)	var(b_3)	var(b_4)	var(b_5)	Condition Index, η
			X3{0,0}			
μ_1	.002	.007	.001	.002	.003	1
μ_2	.016	.750	.000	.009	.041	5
μ_3	.028	.053	.044	.206	.323	7
μ_4	.953	.000	.018	.032	.610	12
μ_5	.001	.190	.937	.751	.023	15
			X3{1,0}			
μ_3	.015	.271	.041	.033	.341	8
μ_4	.952	.009	.012	.006	.585	12
μ_5	.006	.037	.946	.961	.004	30
			X3{1,2}			
μ_4	.995	.962	.094	.117	.998	123
μ_5	.005	.002	.860	.848	.001	30

Exhibit 3.9 Continued

Associated Singular Value	var(b_1)	var(b_2)	var(b_3)	var(b_4)	var(b_5)	Condition Index, η
			X3{3,2}			
μ_4	.967	.936	.000	.000	.979	115
μ_5	.028	.023	1.000	1.000	.020	523
			X3{3,4}			
μ_4	.999	.999	.015	.013	.999	1129
μ_5	.001	.001	.985	.986	.001	517
			X3{4,4}			
μ_4	.698	.705	.633	.631	.699	1357
μ_5	.302	.294	.369	.369	.301	960

The following points are noteworthy:

1. In the X3{0,0} case, three condition indexes are of close magnitude, 7, 12, and 15, and there is some confounding of variate involvement among all three of them.

2. The relation between C3 and C4 is freed from the background in the next case,. X3{1,0}, but there is still some confusion between the two similar condition indexes, 8 and 12.

3. The two relations become more clearly identified as i and j increase, and are strongly separable so long as the condition indexes remain separated by at least one order of magnitude along the 10, 30, 100, 300, 1000 progression.

4. However, no matter how tight the individual relationships, they can be confused when their condition indexes are of similar magnitude, as is seen in the case of X3{4,4}.

Experiment 2: The Y Matrices. Our interest in examining these new experimental data series focuses on several questions. First, does a totally different set of data matrices result in similar generalizations on the behavior of the condition indexes and the variance-decomposition proportions that were beginning to emerge from experiment 1? Second, do "rate-of-change" data series and "rates" series behave differently from the "levels" and "trends" data of experiment 1? The data series Y3 are relevant here. Third, do essential scale problems cause troubles? Data series Y1 and Y2 examine this problem.

Y1 and Y2: The Y1$\{i\}$ series, we recall, consists of a five-column matrix in which $C5 = C1 + C2 + e_i$ as in (3.20a). The variance of C5 introduced by C1 (GNP58) is relatively small, less than 1% of that introduced by C2. Its influence is therefore easily masked. The Y2$\{i\}$ series is exactly the same except that the influence of C1 is made smaller yet. Here $C5 = 0.1C1 + C2 + e_i$ as is clear from (3.20b). These two series allow us to see how sensitive the diagnostic procedure for collinearity is to strong and even severe scaling problems.

For both experimental series Y1 and Y2 we would expect one high condition index associated with high variance-decomposition proportions in $var(b_1)$, $var(b_2)$, and $var(b_5)$. Exhibits 3.10a and 3.11a present these results for Y1 and Y2, respectively, as $i = 0, \ldots, 4$. Exhibits 3.10b and 3.11b present the corresponding supplementary correlations and regressions.

The following points are noteworthy:

1. The results for both data series are in basic accord with expectations.

2. However, the essential scale differences do cause problems in identifying the variates involved in generating the dependency. In the Y1 series, the involvement of the dominated column 1 is not observed at all in

Exhibit 3.10a Variance-decomposition proportions and condition indexes. *Y1 series.* One constructed near dependency (3.20a): $C5 = C1 + C2 + e_i$ ($i = 0, \ldots, 4$)

Associated Singular Value	var(b_1)	var(b_2)	var(b_3)	var(b_4)	var(b_5)	Condition Index, η
			Y1$\{0\}$			
μ_1	.005	.001	.003	.015	.002	1
μ_2	.010	.001	.010	.898	.002	3
μ_3	.782	.036	.001	.010	.081	7
μ_4	.188	.045	.978	.071	.096	10
μ_5	.016	.916	.008	.007	.819	16
			Y1$\{1\}$			
μ_3	.607	.019	.031	.008	.003	8
μ_4	.024	.030	.941	.090	.012	10
μ_5	.360	.950	.014	.036	.985	40

Exhibit 3.10a Continued

Associated Singular Value	var(b_1)	var(b_2)	var(b_3)	var(b_4)	var(b_5)	Condition Index, η
			Y1{2}			
μ_3	.394 ·	.001	.003	.010	.001	7
μ_4	.065	.001	.976	.068	.001	10
μ_5	.534	.998	.008	.020	.999	156
			Y1{3}			
μ_3	.099	.000	.003	.010	.000	7
μ_4	.016	.000	.959	.073	.000	10
μ_5	.883	.997	.024	.003	1.000	397
			Y1{4}			
μ_4	.001	.000	.965	.072	.000	10
μ_5	.993	1.000	.018	.000	1.000	1659

Exhibit 3.10b Regression of C5 on C1, C2, C3, C4

Data Matrix	$r(C5, \widehat{C5})$ $\widehat{C5} = C1 + C2$	C1	C2	C3	C4	R^2
Y1{0}	.776	−0.0264 [−0.01]	0.9262 [5.33]	97.8101 [1.03]	28.3210 [0.34]	.6186
Y1{1}	.939	2.5629 [3.78]	0.8238 [13.20]	35.7769 [1.05]	35.1060 [1.17]	.9116
Y1{2}	.997	0.9505 [5.24]	1.0009 [59.80]	−2.7505 [−0.30]	−4.9108 [−0.61]	.9940
Y1{3}	.999	0.9534 [13.36]	0.9989 [151.80]	2.8747 [0.80]	0.9679 [0.31]	.9991
Y1{4}	1.000	1.0178 [59.57]	0.9975 [633.30]	0.5632 [0.65]	0.0135 [0.0178]	.9999

138

Exhibit 3.11a Variance-decomposition proportions and condition indexes. *Y2 series*. One constructed near dependency (3.20b): $C5 = 0.1C1 + C2 + e_i \; (i = 0, \ldots, 4)$

Associated Singular Value	var(b_1)	var(b_2)	var(b_3)	var(b_4)	var(b_5)	Condition Index, η
			Y2{0}			
μ_1	.005	.003	.003	.014	.003	1
μ_2	.009	.002	.009	.886	.005	3
μ_3	.804	.023	.007	.001	.205	7
μ_4	.157	.366	.201	.000	.781	10
μ_5	.025	.606	.781	.098	.006	11
			Y2{1}			
μ_3	.736	.009	.000	.022	.012	7
μ_4	.248	.005	.887	.072	.012	10
μ_5	.000	.985	.099	.016	.975	42
			Y2{2}			
μ_3	.747	.001	.000	.013	.000	7
μ_4	.190	.001	.880	.068	.001	10
μ_5	.049	.999	.108	.001	.998	153
			Y2{3}			
μ_3	.648	.000	.000	.012	.000	7
μ_4	.157	.000	.867	.069	.000	10
μ_5	.183	1.000	.120	.008	1.000	475
			Y2{4}			
μ_3	.512	.000	.000	.011	.000	7
μ_4	.121	.000	.984	.066	.000	10
μ_5	.357	1.000	.002	.063	1.000	1166

the weakest case Y1{0}, having a condition index of 10 (and a correlation of .78). The involvement of C1 begins to be observed by Y1{1}, but does not show itself completely until Y1{2} and Y1{3}. By contrast, the involvement of the dominant column, C2, along with the generated column 5 is observed from the outset. The same pattern occurs within the supplementary regressions. The regression parameter of C1 is insignificant in case Y1{0}, becomes significant in Y1{1}, and takes the proper order of magnitude (unity) in Y1{2} and Y1{3}.

Exhibit 3.11b   Regression of C5 on C1, C2, C3, and C4

Data Matrix	$r(C5, \widehat{C5})$ $\widehat{C5} = 0.1C1 + C2$	C1	C2	C3	C4	R^2
Y2{0}	.395	0.3789 [0.16]	0.4585 [2.10]	260.8090 [2.18]	47.9654 [0.46]	.1433
Y2{1}	.962	0.0862 [0.12]	1.0613 [16.64]	−35.0793 [−1.00]	27.0102 [0.88]	.9294
Y2{2}	.997	0.2077 [1.15]	1.0179 [60.94]	−13.8609 [−1.52]	−0.4803 [−0.06]	.9943
Y2{3}	.999	0.1331 [2.28]	1.0078 [187.80]	−5.0668 [−1.73]	−1.0565 [−0.41]	.9994
Y2{4}	1.000	0.0845 [3.58]	1.0009 [460.06]	−0.2182 [−0.18]	1.3050 [1.25]	.9999

3. Aggravating the scale problem in the **Y2** series (**C1** now accounts for less than $1/100$ of 1% of the variance in **C5**) has the expected effect. Now the involvement of column 1 is becoming apparent only by the tightest case **Y2{4}** with a condition index of 1166.

4. The several other general patterns noted in experiment 1 seem still to hold: a dependency's effects are beginning to be observed with condition indexes around 10; decreasing the variance of the generated dependency by successive factors of 10 causes the condition index roughly to progress as $10, 30, 100, 300, 1000$.

Y3:   The **Y3**$\{i\}$ series consists of a five-column data matrix in which the fifth column is in a simple relation (3.20c) with the fourth column, $C5 = C4 + e_i$. In this case **C4** is inventory investment, a rate-of-change variate. The results of this series are given in Exhibits 3.12a and 3.12b.

The following points are noteworthy:

1. An interesting phenomenon emerges in case **Y3{0}** that is in need of explanation and provides us with the first opportunity to apply our diagnostic tools to a near dependency that arises naturally in the data, that is, to one not artificially generated. First we note that the generated dependency between **C5** and **C4** is indeed observed—associated with the weak condition index 5. In addition, however, we note that over 80% of var(b_2) and var(b_3) is associated with the larger condition index 11,

Exhibit 3.12a Variance-decomposition proportions and condition indexes. *Y3 series.* One constructed near dependency (3.20c): $C5 = C4 + e_i$ $(i = 0, \ldots, 4)$

Associated Singular Value	var(b_1)	var(b_2)	var(b_3)	var(b_4)	var(b_5)	Condition Index, η
			Y3{0}			
μ_1	.005	.003	.003	.012	.012	1
μ_2	.018	.007	.015	.094	.222	3
μ_3	.953	.180	.112	.003	.001	8
μ_4	.020	.810	.870	.013	.034	11
μ_5	.004	.001	.001	.877	.731	5
			Y3{1}			
μ_3	.706	.133	.118	.027	.019	8
μ_4	.000	.855	.693	.004	.022	11
μ_5	.270	.001	.171	.948	.939	15
			Y3{2}			
μ_3	.953	.171	.102	.000	.000	8
μ_4	.017	.757	.751	.002	.000	11
μ_5	.005	.063	.132	.996	.997	42
			Y3{3}			
μ_3	.956	.180	.115	.000	.000	8
μ_4	.018	.789	.859	.000	.000	11
μ_5	.001	.020	.008	.999	.999	147
			Y3{4}			
μ_3	.890	.179	.107	.000	.000	8
μ_4	.016	.790	.798	.000	.000	11
μ_5	.071	.020	.078	1.000	1.000	416

indicating at least their involvement in a low-level, unintended "background" dependency. The simple correlation between C2 and C3 is very low, .09,[47] so we must look further than a simple dependency between C2 and C3. Further examination of Y3{0} in Exhibit 3.12a shows that there is really not one, but two unintended near dependencies of roughly equal intensity in the basic data matrix associated with the effectively equal condition indexes 11 and 8. Furthermore, these two background

[47]Indeed the highest simple correlation between the four basic columns is −.32.

Exhibit 3.12b   Regression of C5 on C1, C2, C3, and C4

Data Matrix	$r(C5, C4)$	C1	C2	C3	C4	R^2
Y3{0}	.643	0.0015 [0.27]	0.0004 [0.81]	−0.2014 [−0.74]	0.9180 [3.82]	.4231
Y3{1}	.950	−0.0027 [1.93]	0.0001 [0.62]	0.0845 [1.20]	0.9491 [15.38]	.9188
Y3{2}	.995	0.0002 [0.36]	0.0001 [1.23]	−0.0450 [1.80]	1.0048 [45.67]	.9902
Y3{3}	.999	0.0000 [0.17]	−0.0000 [−0.65]	0.0029 [0.40]	1.0004 [161.11]	.9992
Y3{4}	1.000	−0.0000 [−1.31]	−0.0000 [−0.67]	0.0035 [1.42]	1.0013 [455.66]	.9999

dependencies together account for over 95% of $var(b_1)$, $var(b_2)$, and $var(b_3)$, and the three roughly equal condition indexes 5, 8, and 11 account for virtually all of each of the five variances. Applying what we have learned from experiments 1 and 2, we conclude that there are three weak dependencies of roughly equal intensity whose individual effects cannot be separated, a problem we have seen arise when there are several condition indexes of the same order of magnitude. Since we know C5 and C4 are related, we would expect to find two additional near dependencies among the four columns C1, C2, C3, and C4.[48] Indeed, regressing C1 and C3 separately on C2 and C4 gives[49]

$$C1 = 0.0540C_2 + 16.67C_4 \qquad R^2 = .8335$$
$$\quad\;\; [5.86] \qquad\;\; [1.96]$$

$$C3 = 0.0015C_2 + 0.0373C_4 \qquad R^2 = .9045.$$
$$\quad\;\; [8.43] \qquad\;\; [2.27]$$

These background dependencies are, of course, also present in the Y1

[48]Even though $var(b_4)$ is not greatly determined by the condition indexes of 8 and 11, C4's involvement in these two dependencies cannot be ruled out. The variance-decomposition proportions, as we have seen, can be arbitrarily distributed among η's of nearly equal magnitude.

[49]The choice of C1 and C3 on C2 and C4 is arbitrary. Any two of the four variates with a nonvanishing Jacobian could be selected for this descriptive use of least squares. The figures in the square brackets are t's, not standard deviations, and the R^2's are the ratio of predicted sum of squares (not deviations about the mean) to actual sum of squares since there is no constant term.

and **Y2** series (the first four columns being the same in all **Y** series), but their effects there are overshadowed by the presence of the relatively stronger contrived dependency involving **C1**, **C2**, and **C5**. The experience we have gained from these experiments in the use of these diagnostic techniques, however, has clearly led us very much in the right direction.

2. The previously described phenomenon serves to emphasize the point that when two or more condition indexes are of equal or close magnitude, care must be taken in applying the diagnostic test. In such cases the variance-decomposition proportions can be arbitrarily distributed across the roughly equal condition indexes so as to obscure the involvement of a given variate in any of the competing (nearly equal) near dependencies. In **Y3**{0}, for example, the fact that over 80% of var(b_2) and var(b_3) is associated with the single condition index of 11 need not imply only a simple relation between **C2** and **C3**. Other variates (here **C1** and **C4**) associated with competing condition indexes (8 and 5) can be involved as well. Furthermore, when there are competing condition indexes, the fact that a single condition index (like 8 in **Y3**{0}) is associated with only one high variance-decomposition proportion (95% of var(b_1)), need not imply, as it otherwise could,[50] that the corresponding variate (**C1**) is free from involvement in any near dependency. Its interrelation with the variates involved in competing dependencies must also be investigated.

In sum, when there are *competing dependencies* (condition indexes of similar value), they must be treated together in the application of the diagnostic test. That is, the variance-decomposition proportions for each coefficient should be aggregated across the competing condition indexes, and high variance-decomposition aggregate proportions for two or more variances associated with the set of competing high indexes are to be interpreted as evidence of degrading collinearity. The exact involvement of specific variates in specific dependencies cannot be learned in this case, but it is still possible to learn (1) which variates are degraded (those with high *aggregate* variance-decomposition proportions) and (2) the number of near dependencies present (the number of competing indexes).

3. Another, quite different, form of confounded involvement is also exemplified by the foregoing: the *dominant dependency*. Column 4 is apparently involved simultaneously in several near dependencies: weakly, and with scaling problems, in the dependencies associated with η's of 8 and 11; and without scaling problems in the contrived dependency between **C4** and **C5**. In all cases, but particularly as it becomes tighter, this latter dependency dominates the determination of var(b_4), thereby obscuring the weak in-

---

[50]See, however, point 3 following.

volvement of C4 in the other dependencies. Dominant dependencies (condition indexes of higher magnitude), then, can mask the simultaneous involvement of a single variate in weaker dependencies. That is, the possibility always exists that a variate having most or all of its variance determined by a dependency with a high condition index is also involved in dependencies with lower condition indexes, unless, of course, that variate is known to be buffered from the other dependencies through near orthogonality.

4. From within the intricacies of the foregoing points, however, one must not lose sight of the fact that the test for potentially damaging collinearity requires the *joint* condition of (1) two or more variances with high decomposition proportions associated with (2) a single *high* condition index;[51] condition 1 by itself is not enough. It is true in the $Y3\{0\}$ case, for example, that the three condition indexes 5, 8, and 11 account for most of the variance of all five estimates, but by very rough standards, these condition indexes are not high, and the data matrix $Y3\{0\}$, quite likely, could be suitable for many econometric applications. Let us examine this condition further. In our prior examples we noted that contrived dependencies began to be observed when their "tightness" resulted in condition indexes of around 10. We were also able to calculate the correlations that correspond to these relations, so we can associate the magnitudes of condition indexes with this more well-known measure of tightness. A glance through Exhibits 3.5 to 3.12 shows that condition indexes of magnitude 10 result from underlying dependencies whose correlations are in the range of .4 to .6, relatively loose relations by much econometric experience. It is not until condition indexes climb to a level of 15–30 that the underlying relations have correlations of .9, a level that much experience suggests is high.[52] Further insight is afforded by examining the actual variances whose decomposition proportions are given in Exhibit 3.12a; these are presented in Exhibit 3.13.

In the case of $Y3\{0\}$, all variance magnitudes are relatively small, certainly in comparison to the size attained by $\mathrm{var}(b_4)$ and $\mathrm{var}(b_5)$ in the cases of $Y3\{2\}$ and $Y3\{3\}$, when the contrived dependency between them becomes tighter. In short, high variance-decomposition *proportions* surely need not imply large *component values*. This merely restates the notion of

---

[51]As we have just seen in points 2 and 3 above, condition (1) requires some modifications when there are either competing or dominating dependencies. These modifications are treated fully in Section 3.4.

[52]Indeed the experience so far indicates that the condition index goes one further step in the 10, 30, 100, 300, 1000 progression as successive "9's" are added to the underlying correlation. For example, .5→10, .9→30, .99→100, .999→300, and so on.

Exhibit 3.13 Underlying regression variances*

Data Matrix	var(b_1)	var(b_2)	var(b_3)	var(b_4)	var(b_5)
Y3{0}	9.87	16.94	16.32	3.71	3.04
Y3{1}	11.43	16.74	16.94	25.58	26.38
Y3{2}	9.90	17.56	18.18	207.76	204.17
Y3{3}	9.86	16.78	16.06	2559.53	2547.85
Y3{4}	10.59	16.79	17.28	20389.78	20457.85

*Representing the respective Y3{i} matrices by **X**, the figures reported here are diagonal elements of $(\mathbf{X}^T\mathbf{X})^{-1}$—the ϕ_{kk}'s of (3.10)—and do not include the constant factor of s^2, the estimated error variance. Of course, s^2 can only be calculated once a specific **y** has been regressed on **X**.

Section 3.2 that degraded estimates (capable of being improved if calculated from better conditioned data), which apparently can result from even low-level dependencies, need not be harmful; harmfulness depends, in addition, on the specific regression model employing the given data matrix, on the variance σ^2, and on the statistical use to which the results are to be put. For greater details, see Appendix 3D.

5. The contrived dependency between C4 and C5, both rates-of-change variates, seems to behave somewhat differently from previous experience, based on "levels" data; namely, its condition index is lower for comparable tightness in the underlying relation as measured by correlations. Perusal of Exhibits 3.5 to 3.11 indicates that, quite roughly, the condition index jumps one step along the 10, 30, 100, 300, 1000 progression each time another "9" digit is added to the correlation of the underlying dependency. That is, an η of 10 has a corresponding correlation of about .5; an $\eta \simeq 30$, correlation .9; $\eta \simeq 100$, correlation .99; $\eta \simeq 300$, correlation .999. Exhibit 3.12 indicates the rate-of-change data to be one step lower, with $\eta \leqslant 10$, correlation .6; $\eta \simeq 10$, correlation .9; $\eta \simeq 30$, correlation .99, and so on. It may be, therefore, that there is no simple pairing of the level of the strength of a relation as measured by a condition index with that of the same relation as measured by a correlation. There does, however, seem to be stability in the relative magnitudes of these two measures along the progressions noted above.

6. In all of the foregoing, one should not lose sight of the fact that, basically, the diagnostic procedure works in accord with expectations. The contrived relation between C4 and C5 is observed from the outset and takes on unmistakable form once it is removed from the background, in case Y3{1} or Y3{2}.

Exhibit 3.14 Variance-decomposition proportions and condition indexes. *Y4 series.* Two constructed near dependencies $(3.20a)$ and $(3.20c)$: $C5 = C1 + C2 + v_i$ (selected values); $C6 = C4 + v_k$ (selected values)

Associated Singular Value	var(b_1)	var(b_2)	var(b_3)	var(b_4)	var(b_5)	var(b_6)	Condition Index, η
			Y4{0,0}				
μ_1	.004	.001	.002	.008	.001	.008	1
μ_2	.010	.002	.009	.105	.002	.223	3
μ_3	.788	.036	.001	.001	.079	.004	7
μ_4	.181	.051	.981	.004	.100	.044	11
μ_5	.017	.910	.006	.008	.817	.001	17
μ_6	.001	.000	.000	.875	.001	.720	5
			Y4{0,2}				
μ_1	.003	.001	.002	.000	.001	.000	1
μ_2	.011	.002	.008	.002	.003	.002	3
μ_3	.778	.036	.001	.000	.077	.000	8
μ_4	.186	.044	.822	.001	.092	.000	11
μ_5	.015	.916	.009	.000	.772	.000	17
μ_6	.006	.000	.159	.996	.055	.997	47
			Y4{1,1}				
μ_3	.477	.015	.045	.024	.002	.014	8
μ_4	.001	.035	.777	.006	.013	.025	11
μ_5	.346	.948	.010	.000	.984	.005	43
μ_6	.167	.006	.158	.948	.000	.936	17
			Y4{1,2}				
μ_3	.587	.018	.027	.000	.003	.000	8
μ_4	.023	.028	.847	.001	.011	.000	11
μ_5	.173	.453	.068	.234	.543	.250	39
μ_6	.209	.501	.048	.762	.442	.747	51
			Y4{3,3}				
μ_4	.016	.000	.954	.000	.000	.000	11
μ_5	.884	1.000	.025	.003	1.000	.003	428
μ_6	.000	.000	.007	.997	.000	.997	162

Y4:   In the $Y4\{i, k\}$ series the two dependencies of Y1 and Y3 occur simultaneously. Here $C5 = C1 + C2 + v_i$ according to (3.20*a*) and $C6 = C4 + v_k$ as in (3.20*c*). What is new to be learned from this experimental series can be seen from a very few selected variance-decomposition proportion matrices. These are reported in Exhibit 3.14.

The following points are noteworthy:

1. Both relations are observable even in the weakest instance of $Y4\{0,0\}$, one with condition index 17, the other with condition index 5. The presence of the contrived relation between C1, C2, and C5 has somewhat masked the background relation among C1, C2, and C3 that was observed in the Y3 series (although var(b_1) is still being distributed among these relations).

2. Dependencies with differing condition indexes tend to be separately identified, as in the cases of $Y4\{0,2\}$, $Y4\{1,1\}$, and $Y4\{3,3\}$. When the condition indexes are nearly equal, however, as in the case $Y4\{1,2\}$, the involvement of separate variates is confounded between the two dependencies. This fact, observed frequently before, is particularly important in this instance. In earlier experiments, roughly equal condition indexes corresponded to roughly equal underlying correlations. In this case, however, the relation between the "rate-of-change" variates C4 and C6 is .9 while that underlying the relation among C1, C2, and C5 is .99, one 9 stronger. Thus the problem of confounding of relations results from relations of nearly equal tightness as judged by condition indexes, not as judged by correlations.

3. In general, however, the two constructed relations behave together quite independently, and much as they did separately. This was true in experiment 1 as well; the individual behavior of the dependencies in the X1 and X2 series was carried over to their simultaneous behavior in the X3 series. Thus, with the exception of the minor problem of confounded proportions that results from the presence of near dependencies with competing or dominating condition indexes, it seems fair to conclude that the simultaneous presence of several near dependencies poses no critical problems to the analysis.

Experiment 3: The Z Matrices.   The purposes of experiment 3 are (1) to analyze slightly larger data matrices (up to eight columns) to see if size has any notable effect on the procedure; (2) to allow up to three coexisting near dependencies, again to see if new complications arise; (3) to recast some previous experimental series in a slightly different setting to see if their behavior remains stable; (4) to create cases where near orthogonality among data series exists in order to observe its buffering effect against dependencies within nearly orthogonal subgroups. Toward this last objective, columns 4 and 5 of the basic data matrix were generated having correlations with columns 1–3 of no more than .18. Columns 1–3 here are the same as columns 1–3 in the previous experiment 2. Four dependency relations are contrived according to (3.22a–d). Equations (3.22a and b) generate dependencies between the two columns 4 and 5 which were constructed to have low intercorrelations with columns 1–3. Equation (3.22c) generates a dependency using only C1, C2, and C3, which is thereby buffered from C4 and C5. These two data groups are bridged by (3.22d). There are scaling problems built into the generated dependencies. The following are dominated: DUM1 [<.1% of the variance in (3.22b)]; *LHTUR [<.1% of the variance in (3.22c)]; and DUM2 [<.1% of the variance in (3.22d)].

Many of the Z series experiments were designed to duplicate previous experiments with different data in order to observe whether the process exhibits some degree of stability. In those cases where such stability exists, such as Z1 below, and the experiment merely becomes repetitive, it will be reported as such without additional and unnecessary tabulations.

Z1: In this series $C6 = C4 + e_i$. In this basic data matrix **Z**, C4 (DUM1) is generated to have the same mean and variance as column 4, the rate-of-change variate (GV58), of the basic data matrix **Y** of experiment 2. Hence the Z1 series is quite similar to the Y3 series of experiment 2, and we had hoped that this experimental series would exhibit similar properties. This expectation was met in full.

Z2: In this series the dependency is generated by the two "isolated" columns, 4 and 5, by $C6 = C5 - C4$. It also mixes a rate-of-change variate, C4, and a levels variate, C5, and, as noted, has a scaling problem. Exhibit 3.15 presents two Π matrices for cases Z2{2} and Z2{4}.

The following points are noteworthy:

1. The "background" relations with condition indexes 8 and 11 are still present (the first three columns here are the same as in experiment 2).

2. The generated dependency is quite observable, but the scaling problem is evident. Even in Z2{2} with a condition index of 104, the involvement of C4 is not clearly observed, and does not become strongly evident until the condition index increases to the very high value of 1000.

3. The isolation of C1–C3 from C4–C6 is very evident. Even in the case of

Exhibit 3.15 Variance-decomposition proportions and condition indexes. *Z2 series.* One constructed near dependency (3.22*b*): $C6 = C5 - C4 + e_i$ (selected values)

Associated Singular Value	var(b_1)	var(b_2)	var(b_3)	var(b_4)	var(b_5)	var(b_6)	Condition Index, η
			Z2{2}				
μ_1	.004	.003	.002	.000	.000	.000	1
μ_2	.000	.000	.000	.825	.000	.000	2
μ_3	.047	.114	.043	.019	.002	.002	6
μ_4	.944	.162	.085	.016	.000	.000	8
μ_5	.001	.721	.836	.000	.000	.000	11
μ_6	.004	.000	.034	.140	.998	.998	104
			Z2{4}				
μ_5	.001	.661	.839	.000	.000	.000	11
μ_6	.003	.088	.029	.817	1.000	1.000	1039

Z3{4}, the high condition index of 1000 does not add any significant degradation to var(b_1), var(b_2), or var(b_3).

Z3: This series, in which $C6 = 3C1 + 1.5C3 + e_i$, is very similar to the Y1 series and shows effectively identical behavior. The scaling problem here is severe, and the involvement of the dominated variate C3 is not strong even in the Z3{4} case, as is seen by the one relevant row of the Π matrix:

Associated Singular Value	var(b_1)	var(b_2)	var(b_3)	var(b_4)	var(b_5)	var(b_6)	Condition Index, η
			Z3{4}				
μ_6	1.000	.027	.451	.072	.000	1.000	980

Z4: In this series there is the single contrived relation $C6 = C2 + 0.7C5 + e_i$. The behavior is as according to expectation, paralleling that of the qualitatively similar Y1 and X2 series.

Z5: This series possesses two simultaneous dependencies, each isolated from the other by low intercorrelations among the C1–C3 and C4–C5 columns of the basic data matrix Z. Here $C6 = C5 - C4 + e_i$ and $C7 = 3C1 + 1.5C3 + u_j$. A typical Π matrix for this series is given by Exhibit 3.16.

Exhibit 3.16 Variance-decomposition proportions and condition indexes. *Z5 series*. Two constructed near dependencies $(3.22b)$ and $(3.22c)$: $C6 = C5 - C4 + e_i$ $(i=2)$; $C7 = 3C1 + 1.5C3 + u_j$ $(j=3)$

Associated Singular Value	$\text{var}(b_1)$	$\text{var}(b_2)$	$\text{var}(b_3)$	$\text{var}(b_4)$	$\text{var}(b_5)$	$\text{var}(b_6)$	$\text{var}(b_7)$	Condition Index, η
				$Z5\{2,3\}$				
μ_1	.000	.002	.001	.000	.000	.000	.000	1
μ_2	.000	.000	.000	.742	.000	.000	.000	2
μ_3	.000	.029	.005	.029	.002	.002	.000	6
μ_4	.000	.243	.090	.005	.000	.000	.000	8
μ_5	.000	.711	.602	.000	.000	.000	.000	12
μ_6	.000	.000	.021	.121	.950	.949	.000	114
μ_7	.999	.015	.280	.102	.048	.049	.999	368

The following points are noteworthy:

1. The presence of the two relations is clear, and the scaling problems that beset the two relations are observed.

2. Of principal interest is the verification of the expected simultaneous isolation of the relation among C7, C1, and C3 from that among C4, C5, and C6. The low intercorrelations of these two sets of columns allows the variances within each group to be unaffected by the relation among the other group.

3. Although not shown, it should be noted that the usual confounding of relations occurs in this series when the condition numbers are of equal magnitude.

Z6: This case has two contrived near dependencies: $C6 = C4 + e_i$ and $C7 = C2 + 0.7C5 + u_j$. The results are fully in accord with expectations.

Z7: This case presents the first occurrence of three simultaneous relations $C6 = C4 + e_i$, $C7 = 3C1 + 1.5C3 + u_j$, and $C8 = C2 + 0.7C5 + v_k$. Exhibit 3.17 displays three selected cases.

The following points are noteworthy:

1. The presence of three simultaneous near dependencies causes no special problems, each behaving essentially as it did separately.

2. The $Z7\{2,2,3\}$ case illustrates the separate identification of all three

Exhibit 3.17 Variance-decomposition proportions and condition indexes. *Z7 series.* Three near dependencies (3.22a), (3.22c), and (3.22d): C6 = C4 + e_i (selected values); C7 = 3C1 + 1.5C3 + u_j (selected values); C8 = C2 + 0.7C5 + v_k (selected values)

Associated Singular Value	var(b_1)	var(b_2)	var(b_3)	var(b_4)	var(b_5)	var(b_6)	var(b_7)	var(b_8)	Condition Index, η
				Z7{2,2,3}					
μ_5	.000	.000	.657	.001	.020	.001	.003	.000	11
μ_6	.000	.000	.076	.835	.007	.852	.000	.000	35
μ_7	.970	.001	.071	.083	.000	.073	.973	.000	153
μ_8	.028	.999	.177	.078	.829	.072	.025	1.000	455
				Z7{2,3,3}					
μ_6	.000	.000	.071	.886	.007	.900	.000	.000	35
μ_7	.714	.105	.332	.104	.098	.089	.718	.108	345
μ_8	.286	.895	.003	.006	.734	.007	.282	.892	482
				Z7{3,2,2}					
μ_6	.401	.370	.017	.818	.065	.815	.380	.388	221
μ_7	.161	.113	.020	.179	.048	.182	.169	.094	116
μ_8	.436	.506	.068	.003	.124	.003	.451	.516	164

relationships, although the severe scaling problem of **C**3 is masking its influence in the relation associated with μ_7.

3. The other two cases exemplify the problem of separating the individual relationships when the condition indexes are of the same order of magnitude. In $\mathbf{Z}7\{2,3,3\}$ the two relations with similar condition indexes 345 and 482 are confounded; while in $\mathbf{Z}7\{3,2,2\}$, the involved variates have the variance of their estimated regression parameter distributed over the three dependencies with roughly equal condition indexes 221, 116, and 164.

One final conclusion may be drawn rather generally from experiment 3; namely, those Z series that are qualitatively similar to previous X and Y series, result in quantitatively similar Π matrices and condition indexes, attesting to a degree of stability in the diagnostic procedure.

3.4 SUMMARY, INTERPRETATION, AND EXAMPLES OF DIAGNOSING ACTUAL DATA FOR COLLINEARITY

In Section 3.2 a test was suggested for diagnosing the presence of collinearity in data matrices and for assessing the degree to which such near dependencies degrade ordinary least-squares regression estimates. Section 3.3, recognizing the empirical element to this diagnostic procedure, reported a set of experiments designed to provide experience in its use and interpretation. This section summarizes and exemplifies the foregoing. First, the experimental evidence of Section 3.3 is distilled and summarized. A summary set of steps to be followed in employing the diagnostic procedure on actual data sets is then provided, and, finally, four examples of the use of the diagnostic procedure on actual data are given.

Interpreting the Diagnostic Results: A Summary of the Experimental Evidence

Before proceeding with a summary of the evidence, it is worth noting that the experiments of Section 3.3 are necessarily limited in scope and cannot hope to illuminate all that is to be known of the behavior of the proposed diagnostic procedure in all econometric and statistical applications. Indeed, it is to be expected, as experience is gained from future application of these techniques to actual data, that the conclusions presented here will be refined and expanded. For the moment, however, the experimental evidence is gratifyingly stable and provides an excellent point of departure.

This summary begins with a presentation of the experience gained from experiments having a single contrived near dependency. We then summarize the modifications and extensions that arise when analyzing data matrices in which two or more near dependencies coexist.

Experience With a Single Near Dependency

1. *The Diagnostic Procedure Works.* The diagnostic test suggested in Section 3.2 works well and in accord with expectations for a variety of data matrices with contrived dependencies. It is possible not only to determine the presence of the near dependency, but also, subject to the qualifications given below, to determine the variates involved in it.

2. *The Progression of Tightness.* The tighter the underlying dependency (as measured either by its correlation or relevant multiple correlation), the higher the condition index. Indeed, as the underlying correlations or R^2's increase along the progression $<.9, .9, .99, .999, .9999$, and so on, the condition indexes increase roughly along the progression $3, 10, 30, 100, 300, 1000, 3000$, and so on. The correspondence between these two progressions, however, is not constant and depends on the type of data. A given correlation, for example, among rates-of-change data appears to be translated into a lower condition index than for levels data. Some rough generalizations do, however, seem warranted, and these are given next.

3. *Interpreting the Magnitude of the Condition Index.* Most of the experimental evidence shows that weak dependencies (correlations of less than .9) begin to exhibit themselves with condition indexes around 10, and in some cases as low as 5. An index in the neighborhood of 15–30 tends to result from an underlying near dependency with an associated correlation of .9, usually considered to be the borderline of "tightness" in informal econometric practice. Condition indexes of 100 or more appear to be large indeed, causing substantial variance inflation and great potential harm to regression estimates.

4. *Variance-Decomposition Proportions.* The rule of thumb proposed at the end of Section 3.3, that estimates shall be deemed degraded when more than 50% of the variance of two or more coefficients is associated with a single high condition index, still seems good. Future experience may suggest a more appropriate or a more sophisticated rule of thumb, but the 50% rule allows the involved variates to be identified in most instances even when the underlying dependency is reasonably weak (associated correlations of .4 to .7). Indeed, most evidence indicated proportions of over 80% were attained quite early.

5. *Scaling Problems.* Essential scaling imbalance causes the involvement of the dominated variates to be masked and more difficult to detect. Essential scaling imbalance occurs when several variates are interrelated so that the variance introduced by some is very much smaller than that introduced by others. Variates introducing less than 1% of the total variation are dominated, and their involvement can be completely overlooked by this procedure until the condition index rises to 30 or more. Very strongly dominated variates (<.01%) can be masked even with condition index in excess of 300.

6. *Data Type Matters.* As already noted in 2 above, near dependencies among rates-of-change data seem to behave slightly differently from those involving levels type data.

Experience With Coexisting Near Dependencies

7. *Retention of Individuality.* While some new problems of diagnosis and interpretation are introduced, in general it can be concluded that coexisting near dependencies cause the diagnostic procedure no critical problems. Subject to the modifications given in 11–13 below, the several underlying near dependencies behave together much as they did separately. In particular they remain *countable* (8 below) and to a great degree *separable* (9 below).

8. *Countability.* The number of coexisting near dependencies is correctly assessed in all cases by the number of high condition indexes. The presence of a very strong ($\eta > 30$) near dependency, for example, does not obsecure detection of a much weaker coexisting near dependency.

9. *Separability.* The near dependencies remain separable in the following two senses. First, near dependencies which, when existing alone, have a given condition index, retain roughly the same condition index when made to coexist with other near dependencies, regardless of their relative condition indexes. Second, subject to the qualifications given below, the individual involvement of specific variates in specific near dependencies remains observable.

10. *Isolation through Near Orthogonality.* As the theory of Section 3.2 would have it, near orthogonality does indeed buffer the regression estimates of one set of variates from the deleterious effects of near dependencies among the nearly orthogonal variates.

11. *Confounding of Effects with Competing Dependencies.* When two or more near dependencies are competing, that is, have condition indexes of the same order of magnitude, the high variance-decomposition proportions of the variates involved in the separate competing dependencies can be

arbitrarily distributed among them, thus confounding their true involvement. The number of coexisting dependencies is, however, not obscured by this situation, nor is the identification of the variates that are involved in at least one of the competing dependencies. It remains possible, therefore, to diagnose how many dependencies are present and which variates are being degraded by the joint presence of those dependencies. Only information on the separate involvement of specific variates in specific competing dependencies is lost. In this case the test procedure is trivially modified to examine those variates which have high variance-decomposition proportions aggregated over the competing high condition indexes.

12. *Dominating Dependencies.* A dominating dependency, one with a condition index of higher order of magnitude, can become the prime determinant of the variance of a given coefficient and thus obscure information about its simultaneous involvement in a weaker dependency. Consider the example of Exhibit 3.18.

Here there are two dependencies with high condition indexes, 30 and 300; and 300 dominates. The involvement of C3 and C4 in this dominant dependency is clear; however, equally clearly, we cannot rule out the potential involvement of C3 and C4 along with C1 in the dependency associated with $\eta = 30$. Thus, when a dependency is dominated, such as $\eta = 30$ above, it is quite possible for only one high variance-decomposition proportion to be associated with it and still give indication of degradation —the possible involvement of other variate(s) being obscured by the dominant relation. In this case our diagnostic procedure must once again be qualified: two or more high variance-decomposition proportions associated with a single high condition index—*unless that high condition index is dominated by an even larger one, in which case further investigations may be required.* One reasonable procedure to adopt in such cases would be to run an auxiliary regression among the potentially involved variates (C1 on C3 and C4 in the above example) to verify their roles, if such

Exhibit 3.18 Illustrative variance-decomposition proportions and condition indexes

Associated Singular Value	var(b_1)	var(b_2)	var(b_3)	var(b_4)	Condition Index, η
μ_1	.00	.01	.00	.00	1
μ_2	.01	.99	.00	.00	3
μ_3	.99	.00	.01	.01	30
μ_4	.00	.00	.99	.99	300

Exhibit 3.19 Illustrative variance-decomposition proportions and condition indexes

Associated Singular Value	var(b_1)	var(b_2)	var(b_3)	var(b_4)	Condition Index, η
μ_3	.99	.99	.01	.01	30
μ_4	.00	.00	.99	.99	300

information were required. In this example such additional information would be needed to demonstrate the degradation of var(b_1). There is no question that var(b_3) and var(b_4) are degraded—not just by their presence in one, but possibly two, dependencies. However, var(b_1) cannot be said to be degraded unless C1 can be shown to be involved in a linear dependency with C3 and/or C4.

By way of contrast, had the last two rows of the above example read as in Exhibit 3.19, the degradation of all variances would be apparent without further analysis, and auxiliary regressions would not be required unless it was explicitly desired to know whether C3 and C4 entered along with C1 and C2 in the dependency with $\eta = 30$.

13. *Nondegraded Estimates.* On occasion it is also possible to identify those coefficients whose estimates show no evidence of being degraded by the presence of near dependencies. In the example given by Exhibit 3.19, all four variances show degradation due to the two near dependencies with η's of 30 and 300. In the example given by Exhibit 3.18, however, var(b_2) has virtually all of its variance determined in association with the relatively small condition index 3 and is not adversely affected by the two tighter dependencies with η's of 30 and 300. The same situation arises in Section 3.3, for example, for var(b_2) of the X1 or X2 series given in Exhibits 3.5a and 3.6a.

Just where the dividing line between small and large is to be set is a matter that can be answered only with greater practical experience in the use of these techniques. The evidence of the experiments suggests that η's of 10 to 30 are good starting points.[53] We maintain that 10 is a bit on the weak side; 30 seems quite reasonable in almost all instances.

Employing the Diagnostic Procedure

Diagnosing any given data matrix for the presence of near dependencies and assessing the potential harm that their presence may cause regression

---

[53]Compare, however, point 6, page 154.

estimates is effected by a rather straightforward series of steps, the only problems of interpretation arising when there are competing or dominating near dependencies. Two thresholds must be determined at the outset, a condition-index cutoff η^* and a variance-decomposition proportion cutoff π^*, as is seen in steps 3 and 5.

The Steps. It is assumed in the following that the user has a specific parameterization β in mind and has transformed the data (if need be) to conform, so that the model becomes $y = X\beta + \varepsilon$ (see Appendix 3B). Also, an intercept term, if relevant to the model, should remain explicit so that X has a column of ones. Centering the data in this case can mask the role of the constant in any underlying near dependencies and produce misleading diagnostic results.

STEP 1. Scale the data matrix X to have unit column length.[54]

STEP 2. Obtain the singular-value decomposition[55] of X, and from this calculate:
   a. the condition indexes η_k as in (3.7) and
   b. the Π matrix of variance-decomposition proportions as in Exhibit 3.2.

STEP 3. Determine the number and relative strengths of the near dependencies by the condition indexes exceeding some chosen threshold η^*, such as $\eta^* = 10$, or 15, or 30.[56]

STEP 4. Examine the condition indexes for the presence of competing dependencies (roughly equal condition indexes) and dominating dependencies (high condition indexes—exceeding the threshold determined for step 3—coexisting with even larger indexes.)

STEP 5. Determine the involvement (and the resulting degradation to the regression estimates) of the variates in the near dependencies. For this step, some threshold variance-decomposition proportion, π^*, must be chosen ($\pi^* = 0.5$ has worked well in practice). Three cases are to be considered.
   Case 1. Only one near dependency present. A variate is involved in, and its estimated coefficient degraded by, the single near dependency if it is one of two or more variates with variance-decomposition proportions in excess of some threshold value π^*,

---

[54]The need for this is discussed in Section 3.3 and, more rigorously, in Appendix 3B.

[55]Programs effecting this decomposition are discussed in the text below.

[56]Choosing this threshold is akin to choosing a test size (α) in standard statistical hypothesis testing—and only practical experience will help determine a useful rule of thumb. $\eta^* = 15$ or 30 seems a good start. As a matter of practice it seems reasonable to ignore all condition indexes below the threshold as being too weak for further consideration, regardless of what patterns of variance-decomposition proportions may be associated with them.

such as 0.50. Presumably, if only one high variance-decomposition proportion is associated with this single highest condition index, no degradation is exhibited.[57]

Case 2. Competing dependencies. Here involvement is determined by aggregating the variance-decomposition proportions over the competing condition indexes (see point 11 above). Those variates whose *aggregate* proportions exceed the threshold π^* are involved in at least one of the competing dependencies, and therefore have degraded coefficient estimates. In this case, it is not possible exactly to determine in which of the competing near dependencies the variates are involved.

Case 3. Dominating dependencies.[58] In this case (1) we cannot rule out the involvement of a given variate in a dominated dependency if its variance is being greatly determined by a dominating dependency, and (2) we cannot assume the noninvolvement of a variate even if it is the only one with a high proportion of its variance associated with the dominated condition index—other variates can well have their joint involvement obscured by the dominating near dependency. In this case additional analysis, such as auxiliary regressions, is warranted, directly to investigate the descriptive relations among all of the variates potentially involved. See point 12 above.

STEP 6. Form the auxiliary regressions. Once the number of near dependencies has been determined, auxiliary regressions among the indicated variates can be run to display the relations. A simple procedure for forming the auxiliary regressions is described below.

STEP 7. Determine those variates that remain unaffected by the presence of the collinear relations. See point 13 above.

Once the **X** matrix has been analyzed and the potential harm to regression estimates has been assessed, it is possible to analyze the quality of an actual regression based on those data. In particular, one can often learn the following:

1. How many near dependencies plague a given data set and what they are.

---

[57]This situation has, as yet, not occurred in practice, and as long as the data matrix has been properly scaled, as in step 1, it does not seem likely that it will (cf. footnote 40, page 120).
[58]The joint occurrence of dominating and competing dependencies causes no additional difficulties. The competing dependencies, whether dominated or dominating, are merely treated as one in association with their aggregate variance-decomposition proportions.

2. Which variates have coefficient estimates adversely affected by the presence of those dependencies.

3. Whether estimates of interest are included among those with inflated confidence intervals, and therefore whether corrective action (obtaining better conditioned data or applying Bayesian techniques) is warranted.

4. Whether, rather generally, prediction intervals based on the estimated model are greatly inflated by the presence of ill-conditioned data.

5. Whether specific coefficient estimates of interest are relatively isolated from the ill effects of collinearity and therefore trustworthy in spite of ill-conditioned data.

Forming the Auxiliary Regressions. Once the number of near dependencies has been determined, a simple procedure can be used to form the auxiliary regressions that will display them. This is exemplified in the following hypothetical variance-decomposition proportions matrix of a seven-column data matrix, **X**.

Variance-decomposition proportions and condition indexes

Associated Singular Value	var(b_1)	var(b_2)	var(b_3)	var(b_4)	var(b_5)	var(b_6)	var(b_7)	Condition Index, η
μ_1	.0	.0	.0	.0	.0	.0	.0	1
μ_2	.0	.0	.0	.0	.0	.1	.0	2
μ_3	.0	.0	.0	.0	.0	.0	.0	3
μ_4	.0	.0	.1	.0	.0	.6	.0	5
μ_5	.0	.1	⑥	.0	.1	.1	.1	30
μ_6	.5	⑧	.3	.1	.1	.1	.0	50
μ_7	.5	.1	.0	.9	.8	.1	⑨	100

Here we see that there are three near dependencies among the seven variates, associated with condition indexes 30, 50, and 100. Hence, in a sort of reduced-form, we can express three of the seven variates in terms of the remaining four. Choose as the three variates for which to "solve" those three (associated with the circles in the above **Π** matrix) which are most obviously associated with one each of the three separate dependencies. Beginning with the strongest near dependency ($\eta = 100$), either **C4** or **C7** can be picked. Since C7 has the remainder of its variance determined in a more removed dependency, problems from competing relations are minimized if it is picked rather than **C4**. Similarly, it is seen that **C2** and **C3** are most obviously associated with the second ($\eta = 50$) and third ($\eta = 30$) near dependencies, respectively. Hence we pick C7, C2, and C3 as the "pivots" to regress separately on the remaining variates C1, C4, C5, and C6.

There are, of course, many other ways in which such auxiliary regressions could be constructed. Indeed, the specific context of any given data set may suggest a natural set of pivots. In the absence of any other considerations, however, the procedure described above has the advantages that (1) it is simple to employ, (2) it picks as a "dependent" variate for each auxiliary regression one that is known to be strongly involved in the underlying near dependency, and (3) by the procedure itself, the right-hand variates (the remaining set of "regressors") of the auxiliary regressions will be relatively well conditioned.

Software. The computational foundation of the diagnostic procedure reported here is the singular-value decomposition[59] of step 2, a computational routine whose accessibility would seem to be somewhat limited. However, a library called *EISPACK—Release 2* contains a very efficient SVD algorithm and already has been made available to over 200 university computer facilities.[60] Furthermore, an interactive routine has been designed specifically to effect the computational steps 1, 2, and 6 and exists as part of the TROLL system at the MIT Center for Computational Research in Economics and Management Science.[61]

Applications with Actual Data

With one very interesting exception,[62] we have, until now, employed the proposed diagnostic procedure just summarized only on data matrices with contrived near dependencies. We turn now to analyses of four matrices of actual data to see how the procedure fares when dealing with examples of naturally occurring, uncontrived near dependencies. The first example

---

[59]As noted in Section 3.2, the eigenvectors of $X^T X$ and the positive square roots of its eigenvalues provide identical information as the SVD of X, but this is not recommended, for calculations based on $X^T X$ are computationally very much less stable than those based on X when X is ill conditioned—the case that is central to this analysis. (Cf. footnote 32, page 114.)

[60]Copies of EISPACK-Release 2 and further information on it may be obtained from Dr. Wayne Cowell, Argonne Code Center, Argonne National Laboratories, Argonne, Illinois 60439.

[61]The software, known as VARDCOM, is currently incorporated as part of the SENSSYS system at the MIT Center for Computational Research in Economics and Management Science. Information available from Information Processing Services, Publications Office, Room 39–484, MIT, Cambridge, MA 02139.

[62]The unexpected weak, "background" dependency that we discovered when we examined the experimental series Y3.

utilizes the data of the Bauer matrix, introduced in a different context in Section 3.2. The second example examines data familiar to all econometricians, those relevant to an annual, aggregate consumption function. The third example provides diagnostics of the conditioning of the Friedman data that are used in Section 4.3 in the analysis of a monetary equation and will be more fully motivated there. The final example uses data from an equation of the IBM econometric model.

The Bauer Matrix. The modified Bauer matrix, we recall from Section 3.2, had an exact contrived dependency ($C4 = 0.5C5$) between its last two columns, which were in turn orthogonal to the first three. Its purpose there was to exemplify the isolation from collinearity that is afforded those variates that are orthogonal, or nearly so, to the variates involved in the offending near dependencies. In examining the Π matrix, given in Exhibit 3.4, of the Bauer matrix, the involvement of $\mathrm{var}(b_4)$ and $\mathrm{var}(b_5)$ in the exact contrived dependency was clearly observed as well as the isolation of the first three variances from it. But, in addition, there appeared an unexpected occurrence: over 97% of $\mathrm{var}(b_1)$, $\mathrm{var}(b_2)$, and $\mathrm{var}(b_3)$ was associated with the singular value μ_3. We were not prepared at that time to pursue this naturally arising phenomenon, but now we are.

First we note that the variance-decomposition proportions of Exhibit 3.4 and the corresponding singular values cannot be given a wholly meaningful interpretation since they are based on data that have not been column scaled as is required in step 1. Hence, in Exhibit 3.20 we present the Π matrix and condition indexes for the column-scaled Bauer matrix.

In analyzing Exhibit 3.20, it proves instructive to feign ignorance of any prior knowledge we have of the properties of the Bauer matrix to see how well the mechanism discovers what there is to know.

The first and obvious fact is that there are two near dependencies with condition indexes greater than 10. One is dominating; none is competing.

Exhibit 3.20 Variance-decomposition proportions and condition indexes, scaled Bauer matrix

Associated Singular Value	$\mathrm{var}(b_1)$	$\mathrm{var}(b_2)$	$\mathrm{var}(b_3)$	$\mathrm{var}(b_4)$	$\mathrm{var}(b_5)$	Condition Index, η
μ_1	.000	.000	.000	.000	.000	1.0
μ_2	.005	.005	.000	.000	.000	1.0
μ_3	.001	.001	.047	.000	.000	1.3
μ_4	.994	.994	.953	.000	.000	16.0
μ_5	.000	.000	.000	1.000	1.000	2×10^{16}

The dominating dependency is clearly very tight, having the astronomically large condition index of 2×10^{16} and involving only columns 4 and 5. Hence it is safe to conclude that the involvement of C1, C2, and C3 in this dependency is minimal, if any.

The second dependency (and the one that we are really interested in here) possesses the weak to moderate condition index of 16. Clearly, at least the first three columns, C1, C2, and C3, are involved in this dependency, but one cannot rule out the potential involvement of C4 and C5, their roles being masked by their involvement in the dominant dependency.

We may display these two dependencies through auxiliary regressions; we need only to choose the two variates to act as dependent variates, the three remaining being independent. In this case, choosing one of C1, C2, or C3 and one of C4 or C5 is clearly appropriate. Exhibit 3.21 presents auxiliary regression results with C1 and C4 chosen as the two dependent variates to be regressed on C2, C3, and C5. The regressions are based on unscaled data, so that the dependencies are displayed in terms of the original data relationships.

Both near dependencies are clearly displayed. In the first, we see the dominant, essentially perfect relation true of the Bauer data given in Section 3.2 in which C4=0.5C5, exactly. The noninvolvement of C2 and C3 in this relation is also discovered. In the second, we see a weak to moderate ($R^2 = .98$) relation involving C1, C2, and C3, but not C5. This is the naturally occurring dependency whose presence was first suggested in Section 3.2 and is now verified. One can now conclude that all five regression estimates based on this matrix are degraded to varying degrees by the presence of two collinear relations. The variances for the coefficients of C4 and C5 are obviously very seriously degraded, while those for C1, C2, and C3 are considerably less so.

Exhibit 3.21   Auxiliary regressions,* Bauer data (unscaled)

	Coefficient of			
	C2	C3	C5	R^2
C4	0.0000	0.0000	0.5000	1.000
	[0.0]	[0.0]	[∞]	
C1	−0.7008	−1.2693	0.0000	.9820
	[−14.4]	[−7.5]	[0.0]	

*Figures in square brackets are t-statistics.

It is fair to conclude that the diagnostic procedure, when applied to the Bauer matrix, has been very successful in uncovering all relevant properties of the near dependencies contained in it.

The Consumption Function. A relation of great importance in economics is the annual, aggregate consumption function, and so we analyze the following matrix of consumption-function data:

$$\mathbf{X} = [\text{CONST}, \text{C}(T-1), \text{DPI}(T), r(T), \Delta\text{DPI}(T)],$$

where CONST is a column of ones (the constant term),
$\qquad$ C is total consumption, 1958 dollars,
$\qquad$ DPI is disposable personal income, 1958 dollars,
$\qquad$ r is the interest rate (Moody's Aaa).
and all series are annual, 1948–1974.

It must be emphasized that no attempt is being made here to analyze the consumption function itself. There are many well-known, sophisticated alterations to basic consumption data involving, for example, per-capita weightings, disaggregations, wealth effects, and recognition of simultaneity. Our interest here necessarily centers on analysis of one fundamental variant without regard to additional econometric refinements; namely,

$$\text{C}(T) = \beta_1 + \beta_2\text{C}(T-1) + \beta_3\text{DPI}(T) + \beta_4 r(T) + \beta_5\Delta\text{DPI}(T) + \varepsilon(T).$$

$$(3.24)$$

Estimation of (3.24) with ordinary least squares results in

$$\text{C}(T) = \underset{(3.827)^*}{6.7242} + \underset{(0.2374)}{0.2454} \ \text{C}(T-1) \tag{3.25}$$

$$+ \underset{(0.2076)}{0.6984} \ \text{DPI}(T) - \underset{(1.838)}{2.2097} \ r(T) + \underset{(0.1834)}{0.1608} \ \Delta\text{DPI}(T).$$

$$R^2 = .9991 \quad \text{SER} = 3.557 \quad \kappa(\mathbf{X}) = 376 \quad \text{DW} = 1.87$$

*Numbers in parentheses are standard errors.

Only one of these parameter estimates, that of DPI, is significant by a standard t-test; however, few econometricians would be willing to reject the hypotheses that the other β's, either jointly or singly, are significantly different from zero. Furthermore, few econometricians would be happy with the prediction intervals that would result from such a regression. This dissatisfaction stems from the widely held belief that the consumption-function data are highly ill conditioned and that estimates

Exhibit 3.22 Correlation matrix for consumption-function data

	$C(T-1)$	DPI(T)	$r(T)$	ΔDPI(T)
$C(T-1)$	1.000			
DPI(T)	.997	1.000		
$r(T)$	.975	.967	1.000	
ΔDPI(T)	.314	.377	.229	1.000

based on them are too noisy to prove conclusive or useful.[63] A mere glance at the simple correlation matrix for these data, given in Exhibit 3.22, partially confirms this belief. But how ill conditioned are these data? How many near dependencies exist among them and how strong are they? Which variates are involved in them giving evidence of degradation? Which estimates might benefit most from obtaining better conditioned data or from the introduction of appropriate information through a Bayesian prior? Answers to these questions, of course, cannot be obtained from Exhibit 3.22 alone, but can be obtained from an analysis of the Π matrix and condition indexes for the consumption-function data. For this analysis, we are interested only in moderate to strong near dependencies, and so we set the condition-index threshold to $\eta^* = 30$. We continue to employ a variance-decomposition proportion threshold of $\pi^* = 0.50$. Steps 1 and 2 of the diagnostic procedure applied to the consumption-function data result in the Π matrix given in Exhibit 3.23.

Exhibit 3.23 shows the existence of two near dependencies, one dominant with a large condition index of 376 and one strong with a condition index of 39. The dominant relation involves $C(T-1)$, DPI(T),

Exhibit 3.23 Variance-decomposition proportions and condition indexes, consumption-function data

Associated Singular Value	CONST var(b_1)	$C(T-1)$ var(b_2)	DPI(T) var(b_3)	$r(T)$ var(b_4)	ΔDPI(T) var(b_5)	Condition Index, η
μ_1	.001	.000	.000	.000	.001	1
μ_2	.004	.000	.000	.002	.136	4
μ_3	.310	.000	.000	.013	.000	8
μ_4	.264	.004	.004	.984	.048	39
μ_5	.420	.995	.995	.000	.814	376

[63]Indeed, few functions have received greater attention than the consumption function in efforts made to overcome the ill-conditioned data and refine its estimation.

and ΔDPI(T). The variable $r(T)$ does not seem to be involved in this dependency, but it is likely that the constant term, CONST, is being shared in both. The weaker dependency definitely includes $r(T)$; all other variates are potentially involved, their effects clearly being dominated by their involvement in the stronger dependency with $\eta = 376$.

Auxiliary regressions are required in this case to determine those variates involved in the weaker of the two dependencies. One possible choice for the two dependent variates of these auxiliary regressions would be DPI(T) and $r(T)$. Exhibit 3.24 reports these results.

We verify that the dominant relation does involve CONST, C($T-1$), DPI(T), and ΔDPI(T), and note that the weaker involves at least CONST, C($T-1$), and $r(T)$.

Quite generally, then, we may conclude that the data on which the consumption-function regression (3.25) is based possess two strong near dependencies (one very strong). Furthermore, each variate is involved in one or both of these near dependencies, and each is degraded to some degree by their presence. It would appear that the estimates of coefficients of C($T-1$) and DPI(T) are most seriously affected, followed closely by that for ΔDPI(T), these variates being strongly involved in either the tighter of the two dependencies or both. The estimate of the coefficient of $r(T)$ is adversely affected by its strong involvement in the weaker of the two dependencies, but, in our experimental experience, we found η's of 39 to be large, and the R^2 in Exhibit 3.24 confirms this here. Thus we see that all parameter estimates in (3.25), and their estimated standard errors, show great potential for refinement through better conditioning of the estimation problem, either from more appropriate modeling or the introduction of better conditioned data or appropriate prior information. One would be loath to reject, for example, the role of interest rates in the aggregate consumption function on the basis of the estimates of (3.25); and one would feel even more helpless in predicting the effects of a change in r on

Exhibit 3.24 Auxiliary regressions,* consumption-function data (unscaled)

	Coefficients of				
	CONST	C($T-1$)	ΔDPI(T)	R^2	η
DPI(T)	-11.5472	1.1384	.8044	.9999	376
	[-4.9]	[164.9]	[11.9]		
$r(T)$	-1.0244	0.0174	-0.0145	.9945	39
	[-3.9]	[22.3]	[-1.9]		

*Figures in square brackets are t-statistics.

aggregate consumption from a regression equation like (3.25). Thus the econometrician's intuitive dissatisfaction with estimates of the aggregate consumption function, and his seemingly never-ending efforts to refine them, seem fully justified.

Several additional points of interest arise from this example, some of which suggest future directions for research. First, it is not surprising that the estimated coefficient of DPI(T) demonstrates statistical significance even in the presence of the extreme ill conditioning of the consumption-function data, for C(T) and DPI(T) are phenomenally highly correlated (.9999). Indeed, it is in light of this high correlation that the seriousness of the degradation of the estimate of this parameter can be seen, for its standard error is nevertheless quite large, resulting in the very broad 95% confidence interval of [.28, 1.11]. Second, as seen from Exhibit 3.23, no one near dependency dominates the determination of the variance of the estimate of the constant term. This estimate is nevertheless degraded since nearly 70% of the variance is associated with the two near dependencies, as is verified by the auxiliary regressions in Exhibit 3.24. This lack of dominance is to be contrasted with the estimates of the coefficients of C($T-1$) and DPI(T), which also clearly enter both near dependencies but are greatly dominated by the stronger of the two. This situation suggests, in accord with intuition, that it is possible for a variate that is weakly involved in a strong near dependency to be confounded with one that is more strongly involved in a weaker near dependency. Similar results occur in the experiments of Section 3.3, but not in such a way that any definite conclusions can be drawn. Further experimentation will be needed directly to test this suggestion. Third, within a given near dependency, there appears to be a strong rank correlation between the relative size of the variance-decomposition proportions of the variates involved and their t-statistics in the corresponding auxiliary regressions. Comparing the variance-decomposition proportions for the near dependency with $\eta = 376$ in Exhibit 3.23 and the corresponding t's for the DPI(T) regression in Exhibit 3.24 exemplifies the point. Of course, allowance must be made for relations that are dominated (such as the one with $\eta = 39$) or are competing, but again there is considerable support, but no substantiation, for such a hypothesis from the experiments of Section 3.3, and further experiments aimed directly to this point are suggested. Fourth, even with this "real-world" data, the relative progression between correlations and condition indexes summarized in point 2 above continues to hold. The near dependencies of the consumption data are of orders of magnitudes 30 and 300, two steps apart along the progression 3, 10, 30, 300, and so on. Similarly, the R^2's of the auxiliary regressions reported in Exhibit 3.24 are .99 and .9999, two steps apart along the 9's progression .9,

.99, .999, .9999, and so on. Fifth, we once again note the ability of these diagnostic tools to uncover complex relations among three or more variates that are overlooked by simple correlation analysis, a problem first raised in the introduction. The simple correlation matrix in Exhibit 3.22 surely tells us that $DPI(T)$ and $C(T-1)$ are closely related; but the role of $\Delta DPI(T)$ (or, equivalently, the role of $DPI(T-1)$) is not at all observable from this information. The largest simple correlation with $\Delta DPI(T)$ is under .4. The role of $\Delta DPI(T)$ in a near dependency along with $C(T-1)$ and $DPI(T)$, however, is readily apparent from the variance-decomposition proportions matrix of Exhibit 3.23.

The Friedman Data. As a further example of the use of the collinearity diagnostics, we analyze the conditioning of a body of monetary data. These data, relevant to the equation for the household demand for corporate bonds in the Friedman (1977) model, are introduced in greater detail in Chapter 4. We treat these data rather clinically here, without concern for their economic meaning, deferring definitions of the variable names and motivation for the model to Section 4.3. The reader may find it of interest to return to this example once that section has been read. We also see in Section 4.3 that it is possible to use ridge regression to ameliorate somewhat the collinearity that we soon observe to beset the Friedman data, with the result that the instability in the estimated coefficients is greatly reduced.

The Friedman data consist of 56 observations on seven variables, the first of which is a constant variate. We continue to set η^* at 30 and π^* at 0.50. Steps 1 and 2 applied to these data result in the Π matrix given in Exhibit 3.25.

Exhibit 3.25 Variance-decomposition proportions and condition indexes, Friedman monetary data*

Associated Singular Value	var(b_1)	var(b_2)	var(b_3)	var(b_4)	var(b_5)	var(b_6)	var(b_7)	Condition Index, η
μ_1	.000	.000	.000	.000	.000	.000	.000	1
μ_2	.004	.002	.000	.023	.000	.001	.008	6
μ_3	.001	.004	.003	.014	.096	.000	.007	9
μ_4	.024	.028	.008	.200	.022	.000	.005	15
μ_5	.091	.046	.018	.024	.114	.007	.009	21
μ_6	.001	.224	.151	.596	.532	.028	.468	48
μ_7	.878	.695	.819	.142	.235	.963	.502	112

*For definitions of the variables and motivation for the model, see Section 4.3.

Two near dependencies with $\eta > \eta^* = 30$ are observed, one dominant with an η of 112. Clearly columns 1, 2, 3, 6, and 7 are involved in this stronger dependency, while columns 4, 5, and 7 (marginally) are involved in the weaker near dependency with $\eta = 48$. Further, columns 1, 2, 3, and 6 could conceivably be involved in this weaker dependency, their effects possibly being masked by the dominant dependency associated with $\eta = 112$. Auxiliary regressions are required to obtain more detailed information on the exact makeup of these two near dependencies. From among the many ways two of these variables could be chosen to be written as a linear combination of the remaining five, examination of Exhibit 3.25 would suggest pivoting on C4 and C6 to form these auxiliary regressions, since these two columns show simultaneously maximum involvement in one near dependency and minimum involvement in the other. Thus we regress C6 and C4 on the remaining columns C1, C2, C3, C5, and C7 to obtain Exhibit 3.26.

From Exhibit 3.26 we see that all the variates C1, C2, C3, C5, and C7 are strongly involved in the dominant near dependency along with C6, while only C2, C3, C5, and C7 enter into the weaker dependency along with C4. The usual progression of R^2 versus η is again evident, although the level is different here from that for the consumption-function data. The η of 48 here corresponds to a slightly lower R^2 of .98 than was the case for the consumption function, where an η of 39 corresponded to a near dependency having an R^2 of .99. This highlights the fact, noted earlier, that condition indexes and correlations (multiple correlations) provide similar information about the relative tightness of a linear relationship, but do not convey the same information regarding absolute levels of tightness. It remains an open and interesting question which measure, if either, provides the better information regarding the potential harm to regression

Exhibit 3.26 Auxiliary Regressions,* Friedman monetary data†

	Coefficient of						
	C1	C2	C3	C5	C7	R^2	η
C6	−1.818	0.0158	2.3506	3.6033	0.3614	.9989	112
	[−14.0]	[9.6]	[13.2]	[7.5]	[16.8]		
C4	0.0146	0.0103	1.3364	−2.1861	−0.1458	.9835	48
	[0.1]	[7.5]	[9.0]	[−5.5]	[−8.1]		

*Figures in square brackets are t-statistics.
†For definitions of variables and motivation for the model, see Section 4.3.

estimates due to a given near dependency. In any event, it is clear that the information from auxiliary regressions is quite complementary with that obtained from the condition indexes and the variance-decomposition proportions; the two together provide a powerful and efficient tool for uncovering and analyzing the presence, degree, and content of linear near dependencies among data series.

An Equation from the IBM Econometric Model. The next example of the collinearity diagnostics makes use of a data set brought to our attention by Harry Eisenpress of IBM. It serves well both to expand our understanding of the means by which this diagnostic technique can distinguish between two seemingly intertwined linear relations within a data set and to provide an excellent example of the practical distinction between harmful and degrading collinearity made in Section 3.2.

The basic regression model employed in this analysis is of the form

$$\text{NONDUR}(T) = \beta_1 + \beta_2 \text{RATINC}(T) + \beta_3 \text{NONDUR}(T-1) + \varepsilon(T),$$

$$(3.26)$$

where

$$\begin{aligned}
\text{NONDUR} =\ & \text{the ratio of nondurables-and-services consumption} \\
& \text{to deflated discretionary income} \\
\text{RATINC} =\ & \text{the ratio of current deflated discretionary income} \\
& \text{to its lagged value.}
\end{aligned}$$

Employing quarterly data from 1955-1 to 1973-4, this equation is estimated as

$$\text{NONDUR}(T) = \underset{(0.0656)}{0.6906} - \underset{(0.0613)}{0.6653} \text{RATINC}(T) + \underset{(0.0342)}{0.9794} \text{NONDUR}(T-1)$$

$$(3.27)$$

$$R^2 = .9226 \quad \text{SER} = 0.0043 \quad \text{DW} = 2.19 \quad \kappa(\mathbf{X}) = 305.$$

The estimated standard errors are given in parentheses, and on the basis of the t-statistics (all of which are in excess of 10), the three coefficients individually differ significantly from zero. At the same time, the condition number of the scaled $\mathbf{X}$ matrix (of order 76×3 including the constant term) is 305, indicating, according to our previous experience, at least one very strong near dependency among the three columns of $\mathbf{X}$. Thus, while the individual t-statistics are good, there is evidence that a more detailed analysis of the possible sources of collinearity within $\mathbf{X}$ is nevertheless

Exhibit 3.27   Variance-decomposition proportions, IBM equation

Associated Singular Value	CONST var(b_1)	RATINC(T) var(b_2)	NONDUR($T-1$) var(b_3)	η
μ_1	.000	.000	.000	1
μ_2	.045	.085	.975	138
μ_3	.954	.914	.024	305

warranted. Exhibit 3.27 presents the variance-decomposition proportions for this data set.

Exhibit 3.27 reveals not one but two strong near dependencies among the columns of **X**, a dominant relation, with condition index 305, involving CONST and RATINC(T), and a dominated, but nevertheless strong, relation, with condition index 138, involving NONDUR($T-1$) and possibly CONST and/or RATINC(T), the effects of these latter two columns being masked by their involvement in the dominant near dependency. The dominant near dependency between CONST (the constant term) and RATINC(T) is not surprising, for RATINC(T), being a ratio of a relatively smooth time series to its lagged value, is clearly going to take on values around unity. Auxiliary regressions are required to ascertain the nature and extent of the second, dominated near dependency. Special care, however, is required in this case in forming and interpreting the auxiliary regressions.

Since there are two near dependencies among three variates, we can consider forming auxiliary relations between any two of the variates with the third. The two auxiliary regressions that are indicated from an examination of Exhibit 3.27 are between RATINC(T) and CONST on the one hand and between NONDUR($T-1$) and CONST on the other.[64] CONST, of course, is a constant term, so both of these relations involve a regression of a single variate on a constant term alone. While there is nothing to prevent one from running such a regression, it is also clear that R^2, as usually calculated, will necessarily be zero, and this measure will not serve to assess the strength of the auxiliary relations. Rather it will be required to draw on the well-known fact that the cosine of the angle θ between any two n-vectors **u** and **v** can be expressed as $\mathbf{u}^T\mathbf{v}/\|\mathbf{u}\|\|\mathbf{v}\|$. The square of this cosine bears a direct relation to R^2 and serves as an appropriate generalization of that concept when one of the two vectors has

[64]We could just as well have chosen CONST on RATINC(T) and NONDUR($T-1$) on RATINC(T), but we chose instead to put the constant term CONST on the right-hand side.

Exhibit 3.28   Auxiliary regressions,* IBM equation

		$\cos^2\theta$	Associated Condition Index
RATINC(T)	= 1.0092 CONST [1067.0]	.999934	305
NONDUR($T-1$)	= 0.9532 CONST [562.0]	.999763	138

*Figures in square brackets are t-statistics.

constant components.[65] Exhibit 3.28 presents the two auxiliary regressions along with the appropriate squared cosines.

Indeed we note that both near dependencies are very strong, but that the one associated with the dominant condition index 305 is the stronger,[66] and the usual "progression of nines" prevails. The diagnostics are therefore quite correct in indicating two strong near dependencies in the data set, and we have identified them. There does, however, remain one very interesting question: if collinearity is so bad among the columns of $\mathbf{X}$, how did the regression estimates in (3.27) seemingly turn out so well? Each estimated coefficient there, we recall, had a t in excess of 10. But we see that this is only part of the story.

It is appropriate at this point to reiterate the distinction made in Section 3.2 between degrading and harmful collinearity. The presence of collinear dependencies renders tests based on least-squares estimates for a given sample size less powerful than could otherwise be the case; that is, collinearity degrades regression estimates. The degradation need not, however, be great enough actually to cause trouble for some purposes; that

[65]For any two n-vectors $\mathbf{u}$ and $\mathbf{v}$ with angle θ between them we have $\cos(\theta) \equiv \mathbf{u}^T\mathbf{v}/\|\mathbf{u}\|\|\mathbf{v}\|$. If, in addition, $\mathbf{u}$ and $\mathbf{v}$ are centered about their means, this expression becomes r_{uv}, the simple correlation between the components of $\mathbf{u}$ and those of $\mathbf{v}$. The cosine of the angle between two centered vectors is thus directly related to the concept of correlation. Furthermore, it is readily demonstrated that the usual regression R^2 is the simple correlation between the true values of the response variable y and its least-squares fitted values $\hat{y}$. Thus $R^2 = \cos^2(\theta)$, where θ is the angle between $\tilde{y}$ and $\hat{\tilde{y}}$ and where the tilde indicates the vectors have been transformed into deviations about their means. Clearly this latter concept is of no use if either y or $\hat{y}$ has constant components. In this latter case, it is appropriate to consider the angle θ between y and $\hat{y}$ as a measure of their tightness of fit. When $\theta = 0°$, y and $\hat{y}$ lie on the same line and $\cos(\theta) = 1$, indicating this collinearity. When $\theta = 90°$, the two vectors are orthogonal, and $\cos(\theta) = 0$, indicating this lack of collinearity.

[66]It is of interest to note here that the angle between RATINC(T) and CONST is less than $\frac{1}{2}$ degree (about 28 minutes of 1 degree) and that between NONDUR($T-1$) and CONST is less than 0.9 degree (about 53 minutes of 1 degree).

is, it may not actually become harmful.[67] In the estimation of (3.27), for example, we can well assume that all of the regression coefficients are seriously degraded by their involvement in two strong near dependencies, and that our knowledge of all the estimates could be made even more precise if better conditioned data were employed. Such degradation clearly has not been harmful if our interest in (3.27) centers only on tests that the coefficients individually differ significantly from zero, for each coefficient passes this test with flying colors. If, however, our interest were in other tests of hypothesis, we might not be so fortunate. For example, it may well be of interest in a model of this sort with a lagged dependent variable to test the null hypothesis H_0: $\beta_3 = 1$, with an alternative hypothesis H_1: $\beta_3 < 1$. The calculated t for such a test here is .6023, and, on the basis of these data, we may not reject H_0. However, because we know that the estimate b_3 on which this test is based is being degraded by its inclusion in a strong near dependency, we are less willing actually to accept H_0 rather than to feel that the test consequently lacks power and is inconclusive. This test of hypothesis, therefore, is actually being harmed by the presence of degrading collinearity in the sense that there is reason to believe that the introduction of better conditioned data would result in a more refined estimate of β_3, and with it a more conclusive test of H_0: $\beta_3 = 1$. If the test of this hypothesis were truly important to the investigator, the present data set would not be optimal for his needs. He would clearly be better off with a data set in which the effects of NONDUR($T-1$) were not so confounded with those of CONST or RATINC(T). It may very well be the case that, even if such data were available, they would lead to the same outcome, that is, not to reject H_0: $\beta_3 = 1$. However, under these circumstances the investigator would have increased confidence in the conclusiveness of the test of hypothesis, knowing that the acceptance region had not been enlarged by ill-conditioned data. Collinearity is harmful, therefore, only if it is first degrading and then if, in addition, important tests *based on the degraded estimates* are considered inconclusive,[68] for these tests could be refined and made more trustworthy (even if the outcome is the same) when based on better conditioned data.

---

[67]It is recalled that degradation is based on an analysis of the data matrix X alone. The estimated variances, however, depend not only on the elements of $(\mathbf{X}^T\mathbf{X})^{-1}$ but also on the estimated standard error s^2. It could well be that degraded (inflated) elements of $(\mathbf{X}^T\mathbf{X})^{-1}$ are, for specific purposes, counteracted by a sufficiently small estimated regression error variance s^2. The degradation exists, nevertheless, and one would clearly be even better off without it.
[68]This occurs for a test of significance when it fails, and, rather more generally, for a test of hypothesis when one is unable to reject the null hypothesis.

It is important to note that the preceding discussion is not meant to suggest that an investigator should continue to seek out new data sets (should such riches be available) until a given hypothesis achieves a desirable outcome. Rather it is to say that, regardless of the desired outcome, tests of hypothesis of individual parameters[69] which are based on degraded estimates tend to lack power, the confidence intervals of the estimators being enlarged by the ill conditioning and, as such, the investigator is quite justified in viewing an outcome that lies in the "acceptance" region as being inconclusive. Of course, no similar assessment is warranted if the outcome falls in the rejection region, for one cannot be upset when an unpowerful test is nevertheless successful in rejecting a hypothesis.

APPENDIX 3A: THE CONDITION NUMBER AND INVERTIBILITY

In this appendix we examine a means for interpreting the relation that exists between the condition number of a matrix and the "invertibility" of that matrix. We see that the higher the condition number of a matrix, the greater is the potential sensitivity of elements of its inverse to small changes in the elements of the matrix itself. Sensitivity is measured by the economist's familiar notion of elasticity. In particular we see that twice the condition number of a real symmetric matrix A, that is, $2\kappa(A)$, provides an upper bound for the elasticity of the diagonal elements of A^{-1} with respect to elements of A. This result is then particularized to the special case where $A = X^T X$, and it is shown that $2\kappa(X)$ plays a similar role for the elasticity of the diagonal elements of $(X^T X)^{-1}$ with respect to elements of X. This latter result shows how the condition number provides a measure of the potential sensitivity of the estimated standard errors of regression coefficients to small changes in the data.

The elements of any matrix A are denoted by $A = (a_{rs})$ and those of the inverse (if A is square and invertible) by $A^{-1} = (a^{ij})$. The m rows of an $m \times n$ matrix are denoted according to

$$A = \begin{bmatrix} a_1^T \\ \vdots \\ a_m^T \end{bmatrix}.$$

---

[69]As well as most linear combinations of the estimators. As is well known [Theil (1971), pp. 148–152], however, tests based on some, but by no means all, linear combinations of degraded estimates may not be degraded by ill-conditioned data. See also Appendix 3B.

The notation $|\cdot|$ indicates absolute value, and $\|\mathbf{x}\|$ denotes the Euclidean length of a vector $\mathbf{x}$, that is, $(\Sigma_{i=1}^n x_i^2)^{1/2}$.

We first show a result applicable to any nonsingular matrix $\mathbf{A}$ which gains strength when applied to the case where $\mathbf{A}$ is a real symmetric matrix. Here we employ the elasticity notation $\xi_{rs}^{ij} \equiv (\partial a^{ij}/\partial a_{rs})(a_{rs}/a^{ij})$.

Theorem 1.   Let $\mathbf{A}$ be a nonsingular matrix with condition number $\kappa(\mathbf{A})$, then $|\xi_{rs}^{ij}| \leqslant c_{ij}\kappa(\mathbf{A})$ where $c_{ij} \geqslant 1$ for all i and j.

PROOF.   First we recall [Theil (1971), p.33] $\partial a^{ij}/\partial a_{rs} = -a^{ir}a^{sj}$, and from the SVD of $\mathbf{A} = \mathbf{U}\mathbf{D}\mathbf{V}^T$, we note $a_{rs} = \mathbf{u}_r^T \mathbf{D} \mathbf{v}_s$ and $a^{hk} = \mathbf{v}_h^T \mathbf{D}^{-1}\mathbf{u}_k$. Hence we may write

$$\xi_{rs}^{ij} \equiv \frac{\partial a^{ij}}{\partial a_{rs}} \frac{a_{rs}}{a^{ij}} = -\frac{(\mathbf{v}_i^T \mathbf{D}^{-1}\mathbf{u}_r)(\mathbf{v}_s^T \mathbf{D}^{-1}\mathbf{u}_j)(\mathbf{v}_s^T \mathbf{D}\mathbf{u}_r)}{(\mathbf{v}_i^T \mathbf{D}^{-1}\mathbf{u}_j)}. \tag{3A.1}$$

Taking absolute values and applying the Cauchy-Schwartz (C-S) inequality to the numerator produces

$$|\xi_{rs}^{ij}| = \frac{|\mathbf{v}_i^T \mathbf{D}^{-1}\mathbf{u}_r||\mathbf{v}_s^T \mathbf{D}^{-1}\mathbf{u}_j||\mathbf{v}_s^T \mathbf{D}\mathbf{u}_r|}{|\mathbf{v}_i^T \mathbf{D}^{-1}\mathbf{u}_j|} \leqslant \frac{(\mathbf{v}_i^T \mathbf{D}^{-1}\mathbf{v}_i)^{1/2}(\mathbf{u}_j^T \mathbf{D}^{-1}\mathbf{u}_j)^{1/2}}{|\mathbf{v}_i^T \mathbf{D}^{-1}\mathbf{u}_j|}$$

$$\times (\mathbf{u}_r^T \mathbf{D}^{-1}\mathbf{u}_r)^{1/2}(\mathbf{u}_r^T \mathbf{D}\mathbf{u}_r)^{1/2}(\mathbf{v}_s^T \mathbf{D}^{-1}\mathbf{v}_s)^{1/2}(\mathbf{v}_s^T \mathbf{D}\mathbf{v}_s)^{1/2}. \tag{3A.2}$$

Recalling that $\|\mathbf{u}_k\| = \|\mathbf{v}_k\| = 1$ for all k, and that $\mathbf{D}$ is diagonal, we employ the fact [Rao (1973)] that for any positive-definite matrix $\mathbf{B}$ and vector $\boldsymbol{\beta}$ with $\|\boldsymbol{\beta}\| = 1$, $\boldsymbol{\beta}^T \mathbf{B}\boldsymbol{\beta} \leqslant \lambda_{\max}$, where $\lambda_{\max}$ is the maximal eigenvalue of $\mathbf{B}$. Hence we have

$$|\xi_{rs}^{ij}| \leqslant c_{ij}\kappa(\mathbf{A}), \tag{3A.3}$$

where

$$c_{ij} = \frac{(\mathbf{v}_i^T \mathbf{D}^{-1}\mathbf{v}_i)^{1/2}(\mathbf{u}_j^T \mathbf{D}^{-1}\mathbf{u}_j)^{1/2}}{|\mathbf{v}_i^T \mathbf{D}^{-1}\mathbf{u}_j|} \geqslant 1,$$

by the C-S inequality.   ∎

We next examine the case where $\mathbf{A}$ is a nonsingular real symmetric matrix. Recognizing that changing a_{rs} now also changes a_{sr}, the elasticity of an element of $\mathbf{A}^{-1}$ with respect to a_{rs} becomes

$$
\tilde{\xi}_{rs}^{ij} = \begin{cases} \left(\dfrac{\partial a^{ij}}{\partial a_{rs}} + \dfrac{\partial a^{ij}}{\partial a_{sr}} \right) \dfrac{a_{rs}}{a^{ij}} & \text{for } r \neq s \\[3mm] \dfrac{\partial a^{ij}}{\partial a_{rr}} \dfrac{a^{rr}}{a^{ij}} & \text{for } r = s. \end{cases}
\tag{3A.4}
$$

Furthermore, along the diagonal of $\mathbf{A}^{-1}$, where $i = j$, we have

$$
\tilde{\xi}_{rs}^{ii} = \begin{cases} 2 \dfrac{\partial a^{ii}}{\partial a_{rs}} \dfrac{a_{rs}}{a^{ii}} = 2\xi_{rs}^{ii} & \text{for } r \neq s \\[3mm] \dfrac{\partial a^{ii}}{\partial a_{rr}} \dfrac{a_{rr}}{a^{ii}} = \xi_{rr}^{ii} & \text{for } r = s, \end{cases}
\tag{3A.5}
$$

where ξ_{rs}^{ij} is defined as in (3A.1).

Finally we note that the symmetry of $\mathbf{A}$ implies that its SVD takes the form $\mathbf{U}\mathbf{D}\mathbf{V}^T$ where, for all i, $\mathbf{v}_i = \alpha_i \mathbf{u}_i$, $\alpha_i = \pm 1$. Hence, along the diagonal where $i = j$, c_{ij} in Theorem 1 takes the values $c_{ii} = 1$ for all i. Joining this fact and Theorem 1 to (3A.5), we have just proved Theorem 2.

Theorem 2. Let $\mathbf{A}$ be a nonsingular real symmetric matrix with condition number $\kappa(\mathbf{A})$, then, for the diagonal elements $a^{ii} \neq 0$ of $\mathbf{A}^{-1}$, $|\tilde{\xi}_{rs}^{ii}| \leqslant 2\kappa(\mathbf{A})$ for $r \neq s$ and $|\tilde{\xi}_{rr}^{ii}| < \kappa(\mathbf{A})$ for all r.

From the point of view of users of least-squares regression, particular interest is attached to the case where $\mathbf{A} = \mathbf{X}^T\mathbf{X}$, where $\mathbf{X}$ is an $n \times p$ data matrix with condition number $\kappa(\mathbf{X})$. We now prove Theorem 3.

Theorem 3.[70] Let $\mathbf{X} = (x_{tk})$ be an $n \times p$ data matrix, and let $\mathbf{A} = \mathbf{X}^T\mathbf{X}$. Then $|\xi_{tk}^{ii}| \equiv (\partial a^{ii}/\partial x_{tk})(x_{tk}/a^{ii}) \leqslant 2\kappa(\mathbf{X})$ for $i, k = 1, \ldots, p$ and $t = 1, \ldots, n$.

---

[70]We are indebted to R. J. O'Brien of the University of Southampton for providing a proof (employed here, following equation (3A.6)) that substantially tightens our original bounds. In O'Brien (1975) a study of the sensitivity of OLS estimates to perturbations in the data is undertaken.

PROOF.

$$\xi_{tk}^{ij} \equiv \frac{\partial a^{ij}}{\partial x_{tk}} \frac{x_{tk}}{a^{ij}} = \left[\sum_{r,s} \frac{\partial a^{ij}}{\partial a_{rs}} \frac{\partial a_{rs}}{\partial x_{tk}} \right] \frac{x_{tk}}{a^{ij}}$$

$$= 2 \left[\sum_{s=1}^{p} \frac{\partial a^{ij}}{\partial a_{ks}} x_{ts} \right] \frac{x_{tk}}{a^{ij}}$$

$$= -2 \frac{a^{ik}}{a^{ij}} x_{tk} \sum_{s=1}^{p} a^{sj} x_{ts}$$

$$= -2 \frac{a^{ik}}{a^{ij}} x_{tk} \mathbf{x}_t^T \mathbf{a}^j$$

$$\equiv -2 \frac{a^{ik}}{a^{ij}} x_{tk} z_{tj}, \tag{3A.6}$$

where $\mathbf{x}_t^T$ is the tth row of $\mathbf{X}$ and $\mathbf{a}^j$ is the jth column of $\mathbf{A}^{-1}$, and where we note that $\sum_{s=1}^{p} a^{sj} x_{ts} \equiv \mathbf{x}_t^T \mathbf{a}^j$ is the (j,t) element of the $p \times n$ matrix $\mathbf{Z}^T \equiv (\mathbf{X}^T \mathbf{X})^{-1} \mathbf{X}^T$. Letting $\mathbf{z}_j$ be the jth column of $\mathbf{Z}$, we note that $|z_{tj}| \leqslant \|\mathbf{z}_j\| = (a^{jj})^{1/2}$, this latter since $\mathbf{Z}^T \mathbf{Z} = (\mathbf{X}^T \mathbf{X})^{-1}$. Likewise $|x_{tk}| \leqslant \|\mathbf{x}_k\| = (a_{kk})^{1/2}$. Hence for $i=j$, (3A.6) becomes

$$|\xi_{tk}^{ii}| = 2 \frac{|a^{ik}|}{a^{ii}} |x_{tk}||z_{ti}| \leqslant 2 \frac{|a^{ik}|}{a^{ii}} (a_{kk})^{1/2} (a^{ii})^{1/2}. \tag{3A.7}$$

Further, $\mathbf{A}$ positive definite implies $|a^{ik}| \leqslant (a^{ii})^{1/2} (a^{kk})^{1/2}$, resulting in

$$|\xi_{tk}^{ii}| \leqslant 2(a_{kk})^{1/2} (a^{kk})^{1/2}. \tag{3A.8}$$

Now, for any p-vector $\boldsymbol{\alpha}$ such that $\|\boldsymbol{\alpha}\| = 1$, $\boldsymbol{\alpha}^T \mathbf{A} \boldsymbol{\alpha} = \boldsymbol{\alpha}^T \mathbf{V} \mathbf{D}^2 \mathbf{V}^T \boldsymbol{\alpha} \equiv \boldsymbol{\beta}^T \mathbf{D}^2 \boldsymbol{\beta}$, where $\|\boldsymbol{\beta}\| \equiv \|\mathbf{V}^T \boldsymbol{\alpha}\| = 1$, and hence

$$\boldsymbol{\alpha}^T \mathbf{A} \boldsymbol{\alpha} \leqslant \mu_{max}^2. \tag{3A.9}$$

In particular, letting $\boldsymbol{\alpha}$ be the kth-component unit vector, (3A.9) becomes

$$a_{kk} \leqslant \mu_{max}^2. \tag{3A.10}$$

Using $\mathbf{A}^{-1}$ and $\mathbf{D}^{-2}$ in the above results in

$$a^{kk} \leqslant \frac{1}{\mu_{min}^2}, \tag{3A.11}$$

which implies, in conjunction with (3A.10), that

$$(a_{kk})^{1/2}(a^{kk})^{1/2} \leqslant \frac{\mu_{max}}{\mu_{min}} = \kappa(\mathbf{X}). \tag{3A.12}$$

Hence, (3A.8) becomes

$$|\xi_{tk}^{ii}| \leqslant 2\kappa(\mathbf{X}) \tag{3A.13}$$

∎

This last result is directly interpretable in a least-squares context, for it says that the elasticity of the variance of any least-squares estimate with respect to any element of the data matrix $\mathbf{X}$ is bounded by twice the condition number of $\mathbf{X}$. That is, $2\kappa(\mathbf{X})$ provides an upper bound to the possible sensitivity of the parameter variances to changes in $\mathbf{X}$. Since condition numbers in excess of 100 are not uncommon for econometric data matrices (the consumption-function data, we recall, have a κ of 370), a 1% change in any element of $\mathbf{X}$ could result in a $2 \times 100\%$ change in the variance of any estimate or, roughly, a 14% change in its standard error. It is to be emphasized that this result is an inequality, and hence shows the maximum potential sensitivity; it is not an immutable and incontrovertible fact of life. Experience shows that the sensitivity of economic data is usually considerably less than this maximum. Ill-conditioned data, however, obviously have the potential for causing troubles, and the condition index provides a quick measure of the extent of that potential.

APPENDIX 3B: PARAMETERIZATION AND SCALING

This appendix deals with two related issues that arise with respect to the general applicability of the collinearity diagnostics: the effect on the collinearity diagnostics due to linear transformations of the data (the problem of "parameterization") and the validity of and need for column scaling. The second issue has already been introduced in Section 3.3 but is given stronger justification in light of the results of Appendix 3A.

The Effects on the Collinearity Diagnostics Due to Linear Transformations of the Data

The collinearity diagnostics of this chapter analyze the suitability of the conditioning of the $n \times p$ matrix $\mathbf{X}$ for estimating the parameters $\boldsymbol{\beta}$ of the linear model $\mathbf{y} = \mathbf{X}\boldsymbol{\beta} + \boldsymbol{\varepsilon}$ by the technique of linear regression. It may be the case, however, that one is interested in estimating a reparameterized, but equivalent, version of this model in the form of

$$y = (XG^{-1})G\beta + \varepsilon \equiv Z\delta + \varepsilon,$$

where G is a $p \times p$ nonsingular matrix. Since the singular values of Z need not be the same as those of X, the question naturally arises as to whether the collinearity diagnostics when applied to Z can show few or no problems even if they reveal severe problems when applied to X. While no simple answer can be given to this question,[71] we are able to show here that the dependency of the collinearity diagnostics on the choice of parameters (or, equivalently on linear transformations of the data) in no way reduces the validity or usefulness of these diagnostic techniques for any particular application. Furthermore, we see that, in practice, most reparameterizations G leave the collinearity diagnostics little altered, if altered at all. And finally, we see that a linear transformation G^{-1} can "undo" ill conditioning in X only if the transformation G^{-1} is itself ill conditioned in a manner dependent on the nature of the ill conditioning of X. This proposition means (1) that such benignant transformations cannot be presumed to occur in practice, for the parameterization G is chosen on the basis of a priori modeling considerations whereas the ill conditioning of X results from chance outcomes in the data, and (2) that even if such a G were chosen, its ill conditioning would, as a matter of practice, provide unstable computation of G^{-1} and $Z = XG^{-1}$.

Each Parameterization Is a Different Problem. It is well known [Theil (1971), pp. 153–154; Malinvaud (1970), pp. 216–221; Silvey (1969)] that some linear combinations of regression parameters can be precisely estimated even if ill conditioning prevents precise knowledge of the specific parameters estimated. Therefore, should the investigator be interested in such linear combinations of the parameters, reparameterization is a benefit to his cause.[72] If, however, the investigator is not interested in such linear combinations,[73] but rather in estimates of the original parameters, then the fact that such linear combinations exist does him little good indeed. Thus,

---

[71]The solution to this problem depends on knowledge of the relation of the condition indexes of a given matrix X to those of the linear transform, XG^{-1}—and this latter problem stands as an interesting but unsolved problem of numerical analysis.

[72]Of course, testing hypotheses on linear combinations $\delta = G\beta$ of the parameters of a given model $y = X\beta + \varepsilon$ is equivalent to tests of hypotheses on the explicit parameters of an appropriately reparameterized model, $y = Z\delta + \varepsilon$, $Z = XG^{-1}$.

[73]Since the linear combinations of the parameters that can be known with precision depend on the eigenvalues of $X^T X$ [Silvey (1969)] and not on the investigator's model, it is unlikely that one will be interested in such a linear combination in practice.

each parameterization, with its corresponding data matrix, poses a separate problem. The investigator requires a diagnostic procedure that allows him to assess the suitability of the data for estimating the model relevant to his choice of parameterization. The diagnostics of Chapter 3 do just that. If the parameters of interest are the β's of $y = X\beta + \varepsilon$, then the diagnostics are to be applied to X, whereas if the parameters of interest are the δ's of $y = Z\delta + \varepsilon = (XG^{-1})G\beta + \varepsilon$, then the diagnostics should be applied to $Z \equiv XG^{-1}$. As a practical matter, however, we see that the diagnostics often lead to similar basic conclusions in both situations.

A simple example serves well here. Consider a Cobb-Douglas model $Q = AK^{\alpha}L^{\beta}\eta$ estimated as $lnQ = lnA + \alpha lnK + \beta lnL + \varepsilon$. Assume investigator 1 is interested in knowing the individual coefficients, α and β, whereas investigator 2 is interested only in returns-to-scale as measured by $\gamma = \alpha + \beta$. Investigator 1 takes the basic data matrix[74] $X = [\iota\ lnK\ lnL]$, while investigator 2 formulates the model as $lnQ = lnA + \gamma lnL + \phi(lnK - lnL) + \varepsilon$, where reparameterization occurs as[75]

$$\begin{bmatrix} \gamma \\ \phi \end{bmatrix} = \begin{bmatrix} 1 & 1 \\ 1 & 0 \end{bmatrix} \begin{bmatrix} \alpha \\ \beta \end{bmatrix},$$

that is, $\gamma = \alpha + \beta$, $\phi = \alpha$. For illustrative purposes, we assume that investigator 1 finds X ill conditioned on account of a single strong near dependency between lnL and lnK, and, as a result, he is unable to obtain the precise estimates he desires of α and β (although he gets a good estimate of the constant term, lnA). Investigator 2, however, finds, in estimating his model $y = Z\delta + \varepsilon$, where $Z = [\iota\ lnL\ (lnK - lnL)]$ and $\delta = (lnA, \gamma, \phi)^T$, that the estimate of $\gamma = \alpha + \beta$ is quite well determined, and he is happy. This stems from the fact that the single near dependency in X between lnL and lnK (which has wholly foiled investigator 1) has been transformed into a single near dependency in Z between ι (the constant term) and the now relatively constant variate $lnK - lnL$. Investigator 2 therefore finds the Z matrix unsuitable for estimation of lnA and ϕ, but quite useful to his purpose of estimating $\gamma = \alpha + \beta$. Of course, the happiness of investigator 2 in no way diminishes the sorrow of investigator 1; investigator 1's problem is real despite the fact that another parameterization need not suffer the same fate. One should not be surprised, then, by the fact that data that are harmfully ill conditioned for one parameterization need not be so for another. What is important to

---

[74]The term ι is a column of ones.

[75]We are ignoring the constant term here. It is the same in both parameterizations and merely adds an identity component to the transformation.

realize here is that the diagnostics will correctly assess the suitability of the data for each investigator's needs. Investigator 1, in applying the analysis to **X**, will discover the near dependency adversely affecting α and β and will be apprised of the unsuitability of this data set for his needs. Likewise, investigator 2, in analyzing **Z**, will discover the near dependency adversely affecting the estimates of lnA and ϕ, but will also be apprised of the suitability of the data for estimating $\gamma = \alpha + \beta$.

To summarize the foregoing, we see that each parameterization of a model presents an inherently different problem, reflecting different interests of the investigator and requiring different characteristics of the data. In general, once a parameterization β has been decided on, the data should be transformed (if need be) to conform, so that the model becomes $y = X\beta + \varepsilon$. Application of the diagnostics to **X** then assesses the suitability of **X** for estimating the specific parameters β. If the parameterization is to be changed, the data should be appropriately transformed and reanalyzed for their suitability to the new parameterization. In practice, however, analysis of the data in the form of **X** generally also tells a great deal about its suitability for other parameterizations. In particular, we see that reparameterization rarely undoes near dependencies; it merely alters their composition, as occurs in the case considered above.

A More General Analysis. As noted, no fully general analysis of the effect of linear transformations is possible, since there is no known relation, in general, between the condition indexes of **X** and those of **XA**, for **A** nonsingular. The following points, however, can serve to clarify the effects that such linear transformations of the data can have on the collinearity diagnostics:

1. Clearly, in the case of an exact dependency, no reparameterization can undo it, and so change the nature of the diagnostics. If there is a $c \neq 0$ such that $Xc = 0$, then for any $Z = XA$ (**A** nonsingular) there exists a $d = A^{-1}c \neq 0$ such that $Zd = 0$. Hence X^TX has a zero eigenvalue if and only if Z^TZ does, and the diagnostics would detect the dependency whether one analyzes **X** or **Z**.

2. When the dependencies are not exact, however, it is possible for a reparameterization to result in a better conditioned matrix. Consider the matrices

$$\mathbf{A} = \begin{bmatrix} 1 & (1-\alpha) \\ (1-\alpha) & 1 \end{bmatrix} \quad \text{and} \quad \mathbf{B} = \begin{bmatrix} \alpha & 0 \\ 0 & \alpha \end{bmatrix}.$$

As α goes to zero, $\mathbf{A}$ becomes ill conditioned and $\mathbf{B}$ does not. However, $\mathbf{AG}^{-1} = \mathbf{B}$, where

$$\mathbf{G}^{-1} = \frac{1}{2-\alpha}\begin{bmatrix} 1 & -(1-\alpha) \\ -(1-\alpha) & 1 \end{bmatrix}.$$

Hence, there is a transformation $\mathbf{G}^{-1}$ that takes the ill-conditioned matrix $\mathbf{A}$ (for small α) into the well-conditioned matrix $\mathbf{B}$. But this transformation itself becomes ill conditioned (as is obvious) as α goes to zero. Hence, unless the parameterization that is associated with $\mathbf{B}$ is the one desired for estimation, the transformation back to the one associated with $\mathbf{A}$ reintroduces the ill conditioning (you can't get something for nothing).

This result is seen more generally as follows. Consider a data matrix $\mathbf{X}$. The condition number of $\mathbf{X}^T\mathbf{X}$ is $\mu_{X,\max}^2/\mu_{X,\min}^2$ and, as is well known [Wilkinson (1965), p. 57],

$$\mu_{X,\min}^2 = \min_{\mathbf{c}^T\mathbf{c}=1} |\mathbf{c}^T\mathbf{X}^T\mathbf{X}\mathbf{c}| \quad \text{and} \quad \mu_{X,\max}^2 = \max_{\mathbf{c}^T\mathbf{c}=1} |\mathbf{c}^T\mathbf{X}^T\mathbf{X}\mathbf{c}|.$$

Let $\mathbf{c}^*$ be a solution to the min problem, and $\mathbf{c}^\circ$ be a solution to the max problem. Consider the "reparameterization" defined by $\mathbf{Z} = \mathbf{XG}^{-1}$ and let $\mathbf{d}^* = \mathbf{Gc}^*$ and $\mathbf{d}^\circ = \mathbf{Gc}^\circ$. Further normalize $\tilde{\mathbf{d}}^* = \mathbf{d}^*/\|\mathbf{d}^*\|$ and $\tilde{\mathbf{d}}^\circ = \mathbf{d}^\circ/\|\mathbf{d}^\circ\|$. Then we have $|\tilde{\mathbf{d}}^{*T}\mathbf{Z}^T\mathbf{Z}\tilde{\mathbf{d}}^*| - \mu_{X,\min}^2\|\mathbf{d}^*\|^{-2} \geqslant \mu_{Z,\min}^2$ and $|\tilde{\mathbf{d}}^{\circ T}\mathbf{Z}^T\mathbf{Z}\tilde{\mathbf{d}}^\circ| = \mu_{X,\max}^2\|\mathbf{d}^\circ\|^{-2} \leqslant \mu_{Z,\max}^2$, where the $\mu_{Z,i}^2$ are eigenvalues of $\mathbf{Z}^T\mathbf{Z}$. Hence

$$\kappa(\mathbf{Z}) \equiv \frac{\mu_{Z,\max}}{\mu_{Z,\min}} \geqslant \frac{\mu_{X,\max}\|\mathbf{d}^*\|}{\mu_{X,\min}\|\mathbf{d}^\circ\|} = \kappa(\mathbf{X})\frac{\|\mathbf{d}^*\|}{\|\mathbf{d}^\circ\|}. \tag{3B.1}$$

We see from (3B.1) that the condition number of $\mathbf{Z}$ exceeds that of $\mathbf{X}$ by a nonnegative factor $\|\mathbf{d}^*\|/\|\mathbf{d}^\circ\|$. In the case of an orthonormal $\mathbf{G}$, it is clear that $\|\mathbf{d}^\circ\| = \|\mathbf{d}^*\|$ and, as is well known, the equality holds in (3B.1). In this case, ill conditioning in $\mathbf{X}$ is directly reflected in ill conditioning in $\mathbf{Z}$. More generally, (3B.1) shows that reparameterization can make things better, that is, reduce the condition number, only if $\|\mathbf{d}^*\|/\|\mathbf{d}^\circ\| \leqslant 1$, and indeed, if $\kappa(\mathbf{X})$ is very large, $\kappa(\mathbf{Z})$ can be small (but need not be) only if $\|\mathbf{d}^*\|/\|\mathbf{d}^\circ\|$ is very small. But this says that some transform under $\mathbf{G}$ of a unit vector is small in length, namely, $\mathbf{d}^* = \mathbf{Gc}^*$, relative to the length of another, namely, $\mathbf{d}^\circ = \mathbf{Gc}^\circ$. This, of course, can occur only if some eigenvalue of $\mathbf{G}$ is small relative to another, or, equivalently, if **G** is ill conditioned; and the more ill conditioned **X**, the more ill conditioned must be **G**.[76]

3. The above result shows that reparameterization could improve conditioning, but in a way that depends on the singular values of **X** (or the eigenvalues of $\mathbf{X}^T\mathbf{X}$); the transformation must be ill conditioned in a way that just offsets the ill conditioning of the **X** matrix. Thus, any improvement in conditioning brought about by reparameterization depends on aspects of the matrix **X** that are outside the control of the investigator, namely, the singular values of **X**. It can only be by accident that a particular parameterization, chosen on a priori grounds, would just undo the fortuitous ill conditioning of any given data matrix. Therefore, little practical significance attaches to this aspect of parameterization.

4. There is another sense in which the above result shows reparameterization to have little practical value to the improvement of conditioning. Suppose the desired parameterization is as $\mathbf{y}=\mathbf{X}\boldsymbol{\beta}+\boldsymbol{\varepsilon}$, but that **X** is very ill conditioned. Suppose further that it is determined that reparameterization as $\boldsymbol{\delta}=\mathbf{G}\boldsymbol{\beta}$ provides a data matrix $\mathbf{Z}=\mathbf{X}\mathbf{G}^{-1}$ that is well conditioned. At first blush, it would seem reasonable to obtain an estimate **d** of $\boldsymbol{\delta}$ by OLS of **y** on **Z** and then estimate $\boldsymbol{\beta}$ by $\mathbf{b}=\mathbf{G}^{-1}\mathbf{d}$. This maneuver, however, cannot rid the problem of its ill conditioning, for we have seen that the ill conditioning of **X** implies ill conditioning of **G**, and hence of $\mathbf{G}^{-1}$. The ill conditioning of **X** is reintroduced into the solution of $\mathbf{b}=\mathbf{G}^{-1}\mathbf{d}$ through the ill conditioning of $\mathbf{G}^{-1}$. It is true that one can often find a parameterization (usually without a priori interpretation) that takes us from darkness into light, but should we desire to return, the lights must again be dimmed.

5. Finally, as noted above, orthonormal reparameterizations will not change the conditioning of the data matrix; indeed the eigenvalues of $\mathbf{X}^T\mathbf{X}$ remain invariant to such transformations. Likewise, well-conditioned reparameterizations (**G** with nearly equal eigenvalues) will, as a matter of practice, not alter the conditioning much and certainly cannot improve it

---

[76]This is proved as follows: Let **c** by any p-vector with $\|\mathbf{c}\|=1$, and define $\mathbf{d}\equiv\mathbf{Gc}$. Let the SVD of **G** be $\mathbf{UDV}^T$. Then $\|\mathbf{d}\|=\|\mathbf{UDV}^T\mathbf{c}\|=\|\mathbf{Dh}\|$, where $\mathbf{h}\equiv\mathbf{V}^T\mathbf{c}$ and $\|\mathbf{h}\|=1$. We have used the orthogonality of **U** and **V**. Hence $\|\mathbf{d}\|=(\sum_k\mu_k^2h_k^2)^{1/2}$, from which we get $\mu_{min}\leqslant\|\mathbf{d}\|\leqslant\mu_{max}$. Now, from the text, ill conditioning in **X** can be offset by $\mathbf{G}^{-1}$ only if $\|\mathbf{d}^*\|<\|\mathbf{d}^\circ\|$, and hence we have $\mu_{min}\leqslant\|\mathbf{d}^*\|<\|\mathbf{d}^\circ\|\leqslant\mu_{max}$. Thus $\kappa(\mathbf{G})\equiv\mu_{max}/\mu_{min}\geqslant\|\mathbf{d}^\circ\|/\|\mathbf{d}^*\|$. Combining this with (3B.1), we obtain

$$\kappa(\mathbf{Z})\geqslant\kappa(\mathbf{X})\kappa^{-1}(\mathbf{G}),$$

and we see that the larger $\kappa(\mathbf{X})$, the larger must be $\kappa(\mathbf{G})$ to offset.

greatly.[77] Hence, a well-conditioned reparameterization will leave the condition indexes relatively unchanged. As a matter of practice, the diagnostics of this chapter applied to the data matrix **X** will also tell much about the number and relative strengths of linear dependencies that will exist in any well-conditioned linear transformation of **X**.

We may summarize the foregoing by noting that reparameterization (or, equivalently, linear transformations of the data) neither causes problems nor, in general, solves them. On the one hand, one cannot, except by happy accident, make use of linear transformations to relieve ill conditioning in the data. On the other hand, although the collinearity diagnostics are seen to be dependent on the parameterization chosen, it is also seen that such a dependency will not, in the case of well-conditioned transformations, alter the basic story told by the diagnostics, and will not, in any event, invalidate or reduce the usefulness of the collinearity diagnostics for correctly assessing the suitability of the appropriately transformed data for estimating the parameters of the particular parameterization chosen.

Column Scaling

We now turn to an examination of the effects of column scaling on the collinearity diagnostics. In the text, we recall, the data are scaled to have unit column length before being analyzed by the collinearity diagnostics. Some intuitive justification for such scaling is offered in Section 3.3, but a more rigorous justification has awaited the results of Theorem 3 of Appendix 3A.

Of course, column scaling is a special case of the more general linear transformations of the data we have just considered, namely, that case limited to transformations of the form $Z \equiv XB$, where **B** is diagonal and nonsingular. Unlike the general case, however, column scaling does not result in an inherently new parameterization of the regression model; rather, as we have already seen in Section 3.3, it merely changes the units in which the **X** variates are measured. However, it is still true that different column scalings of the same **X** matrix can result in different singular values and, hence, cause the collinearity diagnostics to tell different stories about the conditioning of what are, from a practical point of view, essentially equivalent data sets. In this appendix we see that (1) although there is an optimal scaling that removes this seeming ambiguity, it cannot, in general,

---

[77]This is seen from the inequality (3B.1) above. If $\|\mathbf{d}^*\| \approx \|\mathbf{d}^\circ\|$, so that, in the worst instance, $\|\mathbf{d}^*\| / \|\mathbf{d}^\circ\|$ is not too small, the conditioning of **Z** is bounded away from being much improved over that of **X**.

be simply determined, but (2) scaling for unit length (or, more generally, for equal column length) affords a simple and effective expedient for approximating this optimal scaling.

The optimal scaling for the purpose of the collinearity diagnostics is readily determined from an examination of Theorem 3 of Appendix 3A. There we found the inequality,

$$|\xi_{kt}^{ii}| \leqslant 2\kappa(\mathbf{X}), \tag{3B.2}$$

that relates two measures of the conditioning of $\mathbf{X}$, (1) the condition number of $\mathbf{X}$, $\kappa(\mathbf{X})$, and (2) the sensitivity of the diagonal elements of $(\mathbf{X}^T\mathbf{X})^{-1}$ to small changes in the elements of $\mathbf{X}$, as measured by the elasticities ξ_{kt}^{ii}. It is a simple matter to show that, whereas $\kappa(\mathbf{X})$ can be altered by changing the lengths of the columns of $\mathbf{X}$, the ξ_{kt}^{ii} are invariant to such column scaling. Hence the inequality (3B.2) must be identically true for all such column scalings. That is, for the $n \times p$ data matrix $\mathbf{X}$,

$$|\xi_{kt}^{ii}| \leqslant 2\kappa(\mathbf{XB}) \qquad \text{for all } \mathbf{B} \in \mathcal{B}_p, \tag{3B.3}$$

where ξ_{kt}^{ii} is defined as in (3A.1) and $\mathcal{B}_p$ is the set of all nonsingular diagonal matrices of size p. The bound (3B.3) is obviously tightest when a scale $\mathbf{B}^*$ is chosen such that

$$\kappa(\mathbf{XB}^*) = \min_{\mathbf{B} \in \mathcal{B}_p} \kappa(\mathbf{XB}), \tag{3B.4}$$

and such a scaling thereby becomes most meaningful for an analysis of the extent to which ill conditioning of the data matrix $\mathbf{X}$ can adversely affect linear regression. Unfortunately, the general problem of optimal scaling—column scaling that results in a data matrix $\mathbf{X}$ with minimal condition number $\kappa(\mathbf{X})$—remains unsolved. However, scaling for equal column lengths (which our unit column length is but a simple means for effecting) has known "near-optimal" properties in this regard.

To wit, van der Sluis (1969) has shown that any positive-definite and symmetric matrix $\mathbf{P}$ $(p \times p)$ with all its diagonal elements equal has condition number

$$\kappa(\mathbf{P}) \leqslant p \min_{\mathbf{B} \in \mathcal{B}_p} \kappa(\mathbf{B}^T\mathbf{PB}). \tag{3B.5}$$

Of course, column equilibration (scaling for equal column lengths) of $\mathbf{X}$ results in a $p \times p$ positive-definite, real symmetric matrix $\mathbf{P} = (\mathbf{X}^T\mathbf{X})$ which has all of its diagonal elements equal. Remembering that $\kappa(\mathbf{X}^T\mathbf{X}) = \kappa^2(\mathbf{X})$, we see from the van der Sluis result that column equilibration must result in a data matrix with condition number

$$\kappa(\mathbf{X}) \leqslant \sqrt{p} \min_{\mathbf{B} \in \mathcal{B}_p} \kappa(\mathbf{XB}).$$

Hence, even though column equilibration does not necessarily result in a data matrix with optimal κ, it cannot be off by more than a factor of $\sqrt{p}$. Since p, the number of variates, is usually small, column equilibration is highly desirable in the regression context.[78]

The above fact has been known to numerical analysts for a number of years and has interested them even longer. Forsythe and Straus (1955), for example, had already shown that

$$\kappa(\mathbf{P}) = \min_{\mathbf{B} \in \mathcal{B}_p} (\mathbf{B}^T \mathbf{PB})$$

when $\mathbf{P}$ has "Young's Property A," that is, $\mathbf{P}$ takes the form

$$\begin{bmatrix} \mathbf{I}_r & \mathbf{C} \\ \mathbf{C}^T & \mathbf{I}_q \end{bmatrix}.$$

This result is of limited use in statistics or econometrics, for one cannot assume that any given $\mathbf{P} = \mathbf{X}^T\mathbf{X}$ will possess Young's Property A. It is of interest to note, however, that the Forsythe-Straus result will clearly be true for the bivariate-regression case where $p = 2$, or $r = q = 1$. In summary then, Forsythe and Straus show us that column scaling is optimal for the bivariate regression, and van der Sluis shows us that it is near-optimal in general. Normalizing for unit column length, therefore, is not so arbitrary as it might first appear, and, as a practical matter, it works extremely well. This fact has long been exploited by numerical analysts, and it derives overwhelming verification in the context of this chapter.

APPENDIX 3C. THE WEAKNESS OF CORRELATION MEASURES IN PROVIDING DIAGNOSTIC INFORMATION

Occasionally it has been suggested, informally, that one can make use of the correlation matrix of the estimated parameters as a means of diagnosing collinearity. This is related to the use of the r_{ij} terms advocated

---

[78]Note that the factor of two or three that could normally result here in practice is very small relative to the potential variation in $\kappa(\mathbf{X})$ under different scalings, a variation that can extend to many orders of magnitude.

by Farrar and Glauber (1967) and described in the introduction to this chapter. We show here that these terms cannot be used to indicate whether or not their corresponding variates are involved in a near dependency, since they all approach unity (± 1) as the degree of ill conditioning increases without bound by whatever cause.

Let the SVD of $\mathbf{X}$ be $\mathbf{UDV}^T$, so that the estimated variance-covariance matrix of $\mathbf{b} = (\mathbf{X}^T\mathbf{X})^{-1}\mathbf{X}^T\mathbf{y}$ is $V(\mathbf{b}) = s^2(\mathbf{X}^T\mathbf{X})^{-1} = s^2\mathbf{VD}^{-2}\mathbf{V}^T$, where s^2 is the estimated regression variance. Now the (i, j) term of $(\mathbf{X}^T\mathbf{X})^{-1}$ is seen to be $\mathbf{v}_i^T\mathbf{D}^{-2}\mathbf{v}_j$, where $\mathbf{v}_i$ is the ith column of $\mathbf{V}^T$. Denoting the estimated correlation between b_i and b_j by $\hat{\rho}_{ij}$, we have

$$\hat{\rho}_{ij} = \frac{\mathbf{v}_i^T\mathbf{D}^{-2}\mathbf{v}_j}{\sqrt{\mathbf{v}_i^T\mathbf{D}^{-2}\mathbf{v}_i}\sqrt{\mathbf{v}_j^T\mathbf{D}^{-2}\mathbf{v}_j}} = \frac{\sum\limits_k \dfrac{v_{ik}v_{jk}}{\mu_k^2}}{\sqrt{\sum\limits_k \dfrac{v_{ik}^2}{\mu_k^2}}\sqrt{\sum\limits_k \dfrac{v_{jk}^2}{\mu_k^2}}}, \qquad (3\text{C}.1)$$

where v_{ik} is the kth component of $\mathbf{v}_i$. As any one near dependency gets tighter, some μ_r gets very small, zero in the limit of an exact dependency; and, as long as $v_{ir} \neq 0$ and $v_{jr} \neq 0$ (a condition almost certain to occur in practice), we have

$$\lim_{\mu_r \to 0} \hat{\rho}_{ij} = \frac{v_{ir}v_{jr}}{\sqrt{v_{ir}^2}\sqrt{v_{jr}^2}} = \pm 1. \qquad (3\text{C}.2)$$

In general, then, any one strong near dependency can infect *all* correlations, not just those involved in the specific dependency.

This same problem besets the r_{ij}. partial-correlation terms suggested by Farrar and Glauber (1967) for examining patterns of variable interdependence, and hence seriously limits their usefulness in that context. This is readily seen by noting that, if the $\mathbf{X}$ data are centered and scaled so that $\mathbf{X}^T\mathbf{X}$ is the matrix of correlations, (3C.1) directly reduces to the expression $-r_{ij} = r^{ij}/(\sqrt{r^{ii}}\sqrt{r^{jj}})$ for these partial correlations.

APPENDIX 3D. THE HARM CAUSED BY COLLINEARITY

We note in Chapter 3 that collinearity is a data problem, resulting from ill conditioning in the matrix $\mathbf{X}$ of explanatory variables, and is not a statistical problem. The actual harm caused by collinearity, however, is a statistical problem, as we note in Sections 3.2 and 3.4, where the distinction between degrading collinearity (the potential for harm) and harmful collinearity is discussed. Diagnosing the presence of collinearity and identifying the parameter estimates that are degraded by it requires information only on the data matrix **X**, whereas an examination of the harm that results requires introducing information on the response variable **y** and/or the variance σ^2 of the error term ε. Diagnosing the presence of collinear relations and their potential for harm, therefore, is separable from seeking to determine fully the nature of the harm. We turn now to some of the conceptual issues involved in defining and understanding this latter problem.

The Basic Harm

The essential harm due to collinearity arises from the fact that a collinear relation can readily result in a situation in which some of the observed systematic influence of the explanatory variables on the response variable is swamped by the model's random error term—or in the familiar terminology of electrical engineering, the signal is swamped by the noise. It is intuitively clear that, under these circumstances, estimation can be hindered. A very simple single-variate model, devoid of any collinearity, suffices to illustrate the nature of this problem and provides, as we shall see, a canonical model to which all collinearity problems can be reduced. .

Consider the simple single-variate model (through the origin)

$$y = \beta x + \varepsilon, \tag{3D.1}$$

where **y** and **x** are n-vectors, β a scalar parameter, and ε an n-vector disturbance term with components that are i.i.d. with mean zero and constant variance σ^2. We assume that (3D.1) is the model that generates the observed y's, and, since collinearity is a phenomenon related to a specific data matrix, we assume **x** is fixed and view the generation of **y** as conditional on this given **x**. Of course (3D.1) encompasses the usual "bivariate" model with constant term, $y = \alpha + \beta x + \varepsilon$, simply by interpreting **y**, **x**, and ε in (3D.1) to be variates centered about their respective means.

We note in (3D.1) that the generation of **y** has a systematic (signal) component, βx, and an independent random (noise) component, ε. For a given β, **x**, and ε we can geometrically depict the generation of **y** as in Exhibit 3D.1a. Here we interpret **y**, **x**, and ε as vectors in an n-dimensional space [see Malinvaud (1970), Wonnacott and Wonnacott (1979)]. For purposes of illustration, we set $\beta = \frac{1}{3}$. The plane of the page, then, is the two-dimensional subspace of n-space spanned by **x** and ε.

We can also depict, for given **x**, likely outcomes of ε, and hence **y**. Since

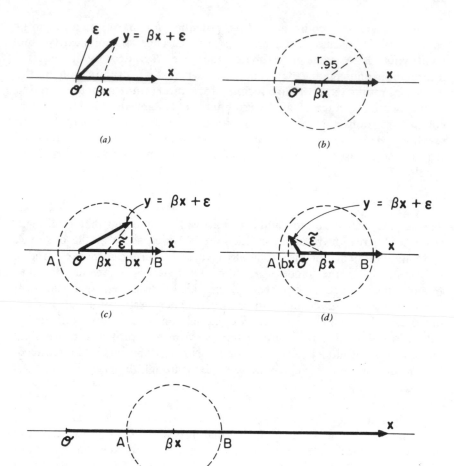

Exhibit 3D.1 Geometry of the statistical problem underlying collinearity. (a) The generation of y given x and $\beta = \frac{1}{3}$. (b) The 95% concentration interval (dotted circle) for y given x—as projected on the plane of the page—x and β as above. The radius of the circle (see text) is proportional to σ, and the case depicted here has a low ratio of signal-to-noise. (c) and (d) Two "likely" outcomes for y (given x, β, and σ as before). Case (c) results in a least-squares estimation $b > 0$ while case (d) has $b < 0$. The 95% concentration interval for b is $\overline{AB}$ and contains θ in this low signal-to-noise situation. (e) The 95% concentration interval for y (dotted circle) and b ($\overline{AB}$) given the same β and σ as before, but longer x. This case has higher ratio of signal-to-noise and $\overline{AB}$ lies wholly above θ.

188

the components of ε are independent and have constant variance, there will be an n-dimensional sphere, centered at the origin, with a given radius such that ε, in repeated trials, will lie within the sphere with probability .95. This is the 95% concentration ellipse for ε, and, should ε be assumed Gaussian (normal), this ellipse would be the n-dimensional sphere with radius $r_{.95} = \sigma\sqrt{\chi^2_{n,.95}}$, where $\chi^2_{n,.95}$ denotes the .95 critical value for a chi-square with n degrees of freedom. The corresponding 95% concentration interval for y is simply this same sphere centered on βx. The intersection of this n-dimensional sphere with the two-dimensional page is depicted by the dotted circle in Exhibit 3D.1b, and from this we can illustrate two "likely" sample outcomes, shown in Exhibits 3D.1c and d. The dotted lines $\tilde{\varepsilon}$ denote the disturbance vector ε shifted to emanate from βx.

It is also possible to see what effect these two different sample outcomes will have on the least-squares estimate b of β. This estimate results from the orthogonal projection of y onto the space spanned by x, depicted as bx in Exhibits 3D.1c and d. We note that b is positive in Exhibit 3D.1c and negative in d, yet either of these cases could readily have been generated by the model. Indeed the 95% concentration interval for bx is seen to be the interval $\overline{AB}$, an interval that includes the origin Θ. It is clear that this variability in the regression estimate is due to the fact that the radius of the 95% concentration sphere for y (which, for given n, is proportional to σ) is large relative to βx; that is, the noise is large relative to the signal, and any systematic elements have little weight in the regression results.

The same degree of variability in the regression results will not arise if the signal becomes relatively greater. Keeping all parameters the same ($\beta = \frac{1}{3}$, σ and n unchanged), we depict this case in Exhibit 3D.1e. Once again the 95% concentration interval for bx is $\overline{AB}$, an interval that now lies wholly above the origin, and the presence of the signal is more readily discernible. Thus, we see that the quality of information derivable from least-squares estimation is diminished the "shorter" is the signal vector βx relative to the noise (which, given n, is proportional to σ)—a situation we may call *hypokinesis*.[79]

[79]The relevance of the length of βx relative to σ—a ratio of signal-to-noise—to this problem is also seen from the following expressions. (1) Conditional on x, we may define an expected (uncentered) R^2 measure derived from $\mathcal{E} y^T y = \beta^T x^T x \beta + n\sigma^2$ (using the independence of x from ε) as $\mathcal{R}^2 \equiv \beta^T x^T x \beta / (\beta^T x^T x \beta + n\sigma^2) \equiv \omega/(\omega + 1)$, where $\omega \equiv \beta^T x^T x \beta / n\sigma^2$ is a measure of signal-to-noise. $\mathcal{R}^2$ measures the conditional expectation (on x) of the "explained" portion of the second moment of y (or the variance of y if the data have been centered—in which case n is replaced by $n-1$). Clearly, when ω is small, so is $\mathcal{R}^2$, and as ω increases, $\mathcal{R}^2$ approaches unity. (2) We also see the relevance of ω from an expression for the variance of b, the least-squares estimate of β, namely $\text{var}(b) = \sigma^2 / x^T x \equiv \beta^2 / n\omega$. Given β and n, the lower ω, the higher the variance of the regression estimate, and inversely.

The Effect of Collinearity

The presence of collinearity among several variates results in the hypokinetic situation just described; that is, it results in a relatively short signal vector relevant to the estimation of some parameters. We can see this by graduating to the two-variate model

$$\mathbf{y} = \beta_1 \mathbf{x}_1 + \beta_2 \mathbf{x}_2 + \varepsilon, \qquad (3D.2)$$

and by showing that the assumption of a collinear relation between $\mathbf{x}_1$ and $\mathbf{x}_2$ allows us to reduce this model to the hypokinetic case just discussed. We continue to view estimation of (3D.2) by regression conditional upon a particular data set $\mathbf{X} = (\mathbf{x}_1 \mathbf{x}_2)$.

If there exists a near dependency between $\mathbf{x}_1$ and $\mathbf{x}_2$, then a mechanical regression of $\mathbf{x}_2$ on $\mathbf{x}_1$ will result in a fit with a relatively small sum-of-squared residuals—a short residual vector. Thus we have

$$\mathbf{x}_2 = a\mathbf{x}_1 + \mathbf{e}_2, \qquad (3D.3)$$

where $a = \mathbf{x}_1^T \mathbf{x}_2 / \mathbf{x}_1^T \mathbf{x}_1$, $\mathbf{x}_1^T \mathbf{e}_2 = 0$, and where $\mathbf{e}_2^T \mathbf{e}_2$ is necessarily small relative to $\mathbf{x}_2^T \mathbf{x}_2$; that is $\mathbf{e}_2^T \mathbf{e}_2 / \mathbf{x}_2^T \mathbf{x}_2 = (1 - R^2)$ is small.[80] Substituting (3D.3) into (3D.2) gives the equivalent model

$$\mathbf{y} = (\beta_1 + \beta_2 a)\mathbf{x}_1 + \beta_2 \mathbf{e}_2 + \varepsilon, \qquad (3D.4)$$

for which the fixed explanatory variables $\mathbf{x}_1$ and $\mathbf{e}_2$ are orthogonal. Thus the regression estimates of $\mathbf{y}$ on $\mathbf{x}_1$ and $\mathbf{e}_2$ together are the same as those obtained by regressing $\mathbf{y}$ on $\mathbf{x}_1$ and $\mathbf{y}$ on $\mathbf{e}_2$ separately, and we may investigate the effectiveness of least squares in estimating β_2 by focusing only on the regression of $\mathbf{y}$ on $\mathbf{e}_2$, a problem identical to that discussed above. It is clear that the effect of the collinear relation has been to result in a short signal variate $\beta_2 \mathbf{e}_2$ associated with the estimation of β_2, while holding σ constant, and hence to make estimation less precise. The greatest part of the signal has been applied to the estimate of the parameter associated with $\mathbf{x}_1$, that is, the linear combination of $\beta_1 + \beta_2 a$ (a known from (3D.3)). There is little signal left over to apply to the independent estimation of β_2. Collinearity results in hypokinesis.

The preceding analysis readily generalizes to p variates containing $p_2 < p$ near dependencies (as diagnosed and identified by the methods of Chapter 3). Here we have

$$\mathbf{y} = \mathbf{X}\beta + \varepsilon = \mathbf{X}_1 \beta_1 + \mathbf{X}_2 \beta_2 + \varepsilon, \qquad (3D.5)$$

---

[80]R^2 is not necessarily based on centered data here.

where $\mathbf{X}$ is $n \times p$, $\mathbf{X}_1$ is $n \times p_1$, $\mathbf{X}_2$ is $n \times p_2$, and $p = p_1 + p_2$. We assume that $\mathbf{X}$ has been partitioned into two blocks, $\mathbf{X}_2$ containing the variates determined by the methods of Chapter 3 to be involved in the p_2 near dependencies with the variates in $\mathbf{X}_1$. Regressing $\mathbf{X}_2$ on $\mathbf{X}_1$ gives

$$\mathbf{X}_2 = \mathbf{X}_1 \mathbf{A} + \mathbf{E}_2, \tag{3D.6}$$

where $\mathbf{A} = (\mathbf{X}_1^T \mathbf{X}_1)^{-1} \mathbf{X}_1^T \mathbf{X}_2$, $\mathbf{X}_1^T \mathbf{E}_2 = 0$, and $\det(\mathbf{E}_2^T \mathbf{E}_2)$ is small relative to $\det(\mathbf{X}_2^T \mathbf{X}_2)$.

Substituting (3D.6) into (3D.5) gives, analogously to the preceding,

$$\mathbf{y} = \mathbf{X}_1(\boldsymbol{\beta}_1 + \mathbf{A}\boldsymbol{\beta}_2) + \mathbf{E}_2 \boldsymbol{\beta}_2 + \boldsymbol{\varepsilon}. \tag{3D.7}$$

Once again, since $\mathbf{X}_1^T \mathbf{E}_2 = 0$, we may estimate $\boldsymbol{\beta}_1 + \mathbf{A}\boldsymbol{\beta}_2$ and $\boldsymbol{\beta}_2$ through separate regressions of $\mathbf{y}$ on $\mathbf{X}_1$ and $\mathbf{y}$ on $\mathbf{E}_2$, respectively. Most of the data information (signal) will be applied to the estimate of the linear combination $\boldsymbol{\beta}_1 + \mathbf{A}\boldsymbol{\beta}_2$ ($\mathbf{A}$ known from (3D.6)), and little information will be available from the hypokinetic variate $\mathbf{E}_2 \boldsymbol{\beta}_2$ for the estimation of $\boldsymbol{\beta}_2$.

Collinearity, then, is a *data weakness* that can manifest itself as a *statistical problem* by creating a situation in which there is insufficient signal relative to the noise to allow estimation with precision. Once again we see that collinearity is not, in itself, a statistical problem, but can result in one;[81] and that a full assessment of the harm of collinearity (as opposed to its potential harm) must be sought in introducing information on the response variable $\mathbf{y}$.

---

[81] It also proves instructive to examine the expression for the variance of the kth regression coefficient, $\mathrm{var}(b_k)$, given in footnote 7, to clarify the relation between the data problem of collinearity and the statistical problem that may result. We recall that $\mathrm{var}(b_k) = \sigma^2 \mathrm{VIF}_k / \mathbf{X}_k^T \mathbf{X}_k = \beta_k^2 \mathrm{VIF}_k / n\omega_k$, where $\mathbf{X}_k$ is the kth column of $\mathbf{X}$, β_k is the kth element of $\boldsymbol{\beta}$, $\mathrm{VIF}_k = (1 - R_k^2)^{-1}$, R_k^2 is the uncentered R^2 from $\mathbf{X}_k$ regressed on the remaining columns of $\mathbf{X}$, and $\omega_k \equiv \beta_k^T \mathbf{X}_k^T \mathbf{X}_k \beta_k / n\sigma^2$. From the first equality above, we see how the three separable elements of signal ($\mathbf{X}_k^T \mathbf{X}_k$—given β_k), noise (σ^2—given n), and collinearity (VIF_k) come together to determine $\mathrm{var}(b_k)$. This interaction is made even clearer by the second equality. Here we note that ω_k is a measure of the signal (relative to noise) associated with the kth regression component $\beta_k \mathbf{X}_k$, which is invariant to the presence of collinearity; and VIF_k is a measure of strength of the collinear relation of $\mathbf{X}_k$ with the remaining data, which is invariant to the relative signal, ω_k. Given β_k and n, then, $\mathrm{var}(b_k)$ will be smaller the greater is the relative signal (ω_k) and the less is the collinearity (VIF_k). Thus, we see that inflated regression variances can, but need not, always attend the presence of collinearity. Indeed, given any VIF_k, $\mathrm{var}(b_k)$ can be small if ω_k is sufficiently large.

CHAPTER 4

Applications and Remedies

Chapters 2 and 3 present two sets of diagnostic techniques for analyzing, respectively, influential observations and collinearity in the linear regression model. Brief examples of the use of these tools are also given there, along with rules of thumb, where possible, for their application. This chapter is devoted to detailed examples, to verification of the usefulness of both of these diagnostic techniques, and to their interpretation in specific economic and statistical contexts. Section 4.1 addresses the problem of remedying ill-conditioned data and applies one specific remedy, that of mixed-estimation, to the ill-conditioned data of the consumption function introduced in Chapter 3. The collinearity diagnostics of Chapter 3 are used alongside the remedial action to show how the improved conditioning that results from the remedy is reflected in the diagnostics. Section 4.2 extends this example, analyzing by means of the influential-data diagnostics of Chapter 2 the effect on estimation of the consumption function due to each piece of prior information that has been introduced.

In Section 4.3 the data used to estimate a monetary equation are subjected to the influential-data analysis of Chapter 2. These diagnostics may point to a possible misspecification of the model. In addition, the conditioning of the data is examined, and ridge regression is employed as an alternative means of remedial action. Its use confirms our belief that collinearity exacerbates the parameter sensitivity to influential data that is observed in this case.

Finally, in Section 4.4, robust regression is employed along with the influential-data techniques to analyze a housing-price model.

4.1   A REMEDY FOR COLLINEARITY: THE CONSUMPTION FUNCTION WITH MIXED-ESTIMATION

As has been indicated on several occasions, it is not the purpose of this book to prescribe or deal with measures for correcting ill-conditioned data (the fix or the remedy). There is, however, a strong temptation to attempt a fix at least once, for such an experiment provides an excellent test of the efficacy of the diagnostic procedure of Chapter 3. That is, an acceptable fix must result in better conditioned data, which should in turn be reflected both in a lower condition number assigned by the diagnostic technique to the "fixed" data and in sharper regression results based on their use.

In this section we turn our attention to the poorly conditioned consumption-function data introduced in Section 3.4. Before applying this corrective action, however, we discuss the range of possible corrective measures and examine in detail one such measure, the Theil-Goldberger (1961) method of mixed-estimation, that subsequently is applied to the consumption function. The resulting regressions and diagnostics are then compared and interpreted.

Corrective Measures

Historically, collinearity was one of the first-encountered and separately named problems of econometric estimation. In fact, however, it is a special case of the more general problem of identification. In its extreme form, some data series are indistinguishable from a linear combination of the others, and hence it becomes impossible to determine uniquely the parameters from knowledge of the conditional distribution of $\mathbf{y}$ given $\mathbf{X}$. Corrective measures for ill conditioning, therefore, must be like corrective measures for identification; namely, some identifying prior or auxiliary information must be introduced. Two major sources currently exist for such condition-identifying information: additional, better conditioned data and Bayes-like methods employing prior information.

Introduction of New Data. The most direct and obvious method of improving data conditioning is through the collection and use of additional data points that provide the needed independent variation relative to the original data. This answer is rarely useful to the econometrician or many other users of least squares who typically have short data series. New data are obtainable in adequate numbers only at substantial cost, either in terms of the time one must wait for new observations to occur or in terms of the collection costs needed to obtain a less collinear sample. Further, even if new data are obtained, there is typically no guarantee that they will be consistent with the original data or that they will indeed provide independent information. Applied statisticians are often unable to control their experiments, whereas nature often closely replicates hers.

The introduction of new data, therefore, is not likely to be a fix of much practical importance in many applications of least squares. When it is possible, however, to provide new data, the corrective action to be taken is straightforward and simple.[1]

Bayesian-type Techniques.   At least three Bayes-like procedures exist for introducing the experimenter's subjective prior information: (1) a pure Bayesian technique, (2) a mixed-estimation technique, and (3) the technique of ridge regression. These procedures are listed in order of decreasing generality and increasing ease of use. We select the mixed-estimation technique in dealing later with the consumption-function data. Ridge regression is used in conjunction with the monetary-equation data in Section 4.3.

Pure Bayes. The use of Bayesian estimation for providing the identifying information that is needed to improve data conditioning is explained in the work of Zellner (1971) and Leamer (1973, 1978). These excellent studies show that the seemingly insoluble problem of collinearity can in fact be dealt with if the investigator possesses (and is willing to use) subjective prior information on the parameters of the model. The drawbacks of this method are severalfold. First, it relies on subjective information that many researchers simply distrust or feel they do not possess. Second, its use requires a rather exact statement of the prior distribution, an apparent precision that many find too exacting to be realistic. Third, it draws on a statistical theory not as widely understood as classical techniques, and fourth, the computer software required for such estimation methods is not widely available. In practice, the first two

---

[1] It is also possible to improve the conditioning of a given body of data to some limited extent through the simple expedient of column scaling [cf. van der Sluis (1969, 1970)]. Such transformations of the data, of course, do not introduce new identifying information; they merely prepare the existing information for optimal processing by computational software that must necessarily perform calculations using finite arithmetic, and therefore should be done as a matter of course. Simple column equilibration (making all columns of **X** the same length, usually unit), for example, can often reduce the condition number of economic data by a factor of 10^3 or more! See also Appendix 3B on scaling.

drawbacks are far more psychological than real, and efforts by Kadane et al. (1977) have produced techniques that allow the user efficiently and straightforwardly to answer questions that reveal his subjective prior. Such research will moderate all four drawbacks of strict Bayesian estimation, and one can expect to see these techniques gain wider acceptance in the near future.

Mixed-Estimation. Theil and Goldberger (1961) and Theil (1963, 1971) have produced a Bayes-like technique which they call mixed-estimation. By this procedure, prior, or auxiliary,[2] information is added directly to the data matrix. Mixed-estimation is simple to employ and need not require a full specification of the prior distribution.

Beginning with the linear model

$$y = X\beta + \varepsilon, \tag{4.1}$$

with $\mathcal{E}\varepsilon = 0$ and $V(\varepsilon) = \Sigma_1$, it is assumed the investigator can construct $r < p$ prior restrictions on the elements of β in the form

$$c = R\beta + \xi, \tag{4.2}$$

with $\mathcal{E}\xi = 0$ and $V(\xi) = \Sigma_2$. Here R is a matrix of rank r of known constants, c is an r-vector of specifiable values, and ξ is a random vector, independent of ε, with mean zero and variance-covariance matrix Σ_2 also assumed to be stipulated by the investigator.

In the method of mixed-estimation as suggested by Theil and Goldberger,[3] estimation of (4.1) subject to (4.2) proceeds by augmenting y and X to give

$$\begin{bmatrix} y \\ c \end{bmatrix} = \begin{bmatrix} X \\ R \end{bmatrix} \beta + \begin{bmatrix} \varepsilon \\ \xi \end{bmatrix}, \tag{4.3}$$

where $V\begin{pmatrix} \varepsilon \\ \xi \end{pmatrix} = \begin{bmatrix} \Sigma_1 & 0 \\ 0 & \Sigma_2 \end{bmatrix} \equiv \Sigma$. If Σ_1 and Σ_2 are known, generalized least squares applied to (4.3) results in the unbiased mixed-estimation estimator

$$b_{ME} = \left(X^T\Sigma_1^{-1}X + R^T\Sigma_2^{-1}R\right)^{-1}\left(X^T\Sigma_1^{-1}y + R^T\Sigma_2^{-1}c\right). \tag{4.4}$$

---

[2] The use of the term "prior" here follows Theil (1971). For an approximation that allows a Bayesian interpretation of mixed estimation, see Theil (1971, pp. 670–672).
[3] See Theil (1971, pp. 347–352).

In practice, an estimate $\mathbf{S}_1$ is substituted for $\mathbf{\Sigma}_1$ (usually in the form $s^2\mathbf{I}$), and $\mathbf{\Sigma}_2$ is specified by the investigator as part of the prior information. The Aitken-type augmented data matrix that results from this procedure is $\begin{bmatrix} \mathbf{\Sigma}_1^{-1/2}\mathbf{X} \\ \mathbf{\Sigma}_2^{-1/2}\mathbf{R} \end{bmatrix}$, and it is the improved conditioning of this matrix that should reflect the usefulness of both the prior information and the diagnostic procedure.[4]

Ridge Regression. Ridge regression is a relatively recent, modified squared-error estimation technique. The reader is directed to Stein (1956), Hoerl and Kennard (1970), Holland (1973), Dempster et al. (1977), and Vinod (1978). The ridge-regression estimator, with the single ridge parameter k, is simply

$$\mathbf{b}_r = (\mathbf{X}^T\mathbf{X} + k\mathbf{I})^{-1}\mathbf{X}^T\mathbf{y}. \tag{4.5}$$

While differing in its philosophical basis, the ridge estimator is seen to be computationally equivalent to the mixed-estimation estimator with $\mathbf{R} = \mathbf{A}$ (where $\mathbf{A}^T\mathbf{A} = \mathbf{I}$), $\mathbf{\Sigma}_1 = \sigma^2\mathbf{I}$, $\mathbf{\Sigma}_2 = \lambda^2\mathbf{I}$, and $\mathbf{c} = \mathbf{0}$. In this event, $k = \sigma^2/\lambda^2$. Of course, in mixed estimation $\mathbf{c}$ is taken to be stochastic with $\mathcal{E}\mathbf{c} = \mathbf{R}\beta$, whereas here $\mathbf{c}$ is taken as a set of constants, which results in a biased estimator. Computationally, therefore, mixed-estimation provides a compromise between the rigors of full Bayes and the somewhat inflexible ridge estimator, whose single parameter k forces the weight attached to all prior constraints to be equal.[5] In practice, mixed-estimation allows prior information to be introduced more naturally and with much greater flexibility than ridge, and with little additional computational effort.

Application to the Consumption-Function Data

We recall from Section 3.4 that standard (scaled) consumption-function data are very poorly conditioned (possessing a κ of 376), contain at least two strong near dependencies (shown in Exhibit 3.24), and result in a regression displaying only one statistically significant coefficient (that of DPI). In this section we employ the mixed-estimation method described

---

[4] In the event that $\mathbf{S}_1 = s^2\mathbf{I}$, this matrix is equivalent to using $\begin{bmatrix} \mathbf{X} \\ \mathbf{DR} \end{bmatrix}$ where $\mathbf{D} = s\mathbf{\Sigma}_2^{-1/2}$.

[5] This restriction is relaxed with the use of the *generalized ridge* estimator, having the form $\mathbf{b}_{GR} \equiv (\mathbf{X}^T\mathbf{X} + \mathbf{\Delta})^{-1}\mathbf{X}^T\mathbf{y}$, where $\mathbf{\Delta}$ is a positive-definite matrix. Yet another near cousin to the ridge estimator, developed in Von Hohenbalken and Riddell (1978), is the *wedge estimator*. This takes the form $\mathbf{b}_w \equiv (\mathbf{Z}^T\mathbf{Z})^{-1}\mathbf{Z}^T\mathbf{y}$, where $\mathbf{Z} = \mathbf{X} + k\mathbf{X}(\mathbf{X}^T\mathbf{X})^{-1}$.

above as one possible means for confronting collinear data. It is also the purpose of this exercise to determine how well the diagnostic techniques presented in Sections 3.2 and 3.3 reflect the better conditioning resulting from the introduction of prior information.

In the experiment that follows, three levels of prior information are specified and examined: a very loose (high variance) set of priors, a middle set, and a tight (relatively low variance) set. Initially these priors are specified as ranges or intervals in which the specific parameters (or linear combinations of them) are assumed to lie with .95 probability. These ranges are then translated into a form usable for mixed-estimation by assuming that the error terms attached to the restrictions in (4.2) are independent Gaussians having means at the center of the specified prior intervals and (implied) variances that cause these prior intervals and the 95% concentration intervals of the error terms to coincide.

An attempt has been made to make these prior restrictions economically reasonable and to allow the alternative levels to have interesting contrasts. But it is not our purpose here to present a highly refined and sophisticated set of priors. It is worth repeating that our objective is mainly to demonstrate how the increased information obtained from the application of successively stronger priors, which should manifest itself in better conditioned augmented data matrices, is in fact reflected in the diagnostics. We nevertheless feel that there is much of economic value to be learned from this study.

The consumption function given by (3.24) was introduced without comment in Section 3.4. We reproduce it here:

$$C(T) = \beta_1 + \beta_2 C(T-1) + \beta_3 \mathrm{DPI}(T) + \beta_4 r(T) + \beta_5 \Delta \mathrm{DPI}(T) + \varepsilon(T).$$

$$(4.6)$$

This consumption function, which allows for Friedmanesque (1957) dynamics in its inclusion of $C(T-1)$ and $\Delta\mathrm{DPI}(T)$, is somewhat unorthodox in its inclusion of interest rates, $r(T)$. We will have more to say on these matters as we proceed.

Prior Restrictions.   Among the five parameters, three prior restrictions are imposed. As noted, the error terms attached to these restrictions are assumed to be independent and normal.

1.   *The marginal propensity to consume, β_3, has an expected value of 0.7.* The 95% prior intervals for the three levels (with their implied variances) are given in Exhibit 4.1.

Exhibit 4.1   Prior: $\beta_3 = 0.7 + \xi_1$. Ranges and implied variances

Prior	95% Range	Implied Variance
Loose	0.50—0.90	0.0104
Medium	0.55—0.85	0.0058
Tight	0.60—0.80	0.0026

The loose prior here is loose indeed, allowing a short-term marginal propensity to consume between 0.5 and 0.9 with high odds. The tight extreme is not excessive.

2. *The long-run marginal propensity to consume* $\beta_3/(1-\beta_2)$, *has expected value 0.9.*[6] One must make an approximation here to mold this prior restriction into the linear framework of (4.2). If $\beta_3/(1-\beta_2) = 0.9 + u$, then $\beta_3 + 0.9\beta_2 = 0.9 + \xi_2$, resulting in an error structure, $\xi_2 \equiv (1-\beta_2)u$, dependent on β_2. To capture the flavor of this result in the necessary linear approximation, we treat β_2 in the error structure as a known constant (but in the error structure only) and set it equal to its OLS estimate from (3.25), that is, 0.24.[7] Thus we approximate $\mathrm{var}(1-\beta_2)u$ by $(1-\beta_2)^2\sigma_u^2 \simeq 0.6\sigma_u^2$. Exhibit 4.2 presents the 95% prior intervals and implied variances for u and ξ_2.

Exhibit 4.2   Prior: $\beta_3/(1-\beta_2) = 0.9 + u$. Incorporated as $\beta_3 + 0.9\beta_2 = 0.9 + \xi_2$. Ranges and implied variances

Prior	95% Range	Implied Variance, σ_u^2	$\sigma_{\xi_2}^2 \simeq 0.6\sigma_u^2$
Loose	0.80—1.00	0.002603	0.0015618
Medium	0.85—0.95	0.000650	0.0003904
Tight	0.89—0.91	0.000026	0.0000156

The prior variances are specified on u and reflect strong prior information even in the loose case, depicting our strong belief in the 0.9 figure. In the tightest case, little variation is allowed.

[6] The long-run marginal propensity to consume is simply the partial derivative of the steady-state solution to (4.6) with respect to income, DPI. In the steady-state, time-dimensioned variables remain constant, and, ignoring the error term, (4.6) becomes

$$C^* = \frac{\cdot \beta_1}{1-\beta_2} + \frac{\beta_3}{1-\beta_2} DPI^* + \frac{\beta_4}{1-\beta_2} r^*,$$

where the asterisks indicate steady-state values.

[7] This procedure is, of course, in violation of a strict application of mixed-estimation, which requires that the prior information be independent of the original data.

3. *Twenty-five cents of every "windfall" dollar is spent; that is, β_5 has expected value 0.25.* We are not too sure of this parameter. If we were Professor Friedman, we would specify $\beta_5 = 0$. To capture this position, we allow the loose case to encompass this possibility with high probability, as well as the possibility that $\beta_5 < 0$. This later case would occur if, instead of consuming part of windfall income, consumers added to windfalls with increased savings in preparation for even larger purchases at a latter date. Exhibit 4.3 presents the three prior levels for this parameter. The middle case allows marginally for Friedman's hypothesis, and the tight case depicts a prior belief that people do indeed spend some of their windfall income.

Exhibit 4.3 Prior: $\beta_5 = 0.25 + \xi_3$. Ranges and implied variances

Prior	95% Range	Implied Variance
Loose	−0.25—0.75	0.0650
Medium	0.00—0.50	0.0162
Tight	0.15—0.35	0.0026

Ignored Information. No prior information is included for either the constant term β_1 or the effect of interest rates β_4. The former merely reflects our doubts of the validity of a linear model at the origin, a data point quite far removed from the center of the **X** data. The latter, the effect of interest rates, is of greater interest. Interest rates are often ignored in simple versions of consumption functions, a characteristic element of most elementary Keynesian models where r is determined in the investment equation, and indeed, most statistical estimation has failed to observe the significance of r in a consumption function. There are, of course, theoretical reasons for its inclusion. The consumption data used here are total consumption figures, total goods (durables plus non-durables) and services, and the relevance of r to decisions on consumer durables has long been recognized. Furthermore, any intertemporal theory of consumption would suggest that, *ceteris paribus*, the greater r, the more expensive current consumption is relative to future consumption. Both of these reasons would argue that r should enter the consumption function with a negative coefficient. Indeed β_4 is estimated as negative in (3.25), but the coefficient is insignificant. We choose to ignore any prior on β_4 here, but a more sophisticated analysis would surely provide at least a one-sided diffuse prior on β_4. We also choose to ignore any prior specification of the possible off-diagonal elements of Σ_2.

Summary of Prior Data. The prior restrictions described above result in the following specification relative to (4.2):

$$
\mathbf{R} = \begin{bmatrix} 0 & 0 & 1 & 0 & 0 \\ 0 & .9 & 1 & 0 & 0 \\ 0 & 0 & 0 & 0 & 1 \end{bmatrix}, \qquad \mathbf{c} = \begin{bmatrix} 0.7 \\ 0.9 \\ 0.25 \end{bmatrix},
$$

and Σ_2 for the three levels of prior,

Σ_2: loose prior,

$$
\begin{bmatrix} 0.0104 & 0 & 0 \\ 0 & 0.0015618 & 0 \\ 0 & 0 & 0.0650 \end{bmatrix}
$$

Σ_2: medium prior,

$$
\begin{bmatrix} 0.0058 & 0 & 0 \\ 0 & 0.0003904 & 0 \\ 0 & 0 & 0.0162 \end{bmatrix}
$$

Σ_2: tight prior,

$$
\begin{bmatrix} 0.0026 & 0 & 0 \\ 0 & 0.0000156 & 0 \\ 0 & 0 & 0.0026 \end{bmatrix}.
$$

Σ_1 was estimated as $s^2 \mathbf{I}$, where s^2 is the estimated variance from (3.25), 12.6534.

Regression Results and Variance-Decomposition Proportions for Mixed-Estimation Consumption-Function Data

Exhibits 4.4–4.6 present, respectively, the regression output, the variance-decomposition proportions, and appropriate auxiliary regressions for the original data and the these data augmented by three degrees of tightness of prior information. These latter variance-decomposition proportions are based on a column-scaled transformation of the augmented data matrix $\begin{bmatrix} \mathbf{X} \\ \mathbf{DR} \end{bmatrix}$, where $\mathbf{D} = s\Sigma_2^{-1/2}$.

Exhibit 4.4 Regression and mixed-estimation results*

	CONST b_1	$C(T-1)$ b_2	DPI(T) b_3	$r(T)$ b_4	ΔDPI(T) b_5
Original data	6.72 [1.76]	0.245 [1.033]	0.698 [3.363]	−2.209 [−1.202]	0.161 [0.877]
Data augmented by					
Loose prior	7.22 [2.51]	0.243 [2.356]	0.693 [7.869]	−1.747 [−1.175]	0.178 [1.860]
Medium prior	7.61 [3.10]	0.241 [3.124]	0.688 [10.351]	−1.354 [−1.260]	0.198 [2.681]
Tight prior	7.77 [3.89]	0.240 [4.821]	0.684 [15.372]	−1.078 [−2.317]	0.229 [5.346]

*Since interest here centers on tests of significance, t-statistics are given in the square brackets.

Exhibit 4.5 Variance-decomposition proportions and condition indexes

Associated Singular Value	CONST var(b_1)	$C(T-1)$ var(b_2)	DPI(T) var(b_3)	$r(T)$ var(b_4)	ΔDPI(T) var(b_5)	Condition Index, η
			Original Consumption Function			
μ_1	.001	.000	.000	.000	.001	1
μ_2	.004	.000	.000	.002	.136	4
μ_3	.310	.000	.000	.013	.000	8
μ_4	.264	.004	.004	.984	.048	39
μ_5	.420	.995	.995	.000	.814	376
		Mixed-Estimation Consumption-Function Data				
			Loose Prior			
μ_1	.002	.000	.000	.000	.006	1
μ_2	.007	.000	.000	.002	.465	4
μ_3	.549	.000	.000	.019	.002	8
μ_4	.270	.014	.017	.967	.092	31
μ_5	.171	.985	.982	.011	.434	160
			Medium Prior			
μ_1	.003	.000	.000	.000	.009	1
μ_2	.008	.000	.000	.003	.648	4
μ_3	.750	.000	.000	.034	.002	8
μ_4	.134	.012	.014	.958	.043	22
μ_5	.104	.987	.985	.004	.297	120

Exhibit 4.5   **Continued**

Associated Singular Value	CONST $\mathrm{var}(b_1)$	C(T−1) $\mathrm{var}(b_2)$	DPI(T) $\mathrm{var}(b_3)$	r(T) $\mathrm{var}(b_4)$	ΔDPI(T) $\mathrm{var}(b_5)$	Condition Index, η
			Tight Prior			
μ_1	.006	.000	.000	.004	.016	1
μ_2	.003	.000	.000	.003	.846	3
μ_3	.285	.001	.001	.066	.039	5
μ_4	.687	.000	.000	.895	.000	8
μ_5	.018	.998	.998	.031	.098	83

Exhibit 4.6   Auxiliary regressions*

	Estimated coefficients of			R^2	η
	CONST	C(T−1)	ΔDPI(T)		
		Original Consumption-Function Data			
DPI(T)	− 11.547 [−4.9]	1.138 [164.9]	0.804 [11.9]	.9999	376
r(T)	− 1.02 [−3.9]	0.018 [22.3]	−0.015 [−1.9]	.9945	39
		Mixed-Estimation Consumption-Function Data			
		Loose Prior			
DPI(T)	− 11.152 [−2.2]	1.139 [75.5]	0.752 [5.2]	.9996	160
r(T)	− 0.899 [−2.9]	0.017 [18.9]	−0.012 [−1.4]	.9917	31
		Medium Prior			
DPI(T)	− 10.029 [−1.5]	1.141 [59.5]	0.631 [3.6]	.9994	120
r(T)	− 0.553 [−1.3]	0.016 [13.2]	−0.007 [−0.68]	.9839	22
		Tight Prior			
DPI(T)	0.209 [0.03]	1.124 [69.1]	0.322 [1.8]	.9987	83
r(T)	2.862 [4.5]	0.005 [3.2]	0.009 [0.52]	.9113	8

*Since interest here centers on tests of significance, *t*-statistics are given in the square brackets.

The following conclusions are indicated:

1. The introduction of prior information greatly sharpens the regression results. On the basis of the original, poorly conditioned data, only the estimate of β_3 was statistically significant. Even the information from the loose prior, however, brings b_1 and b_2 into significance, and the tight prior provides sufficient information to make all estimates significant, even that for b_4, the coefficient of the interest rate!

2. The improved conditioning of the data through the introduction of increasingly strong prior information is clearly and dramatically reflected in the condition indexes. The loose prior reduces the condition index of the dominant relation by more than one-half. The medium prior adds slightly more conditioning information, mainly to the second near dependency, and the tight prior has reduced the dominant condition index to a more nearly manageable 83 and has all but undone the second near dependency, whose condition index is now reduced to 8. There is, therefore, a strong parallel between the diagnostic indications of better conditioning and the sharpened regression estimates that result as increasingly tight prior information is introduced.

3. Examination of the patterns of variance-decomposition proportions reveals two important changes. The first is the ever decreasing proportion of $\text{var}(b_1)$ determined by μ_5, the dominant singular value, and the second is the ever increasing proportion of $\text{var}(b_5)$ determined by μ_2 (and away from μ_5). On the basis of these changes we would expect both of these variates to play a diminished role in the near dependency associated with μ_5, and this is verified by examination of the auxiliary regressions in Exhibit 4.6.

4. We also see in the auxiliary regressions how the decreasing condition indexes are reflected in decreasing R^2, quite in line with the progression we have come to expect.

5. Finally, it is to be noted that the tightest prior information has provided sufficient conditioning information so that even the coefficient of the interest rate, $r(T)$, about which no prior restriction was introduced, has become significant. In essence, the prior information on the other coefficients appears to have "freed up" what independent information there is in the original data on $r(T)$ to allow its coefficient to be more precisely estimated.

This phenomenon is also faithfully reflected in the corresponding diagnostic results (as we hoped it would be). We have already seen in Section 3.4 that $r(T)$ is involved only in the second of the two near dependencies present in the original data. This characteristic is elucidated in the auxiliary regressions relevant to the original data given in Exhibit 4.6. It is also seen clearly there that this second near dependency becomes weaker and weaker as increasingly strong prior information is introduced, finally reduced to having a condition index of 8. Thus the prior information has broken the collinear relation involving $r(T)$, making it possible to test effectively its significance in the estimated consumption function.

4.2 ROW-DELETION DIAGNOSTICS WITH MIXED-ESTIMATION OF THE U.S. CONSUMPTION FUNCTION

We have seen in the previous section that the introduction of prior information can greatly improve the conditioning of a data matrix and result in increased precision in parameter estimates. In this section we show, still using the consumption-function data, that the introduction of prior information also has interesting implications for the analysis of parameter sensitivity to influential data. We first analyze the consumption-function data using single-row diagnostic techniques to highlight possible data problems and then reanalyze the data after they have been augmented by the medium-strength prior introduced in the previous section. This latter exercise allows us both to assess the effect of introducing the prior information and to provide a further illustration of the use of the row-deletion diagnostics.

A Diagnostic Analysis of the Consumption-Function Data

We apply various of the single-row diagnostic measures to the aggregate consumption-function data introduced in Chapter 3 and again in the previous section. For convenience we reproduce in (4.7) the basic least-squares regression for the consumption function in (4.6).

We note that the only statistically significant coefficient is that for personal disposable income. The lack of significance for the other coefficients is quite possibly attributable to collinearity, as evidenced by the very large (scaled) condition number of 376. Our interest at the moment, however, centers on an analysis of the basic data.

$$C(T) = 6.7242 + 0.2454\,C(T-1)$$
$$\quad\quad (3.827)\quad (0.2374)$$

$$+ 0.6984\,\mathrm{DPI}(T) - 2.2097\,r(T) + 0.1608\,\Delta\mathrm{DPI}(T)$$
$$(0.2076)\quad\quad\quad (1.838)\quad\quad (0.1834)$$

$$R^2 = .9991 \quad\quad \mathrm{SER} = 3.557 \quad\quad \mathrm{DW} = 1.89 \quad\quad \kappa(\mathrm{X}) = 376, \quad (4.7)$$

Single-Row Diagnostics. Before proceeding to the single-row diagnostics, a word of warning is in order. The concept of deletion takes on a new meaning when applied to data series appearing along with their own lagged values. In this case deletion of a row does not fully remove the presence of an observation, since lagged values of the variates will occur in the neighboring rows; that is, deletion of a row does not truly correspond to deletion of an observation. We ignore any further examination of this problem here, but note its relevance to future research in Chapter 5.

Residuals. Exhibit 4.7 shows a time-series plot of the studentized residuals. One clear outlier appears in 1973(26) with value -2.44. Two sizable positive residuals, each equal to 1.88, surface in 1969(22) and 1972(25). These magnitudes, along with an examination of Exhibit 4.7, indicate the possibility of intertemporal heteroscedasticity or structural change.

Leverage and Hat-Matrix Diagonals. Exhibit 4.8 tabulates the hat-matrix diagonals in the first column. The approximate cutoff of $2p/n=0.37$ is exceeded only in 1970(23) and 1974(27) with respective values of 0.57 and 0.74. The remaining sizable leverage points, associated with the intervening years 1971–1973, are all less than 0.37.

Coefficient Sensitivity. The DFBETAS are shown in the last five columns of Exhibit 4.8. It is evident from the asterisks, which appear opposite magnitudes in excess of $2/\sqrt{n}=0.38$, that the two worst data points, 23 and 25, have larger DFBETAS for the majority of their

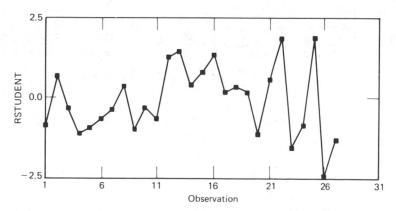

Exhibit 4.7   Time-series plot of studentized residuals: U.S. consumption-function data.

Exhibit 4.8 Some major diagnostics using OLS: U.S. consumption-function data

Index	Year	(1) h_i	(2) RSTUDENT	(3) DFFITS	DFBETAS				
					(4) b_1 CONST	(5) b_2 $C(T-1)$	(6) b_3 DPI(T)	(7) b_4 $r(T)$	(8) b_5 ΔDPI(T)
1	1948	0.1798	−0.812	−0.3805	−0.1609	−0.1261	0.1602	−0.1984	−0.1857
2	1949	0.1340	0.733	0.2883	0.1981	−0.0757	0.0706	0.0094	−0.1132
3	1950	0.1243	−0.272	−0.1027	−0.0545	−0.0009	0.0063	−0.0222	−0.0268
4	1951	0.0886	−1.087	−0.3391	−0.1724	0.0017	0.0003	0.0260	0.0419
5	1952	0.1276	−0.891	−0.3412	−0.2832	0.2086	−0.2035	−0.0089	0.2061
6	1953	0.1586	−0.602	−0.2616	−0.2405	0.1872	−0.1747	−0.0712	0.1429
7	1954	0.1523	−0.310	−0.1314	−0.0412	0.0354	−0.0442	0.0714	0.0711
8	1955	0.1031	−0.398	0.1351	0.0754	−0.0583	0.0599	−0.0270	−0.0380
9	1956	0.0708	−0.954	−0.2634	−0.0058	−0.0914	0.0817	0.1014	−0.0641
10	1957	0.0756	−0.250	−0.0717	−0.0518	0.0301	−0.0279	−0.0129	0.0341
11	1958	0.0905	−0.612	−0.1933	−0.0438	0.0001	−0.0070	0.0563	0.0601
12	1959	0.1565	1.300	0.5602	0.4665*	−0.2857	0.2286	0.4000*	−0.1333
13	1960	0.1140	1.480	0.5311	−0.0177	0.3298	−0.3547	0.1638	0.2798
14	1961	0.1643	0.427	0.1895	−0.0663	0.1602	−0.1626	0.0094	0.1383
15	1962	0.0532	0.840	0.1993	−0.0169	0.0908	−0.0914	−0.0083	0.1014
16	1963	0.1259	1.360	0.5163	−0.2824	0.3434	−0.3049	−0.2968	0.2325
17	1964	0.1795	0.209	0.0979	−0.0514	0.0591	−0.0546	−0.0382	0.0646
18	1965	0.1879	0.369	0.1777	−0.0746	0.0258	−0.0087	−0.1280	0.0232
19	1966	0.1282	0.225	0.0865	−0.0482	0.0287	−0.0213	−0.0523	0.0257
20	1967	0.1315	−1.086	−0.4228	0.2765	−0.1595	0.1171	0.2861	−0.0815
21	1968	0.1182	0.619	0.2270	0.0432	−0.1387	0.1418	0.0006	−0.1081
22	1969	0.1371	1.878	0.7488	−0.1296	0.2023	−0.2443	0.4004*	0.2312
23	1970	0.5741*	−1.517	−1.7624*	−0.0173	−0.5234*	0.7099*	−1.4653*	−0.9079*
24	1971	0.2922	−0.797	−0.5124	−0.2020	0.3800*	−0.3642	−0.1775	0.2918
25	1972	0.2907	1.876	1.2017*	0.1846	−0.8481*	0.8800*	−0.0886	−0.7210*
26	1973	0.2986	−2.438*	−1.5911*	0.7473*	−0.1574	0.1080	0.2315	−0.4604*
27	1974	0.7415*	−1.233	−2.0902*	0.5981*	−0.1711	0.0830	0.3349	0.6329*

*Exceeds cutoff values: $h_i=0.37$; RSTUDENT=2.0; DFFITS=0.86; DFBETAS=0.38.

parameters, while the somewhat less troublesome points, 12, 26, and 27 show a large change in no more than two coefficients, and at lower values of DFBETAS. It is also seen that all coefficients are estimated with some degree of sensitivity to influential data.

The most influential observation based on DFBETAS is 1970(23), where both substantial leverage and a moderately large residual coincide. The next most influential observation occurs two years later in 1972, which is most obviously linked to a moderately large residual. The last two observations are potentially troublesome and have, respectively, a large residual and an extremely large h_i value of 0.74.

Summary. In summary, all coefficients demonstrate sensitivity to influential data, with two data points, 1970(23) and 1972(25), highlighted on account of their large number of sizable DFBETAS. Since the data are plagued by collinearity, we next examine what happens when remedial action is taken.

A Reanalysis after Remedial Action for Ill Conditioning

It is apparent that the row-deletion diagnostics have revealed substantial sensitivities in the estimated coefficients of the consumption function. We shall see, however, that some of this sensitivity disappears with the introduction of prior information, a phenomenon that is clearly discernible from the influential-data diagnostics. For the purposes of this discussion we limit ourselves to an examination of the consumption function estimated by means of mixed-estimation subject to the introduction of the medium-strength prior described in the previous section. The mixed-estimation regression results for the medium-strength prior, taken from Exhibit 4.4, are given in Equation 4.8.[8]

$$C(T) = 7.61 + 0.241\,C(T-1) + 0.688\,\mathrm{DPI}(T) - 1.354\,r(T) + 0.198\,\Delta\mathrm{DPI}(T)$$
$$(2.45)\,(0.077)(0.066)(1.072)(0.074)$$

$$R^2 = .99 \qquad \mathrm{SER} = 4.264 \qquad \mathrm{DW} = 1.97 \qquad \kappa(\mathbf{X}) = 120. \tag{4.8}$$

A reanalysis of the residuals yields little additional information. The data matrix, however, has to be substantially altered, having three additional rows representing the three prior constraints. The row

---

[8] Standard errors in (4.8) are given in parentheses. These, as well as the SER, are calculated on the basis of the full augmented data matrix described in the previous section with 25 degrees of freedom. The Durbin-Watson statistic, however, has been calculated using only the unaugmented data series.

diagnostics should, therefore, reflect the introduction of the prior information, and we turn to them now.

The Row Diagnostics. Let us examine first the hat-matrix diagonals, h_i, which are now very different from those of the unaugmented consumption-function data. The h_i are given in column 1 of Exhibit 4.9, and the leverage points are now concentrated in the priors. The cutoff value for h_i is $2p/n = 0.33$.[9] Despite the fact that the prior variances are not especially tight, the largest h_i's of 0.75 and 0.68 are associated with the rows 28 and 29 corresponding to the prior restrictions on β_3, the DPI coefficient, and the long-run marginal propensity to consume $\beta_3/(1-\beta_2)$. Given the extreme extent of collinearity in the original data, it is not surprising that even moderate prior information possesses so much leverage. The largest previous leverage point of 0.74 (Exhibit 4.8) for 1974 (27) has diminished to 0.59 in the presence of the prior information, while the only other large leverage point of 0.57 in 1970(23) drops to 0.27.

The introduction of the prior information, therefore, has somewhat stabilized the relative leverage of the original data, a fact that should prove reassuring to the investigator who believes in the validity of the prior. It is also seen that these diagnostic techniques allow the investigator to assess just how influential his prior information is in the final estimation process.

The impact of the introduction of prior information on the measures for coefficient changes, DFBETAS, is also substantial, as is shown in Exhibit 4.9. Originally the most sensitive data point was 1970(23) and the next most troublesome 1972(25). Both are dramatically less sensitive in the presence of prior information, while 1974(27), which was marginally disturbing before, is now even less a source of concern. The observation for 1973(26), which still has a large studentized residual, continues to demonstrate some sensitivity.

The three data points 28, 29, and 30, reflecting prior information, tell a complementary story. When the rows corresponding to each prior are deleted one at a time, sizeable coefficient changes occur, particularly for deletions of the priors on short-run and long-run marginal propensities to consume, indexes 28 and 29, respectively. The impact from deleting the prior on income change β_5 (30) is less.

The partial-regression leverage plot for the interest rate coefficient, given in Exhibit 4.10, clearly shows the leverage extended by the prior on $\beta_3/(1-\beta_2)$, the long-run marginal propensity to consume, index 29. Even

---

[9] The proper way to include the prior information in the determination of the cutoffs is by no means clear. We have simply increased n by the number of such constraints here and in the calculations of the cutoffs for all other statistics.

Exhibit 4.9 Some major diagnostics using mixed estimation: consumption-function data with medium prior

Index	Year	(1) h_i	(2) RSTUDENT	(3) DFFITS	DFBETAS				
					(4) b_1 CONST	(5) b_2 $C(T-1)$	(6) b_3 DPI(T)	(7) b_4 $r(T)$	(8) b_5 ΔDPI(T)
1	1948	0.1033	-0.975	-0.3309	-0.2730	-0.0237	0.0605	-0.0671	-0.0604
2	1949	0.1167	0.789	0.2870	0.2254	-0.0035	-0.0043	-0.0241	-0.1065
3	1950	0.1069	-0.388	-0.1342	-0.0963	0.0127	-0.0066	0.0119	-0.0409
4	1951	0.0864	-1.126	-0.3465	-0.2600	-0.0180	0.0223	0.0598	0.0746
5	1952	0.0819	-0.913	-0.2731	-0.2274	0.0575	-0.0512	0.0311	0.0783
6	1953	0.0780	-0.675	-0.1964	-0.1692	0.0718	-0.0592	-0.0128	0.0229
7	1954	0.1052	-0.176	-0.0604	-0.0338	-0.0014	-0.0025	0.0266	0.0265
8	1955	0.0842	0.407	0.1235	0.0741	-0.0257	0.0280	-0.0378	0.0180
9	1956	0.0605	-0.964	-0.2448	-0.1296	-0.0342	0.0249	0.0903	-0.0104
10	1957	0.0585	-0.263	-0.0656	-0.0450	0.0062	-0.0035	-0.0045	0.0230
11	1958	0.0782	-0.553	-0.1613	-0.0777	-0.0210	0.0171	0.0298	0.0727
12	1959	0.0690	1.137	0.3096	0.2120	-0.1052	0.0606	0.1651	0.0076
13	1960	0.0627	1.475	0.3816	0.1618	0.1176	-0.1436	0.0881	-0.0348
14	1961	0.0574	0.415	0.1027	0.0288	0.0502	-0.0528	0.0025	0.0103
15	1962	0.0419	0.864	0.1809	0.0583	0.0214	-0.0211	-0.0191	0.0472
16	1963	0.0567	1.484	0.3640	0.0173	0.1373	-0.0991	-0.1854	0.0320
17	1964	0.0966	0.226	0.0740	-0.0060	0.0130	-0.0076	-0.0287	0.0415
18	1965	0.1169	0.481	0.1752	-0.0283	0.0034	0.0187	-0.0994	0.0599
19	1966	0.0808	0.307	0.0912	-0.0252	0.0091	0.0010	-0.0386	0.0312
20	1967	0.0712	-0.975	-0.2703	0.1068	-0.0556	0.0199	0.1163	-0.0289
21	1968	0.0772	0.640	0.1853	-0.0433	-0.0533	0.0574	0.0205	0.0122
22	1969	0.1107	1.861	0.6569	-0.1781	0.0562	-0.0968	0.3373	0.0072
23	1970	0.2698	-1.597	-0.9714*	0.1502	-0.0336	0.1730	-0.7747*	-0.2646
24	1971	0.1456	-0.836	-0.3451	0.0825	0.1322	-0.1188	-0.1508	0.0210
25	1972	0.1500	1.823	0.7661	-0.3058	-0.2673	0.2987	0.0673	0.0071
26	1973	0.2589	-2.518*	-1.4888*	0.7677*	0.1024	-0.1697	0.0251	-0.6801*
27	1974	0.5987*	-0.568	-0.6945	0.2041	-0.1344	0.1145	-0.0868	0.3842*
28	β_3 Prior	0.7540*	0.329	0.5773	0.1506	0.5612*	0.5773*	-0.0254	-0.3298
29	$\beta_3/(1-\beta_2)$ Prior	0.6838*	-0.479	-0.7057	0.3833*	-0.2124	0.0486	0.6526*	0.0984
30	β_5 Prior	0.3367*	0.515	0.3671	-0.0762	0.1884	-0.2097	0.0486	0.3671*

*Exceeds cutoff values: $h_i = 0.33$; RSTUDENT $= 2.0$; DFFITS $= 0.82$; DFBETAS $= 0.37$.

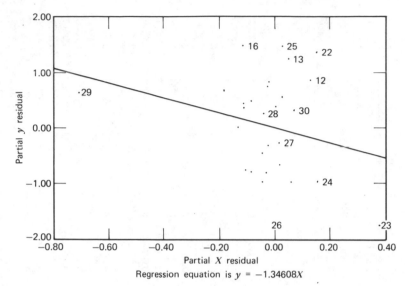

Exhibit 4.10   Partial-regression leverage plot for $b_4(R)$, S.E. = 1.0719 medium strength prior.

though no prior information was provided about $r(T)$, we see that other prior information has had a substantial impact on the estimation of this coefficient.

Exhibit 4.11 brings together the most significant individual coefficient changes associated with row deletion in the presence and absence of prior information. With the exception of 1973(26), where coefficient b_5 deteriorated sharply when the prior was applied, introduction of the priors reduced large coefficient changes.

A Suggested Research Strategy.   The decrease in parameter sensitivity that results from introducing prior information, while far from definitive in the absence of more experimental evidence, does suggest that apparent segments of anomalous data can be readily confounded with problems arising from collinearity. Indeed, collinearity can even disguise anomalous data, as indicated by the increased coefficient sensitivity which became apparent in 1973(26) when data conditioning was improved through the use of prior information. Thus, we provisionally conclude that reduction in collinearity should be a first step for the effective detection of unusual data components. In the absence of additional, better conditioned data, this should be accomplished with Bayesian or mixed-estimation methods, as has been done here. When specific priors are not available, however, more

Exhibit 4.11 DFBETAS with and without prior information:* consumption-function data

| | | \multicolumn{10}{c}{DFBETAS (NP=No Prior, MP=Medium Prior)} | | | | | | | | |
| | | b_1 CONST | | b_2 $C(T-1)$ | | b_3 $DPI(T)$ | | b_4 $r(T)$ | | b_5 $\Delta DPI(T)$ | |
Index	Year	NP	MP	NP	MP	NP	MP	NP	MP	NP	MP
23	1970	-0.0173	0.1502	-0.5234*	-0.0336	0.7099*	0.1730	-1.4653*	-0.7747*	-0.9079*	-0.2646
25	1972	0.1846	-0.3058	-0.8481*	-0.2673	0.8800*	0.2987	-0.0866	0.0673	-0.7210*	0.0071
26	1973	0.7473*	0.7677*	-0.1574	0.1024	0.1080	-0.1697	0.2315	0.0251	-0.4604*	-0.6801*
27	1974	0.5981*	0.2041	-0.1711	-0.1344	0.0830	0.1145	0.3349	-0.0868	0.6329*	0.3842*
28	β_3 Prior	—	0.1506	—	-0.5612*	—	0.5773*	—	-0.0252	—	-0.3298
29	$\beta_3/(1-\beta_2)$ Prior	—	0.3833*	—	-0.2124	—	0.0486	—	0.6526*	—	0.9840*
30	β_3 Prior	—	-0.0762	—	0.1884	—	-0.2097	—	0.0486	—	0.3671*

*Exceeds cutoff values: no prior (NP)=0.38; medium prior (MP)=0.36.

mechanical methods, such as ridge regression, might still improve the sort of diagnostic analysis we have developed in this book.[10]

4.3   AN ANALYSIS OF AN EQUATION DESCRIBING THE HOUSEHOLD DEMAND FOR CORPORATE BONDS

Economic theory and empirical representations of monetary phenomena have come to play a more important, albeit controversial, part in economy-wide econometric models. This sectoral expansion has been accompanied by concern over the issue of the structural stability of estimated relations since financial regulations and institutions change frequently and the impact of monetary controls, actual or anticipated, could readily lead to modifications of underlying behavior. We have therefore selected a recently developed financial market equation as an appropriate context in which to examine possible structural instability and parameter sensitivity through a variety of means.

First we apply the well-known Chow (1960) and the Brown-Durbin-Evans (1975) tests for overall structural stability—tests that, in this case, result in conclusive evidence against instability. We then exploit the diagnostic techniques of Chapter 2 to expose those elements of the monetary equation that show special sensitivity to specific elements of the data. Having thus isolated the potentially troublesome data points, we examine the interaction of the conditioning of the data with the presence of influential observations by means of ridge regression. It is here discovered that there is a reduction in the influence demonstrated by the original data, although the periods of tight money continue to show some influence despite the improved conditioning brought about by ridge regression. This suggests the possibility that the model is not adequately specified.

In the process of building a detailed model of the financial sector, Benjamin Friedman (1977) has devised an equation portraying the household demand for corporate bonds. He has kindly made the underlying data available for our subsequent analysis. The theoretical framework is described in the following excerpt:[11]

In a world in which transactions costs are nontrivial, it is useful to represent investors' portfolio behavior by a model which determines the desired long-run

---

[10]One should be aware, however, that there is a prior implicit in the use of ridge regression that is not always appropriate, namely that all the β's are zero; on this see Holland (1973).
[11]Reprinted with the kind permission of Benjamin Friedman and the *Journal of Political Economy* (Copyright 1977 by The University of Chicago. All rights reserved. ISSN 0022-3808.)

equilibrium portfolio allocation together with a model which determines the short-run adjustment toward the equilibrium allocation.

A familiar model of the selection of desired portfolio allocation, for a given investor or group of investors, is the linear homogeneous form

$$\frac{A_{it}^*}{W_t} = \sum_k^N \beta_{ik} r_{kt} + \sum_h^M \gamma_{ih} X_{ht} + \pi_i, \qquad i = 1, \ldots, N, \tag{4.9}$$

where

A_{it}^*, $i = 1, \ldots, N$ = the investor's desired equilibrium holding of the ith asset at time period t $(\sum_i A_{it}^* = W_t)$;

W_t = the investor's total portfolio size (wealth) at time period t;

r_{kt}, $k = 1, \ldots, N$ = the expected holding-period yield on the kth asset at time period t;

X_{ht}, $h = 1, \ldots, M$ = the values at time period t of additional variables which influence the portfolio allocation;

and the β_{ik}, γ_{ih}, and π_i are fixed coefficients which satisfy $\sum_i \beta_{ik} = 0$ for all k, $\sum_i \gamma_{ih} = 0$ for all h, and $\sum_i \pi_i = 1$. The role of the wealth homogeneity constraint is to require that any shift in an asset's share in the desired equilibrium portfolio be due to movements either of relevant yields (r_k) or of other variables (X_h), rather than to overall growth of the total portfolio itself; particularly for the case of equations representing the behavior of categories of investors, this assumption seems appropriate. ...

Given the desired equilibrium portfolio allocation indicated by model (4.9), the usual description of investor behavior involves a shift of asset holdings which eliminates some, but not all, of the discrepancy between holdings $A_{i,t-1}$ at the end of the previous period and the new desired holdings A_{it}^*. One familiar representation of the resulting portfolio adjustment process is the stock adjustment model

$$\Delta A_{it} = \sum_k^N \theta_{ik} (A_{kt}^* - A_{k,t-1}), \qquad i = 1, \ldots, N, \tag{4.10}$$

where A_{it} = the investor's actual holding of the ith asset at time period t $(\sum_i A_{it} = W_t)$, and the θ_{ik} are fixed coefficients of adjustment such that $0 \leqslant \theta_{ik} \leqslant 1$, $k = i$, and $\sum_i \theta_{ik} = 1$ for all k.

The empirical implementation by Friedman of this hypothesis is the following linear regression:

$$\begin{aligned} \text{HCB}(T) = {}& \beta_1(\text{HAFA}(T) \cdot \text{MR}(T)) + \beta_2(\text{HFA}(T-1) \cdot \text{MR}(T)) \\ & + \beta_3(\text{PER}(T) \cdot \text{HAFA}(T)) + \beta_4(\text{CPR}(T) \cdot \text{HAFA}(T)) \\ & + \beta_5 \text{HCB}(T-1) + \beta_6 \text{HLA}(T-1) + \beta_7 \text{HE}(T-1) + \varepsilon(T). \end{aligned} \tag{4.11}$$

The symbol definitions are the following:

Symbol	Definition
CPR	Commercial Paper Rate
HAFA	Household Net Acquisition of Financial Assets
HCB	Household Stock of Corporate Bonds
HE	Household Stock of Equities
HFA	Household Stock of Financial Assets
HLA	Household Stock of Liquid Assets
MR	Moody AA Utility Bond Rate
PER	Standard and Poor's Price-Earnings Ratio

Equation (4.12) shows least-squares estimates for the regression described above, for the period 1960:1–1973:4 using quarterly data. While Friedman used instrumental-variable estimation, his coefficients differ by at most 5% from those reported here:

$$\text{HCB}(T) = 0.0322 \text{ HAFA}(T) \cdot \text{MR}(T) + 0.000316 \text{ HFA}(T-1) \cdot \text{MR}(T)$$
$$\phantom{\text{HCB}(T) = } (0.00648) \phantom{\text{ HAFA}(T) \cdot \text{MR}(T) + } (0.000091)$$

$$-0.0242 \text{ PER}(T) \cdot \text{HAFA}(T)$$
$$ (0.0125)$$

$$-0.01549 \text{ CPR}(T) \cdot \text{HAFA}(T) + 0.8847 \text{ HCB}(T-1)$$
$$ (0.00454) \phantom{\text{ CPR}(T) \cdot \text{HAFA}(T) + } (0.0161)$$

$$(4.12)$$

$$+0.00894 \text{ HLA}(T-1)$$
$$ (0.00369)$$

$$-0.00619 \text{ HE}(T-1).$$
$$ (0.00157)$$

$$R^2 = .99 \quad \text{SER} = 490.6 \quad \text{DW} = 2.25 \quad \kappa(\mathbf{X}) = 106$$

The multiple correlation coefficient is large, as is to be expected with a smoothly growing dependent variable such as the stock of corporate bonds in an equation which also includes a lagged dependent variable. The signs of the coefficients are correct, the t-statistics are near two or greater in magnitude, and the Durbin-Watson statistic (biased in the presence of a lagged dependent variable towards serial independence) does not indicate autocorrelation problems. The condition number is large and suggests the need to take corrective action.

An Examination of Parameter Instability and Sensitivity

We approach the investigation of parameter instability and sensitivity in two steps. The first step is the application of significance tests for overall stability using the Chow test and the Brown-Durbin-Evans cusum or cusum-of-squares tests based on recursive residuals. The second step relies on row-deletion diagnostics to ascertain the existence and nature of potential sensitivity to influential data.

Tests for Overall Structural Instability. One standard test for structural instability is the Chow (1960) test, which we examine first. To this end the sample period was divided into two equal parts, for which regressions appear in (4.13) and (4.14), and the Chow statistic calculated as $F = 1.20$. The null hypothesis of stability cannot be rejected by the Chow test. Before turning to somewhat more sensitive stability tests, it is worth noting that the coefficients estimated from the second half of the data resemble those for the entire period much more than those estimated from the first half of the data. This finding is consistent with the fact that the influential observations appear only in the last part of the sample.

Regression from first half of data:

$$HCB(T) = 0.00963 \; HAFA(T) \cdot MR(T) + 0.000770 \; HFA(T-1) \cdot MR(T)$$
$$\quad (0.02330) \qquad\qquad\qquad (0.00022)$$

$$+ \; 0.03931 \; PER(T) \cdot HAFA(T)$$
$$\quad (0.02571)$$

$$- \; 0.06589 \; CPR(T) \cdot HAFA(T) + 0.73652 \; HCB(T-1)$$
$$\quad (0.02453) \qquad\qquad\qquad (0.08500)$$

$$- \; 0.00355 \; HLA(T-1)$$
$$\quad (0.00657)$$

$$+ \; 0.00034 \; HE(T-1).$$
$$\quad (0.00243)$$

$$(4.13)$$

$$R^2 = .92 \quad SER = 327.9 \quad DW = 2.25 \quad \kappa(X) = 140$$

Regression from second half of data:

$$HCB(T) = 0.03144\ HAFA(T) \cdot MR(T) + 0.000398\ HFA(T-1) \cdot MR(T)$$
$$\quad\ (0.01144) \qquad\qquad\qquad\qquad (0.000210)$$

$$-0.01957\ PER(T) \cdot HAFA(T)$$
$$\quad (0.02489)$$

$$-0.01830\ CPR(T) \cdot HAFA(T) + 0.87582\ HCB(T-1) \qquad\qquad (4.14)$$
$$\quad (0.00740) \qquad\qquad\qquad (0.02221)$$

$$+0.00905\ HLA(T-1)$$
$$\quad (0.00717)$$

$$-0.00756\ HE(T-1)$$
$$\quad (0.00226)$$

$$R^2 = .99 \quad SER = 600.3 \quad DW = 1.99 \quad \kappa(\mathbf{X}) = 155$$

The Brown-Durbin-Evans test appears to have greater power to detect departures from the null hypothesis than the Chow test. There are four possible versions of this test. The cusum, based on the cumulative sum of one-step ahead prediction residuals, has more power against one-sided departures from the null hypothesis. The cusum-of-squares, on the other hand, has most power against shifts in either direction and hence seems most appropriate for the type of analysis that concerns us. These two test statistics can be cumulated either forward or backward, thus providing four tests altogether. Schweder (1976) has suggested that the backward versions provide locally most powerful tests. Neither cusum test indicates departures at the 10% level. The backward cusum-of-squares, however, shown in Exhibit 4.12, indicates marginal instability at the 10% level, and the forward cusum of squares (not reproduced here) behaves similarly.[12] However, the departure from the null hypothesis is negligible and the cusum-of-squares retracks within the confidence limits.

Thus, on the basis of these two tests, Chow and Brown-Durbin-Evans, we have little reason to conclude that there are substantial instabilities in the overall regression regime.

---

[12]The actual significance level for rejecting the null hypothesis is greater for the combination of points.

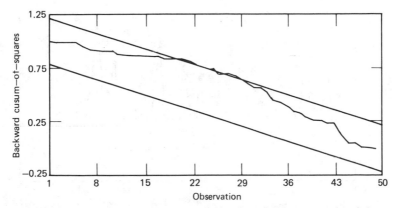

Exhibit 4.12 Backward cusum-of-squares plot: household corporate-bond data.

Sensitivity Diagnostics.

Residuals. Exhibit 4.13 shows the normal probability plot, and Exhibit 4.14 the time-series plot, of the studentized residuals. Three residuals exceed two in magnitude: 1968-1(33) is -2.07, 1971-3(47) is 2.95, and 1973-4(56) is 2.23. The noticeable wiggles in the normal probability plot suggest possible non-normality. There is, furthermore, evidence for some intertemporal heteroscedasticity, since the residuals for the first 21

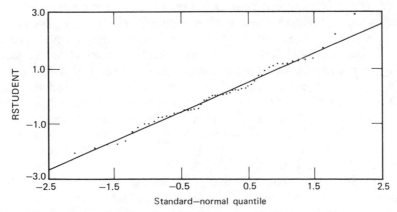

Exhibit 4.13 Normal probability plot of studentized residuals: household corporate-bond data.

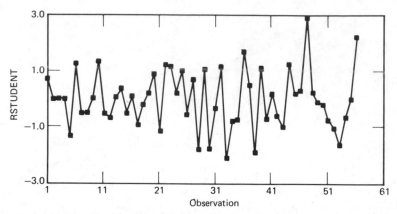

Exhibit 4.14   Time plot of studentized residuals: household corporate-bond data.

data points have less variance than those in the remainder of the data, as is evident in Exhibit 4.14.

Leverage and Coefficient Sensitivity. The diagonal elements of the hat matrix have five values which exceed $2p/n = 0.25$ and appear in two clusters from 1970-2 onward. This shows up clearly in column 1 of Exhibit 4.15, where the large leverage values are denoted by asterisks. Columns 4 to 10 for DFBETAS have 12 rows with at least one entry exceeding the size-adjusted cutoff of $2/\sqrt{n} = 0.27$. The majority occur in the latter half of the periods, and then, for the most part, in clusters.

At this point it is appropriate to examine relations among large coefficient changes and potential sources of sensitivity reflected in leverage or residuals. The following relevant points emerge. By far the largest leverage point is that for 1973-3(55), yet the estimated parameters are impacted negligibly because of a small associated studentized residual. Four changes show up concurrently with high leverage points, of which the three 1970-3(43) and 1970-4(44) and 1973-4(56) appear as potentially the most troublesome. Also, 1970-2(42) has high leverage, yet its DFBETAS, taken as a whole, are not troublesomely large. We also note that, in addition, 1966-4(28), 1971-3(47), and 1973-1(53) have large DFFITS in combination with several large DFBETAS.

We might therefore conclude that 1973-3(55) is helpful leverage, in that it reduces variance without exerting exceptional influence on coefficient estimates. It also appears that certain periods, as noted above, have adverse effects associated with leverage. We combine the preceding diagnostic information with knowledge of the specific economic events of the period to determine whether there is a systematic relation between sensitive coefficients and particular episodes in monetary policy.

The Monetary Background

Exhibit 4.16 shows historical patterns of short-term business borrowing rates, while Exhibit 4.17 shows member-bank excess reserves and member-bank borrowing from the Federal Reserve System. Short-term rates are the most commonly used indexes for conditions of monetary tightness, even though in this instance long-term corporate rates shown in Exhibit 4.18 are of most direct concern.[13] Since long-term rates ordinarily lag both short-term rates and member-bank borrowings from the Federal Reserve System, our principal criteria are short-term rates and member-bank borrowing.

Over the period to which the household demand for corporate bonds has been fitted, 1960-1–1973-4, there were two extended periods of extreme tightness shown by all the relevant series: 1969-1–1970-4 (observations 37–44) and 1972-2 through 1973-4 (observations 50–56). These intervals, while approximate, help to delineate circumstances of particular interest.

From the preceding analysis, large leverage is associated with each of these periods, as is one of the largest residuals, particularly for 1973-4(56). With this equation specification, then, we tentatively conclude that the coefficients are sensitive to periods of tight money. Since it is most difficult to model periods of monetary turbulence and uncertainty, we would not be surprised to find specification problems during periods of tight money.

A Use of Ridge Regression

As suggested at the end of Section 4.2, it is reasonable to make every attempt to remove as much ill conditioning due to collinearity as possible before exploring row-deletion diagnostics. The scaled condition number of 106 suggests that linear dependencies among the variables of the Friedman data are potentially troublesome, despite the "acceptably high" level of reported t-statistics. While mixed-estimation or Bayesian methods are the preferred means for introducing prior information that may improve conditioning, circumstances often arise in which prior information about parameter magnitudes and their variability is deficient or is incorporated only with great difficulty. In the present instance, where many of the variables are cross-products of the more basic variables, priors are especially difficult to establish.

---

[13]These charts have been reproduced from the 1976 *Historical Chart Book*, with the kind permission of the Board of Governors, Federal Reserve Board, Washington, D.C.

Exhibit 4.15 Some major diagnostics using OLS: household corporate-bond demand data

Index	Date	(1) h_i	(2) RSTUDENT	(3) DFFITS	(4) b_1 HAFA(T)·MR(T)	(5) b_2 HFA(T−1)·MR(T)	(6) b_3 PER(T)·HAFA(T)	(7) b_4 CPR(T)·HAFA(T)	(8) b_5 HCB(T−1)	(9) b_6 HLA(T−1)	(10) b_7 HE(T−1)
1	1960 1	0.0252	0.742	0.1194	−0.0602	0.0019	0.0420	0.0058	0.0550	−0.0530	0.0696
2	1960 2	0.0199	0.010	0.0015	0.0003	−0.0006	−0.0003	0.0000	−0.0001	0.0006	−0.0004
3	1960 3	0.0345	0.024	0.0045	0.0014	−0.0025	−0.0018	0.0006	0.0002	0.0025	−0.0012
4	1960 4	0.0772	0.053	0.0154	0.0098	−0.0093	−0.0099	0.0020	−0.0049	0.0137	−0.0103
5	1961 1	0.0832	−1.295	−0.3902	−0.2780*	0.3328*	0.3137*	−0.1692	0.0094	−0.3040*	0.1283
6	1961 2	0.0438	1.299	0.2782	0.1437	−0.2010	−0.1655	0.1056	0.0487	0.1055	0.0381
7	1961 3	0.0585	−0.501	−0.1249	−0.0723	0.1004	0.0866	−0.0597	−0.0265	−0.0563	−0.0120
8	1961 4	0.0658	−0.476	−0.1265	−0.0675	0.1025	0.0841	−0.0679	−0.0349	−0.0466	−0.0257
9	1962 1	0.0732	0.039	0.0111	0.0013	−0.0059	−0.0033	0.0052	0.0060	−0.0013	0.0070
10	1962 2	0.1095	1.354	0.4750	0.2071	−0.3759*	−0.3147*	0.2785*	0.1669	0.1844	0.0772
11	1962 3	0.0907	−0.495	−0.1564	0.0275	−0.0253	−0.0483	0.0853	0.0480	−0.0415	0.0839
12	1962 4	0.0801	−0.629	−0.1858	−0.0511	0.0565	0.0387	0.0357	0.0630	−0.1244	0.1252
13	1963 1	0.0468	0.069	0.0153	0.0023	−0.0051	−0.0018	−0.0020	−0.0023	0.0071	−0.0057
14	1963 2	0.0399	0.390	0.0796	0.0002	−0.0158	0.0042	−0.0181	−0.0082	0.0211	−0.0164
15	1963 3	0.0366	−0.502	−0.0979	0.0065	0.0267	−0.0041	0.0037	−0.0076	−0.0128	−0.0030
16	1963 4	0.0379	0.113	0.0225	0.0039	−0.0111	−0.0053	0.0043	0.0005	0.0069	0.0001
17	1964 1	0.0391	−0.910	−0.1837	−0.0263	0.0862	0.0369	−0.0378	−0.0072	−0.0441	−0.0135
18	1964 2	0.0480	−0.163	−0.0367	0.0122	−0.0017	−0.0134	0.0048	−0.0028	0.0111	−0.0137
19	1964 3	0.0471	0.222	0.0494	−0.0120	0.0009	0.0138	−0.0053	0.0005	−0.0094	0.0142
20	1964 4	0.0485	0.915	0.2068	0.0589	−0.1102	−0.0723	0.0591	−0.0130	0.0719	0.0038
21	1965 1	0.0923	−1.131	−0.3611	−0.2037	0.2853*	0.2543	−0.1893	0.0142	−0.2289	0.0569
22	1965 2	0.0608	1.237	0.3150	0.1168	−0.1895	−0.1492	0.1089	−0.0356	0.1607	−0.0434
23	1965 3	0.0963	1.164	0.3802	−0.2049	0.1582	0.2524	−0.1606	−0.0554	−0.1446	0.0566
24	1965 4	0.1005	0.238	0.0798	−0.0501	0.0325	0.0545	−0.0234	0.0073	−0.0488	0.0388
25	1966 1	0.0798	1.028	0.3028	−0.1834	0.1262	0.1902	−0.0719	0.0130	−0.1799	0.1506
26	1966 2	0.0426	−0.503	−0.1063	0.0357	−0.0236	−0.0331	0.0158	0.0232	0.0152	−0.0116
27	1966 3	0.0718	0.708	0.1970	0.0029	0.0441	0.0103	−0.0669	−0.1301	0.0753	−0.1123
28	1966 4	0.1710	−1.745	−0.7931*	0.4043*	−0.5024*	−0.5275*	0.4762*	0.3680*	0.1247	0.2421
29	1967 1	0.1052	1.084	0.3719	−0.1519	0.1502	0.1973	−0.1437	−0.1496	−0.0389	−0.0754

30	1967 2	0.0468	-1.756	-0.3893	-0.0272	-0.0814	-0.0383	0.1478	0.2442	-0.0774	0.1149
31	1967 3	0.0630	-0.323	-0.0838	-0.0185	-0.0158	-0.0020	0.0236	0.0619	-0.0136	0.0193
32	1967 4	0.0968	1.162	0.3807	-0.0399	0.1851	0.1266	-0.1266	-0.1834	-0.1327	0.0824
33	1968 1	0.0547	-2.069*	-0.4979	-0.1090	0.0116	0.1362	-0.0354	0.1595	-0.1072	0.0307
34	1968 2	0.1296	-0.771	-0.2976	-0.0361	-0.0630	-0.0290	0.0196	0.1697	0.0209	0.0017
35	1968 3	0.0868	-0.723	-0.2229	-0.0038	0.0119	0.0249	-0.0835	-0.0193	0.0695	-0.1349
36	1968 4	0.1461	1.729	0.7153*	-0.1753	0.2209	0.1806	0.0500	0.0471	-0.4413*	0.5029*
37	1969 1	0.1485	0.512	0.2139	-0.0236	0.0250	-0.0113	0.0556	0.0378	-0.0762	0.1188
38	1969 2	0.1251	-1.889	-0.7143*	0.0770	-0.0953	0.0598	-0.2294	-0.0817	0.2098	-0.3274*
39	1969 3	0.1265	1.147	0.4368	-0.0342	0.1038	-0.0423	0.0930	-0.0505	-0.0466	0.0603
40	1969 4	0.1371	-0.702	-0.2800	0.0084	-0.1030	0.0040	-0.0384	0.0940	0.0391	-0.0144
41	1970 1	0.1210	0.224	0.0834	0.0030	0.0224	-0.0123	0.0115	-0.0243	0.0053	-0.0113
42	1970 2	0.3320*	-0.588	-0.4152	0.1496	-0.2713*	-0.1432	0.1915	0.1016	0.0445	0.1134
43	1970 3	0.3669*	-0.989	-0.7533*	-0.3808*	0.0637	0.2364	-0.0445	0.5007*	-0.3759*	0.4470*
44	1970 4	0.2983*	1.258	0.8209*	0.6718*	-0.2743*	-0.5305*	0.0682	-0.4963*	0.6259*	-0.5699*
45	1971 1	0.1547	0.197	0.0845	0.0608	-0.0265	-0.0386	-0.0161	-0.0281	0.0497	-0.0447
46	1971 2	0.0966	0.352	0.1153	0.0527	-0.0176	-0.0394	-0.0109	0.0053	0.0245	-0.0124
47	1971 3	0.0891	2.953*	0.9237*	0.5007*	-0.3416*	-0.4394*	0.1763	0.1250	0.2519	-0.0643
48	1971 4	0.1097	0.263	0.0924	-0.0093	0.0088	0.0237	-0.0278	0.0392	-0.0301	0.0244
49	1972 1	0.2169	-0.066	-0.0348	0.0039	-0.0107	-0.0144	0.0234	-0.0044	0.0096	-0.0019
50	1972 2	0.2129	-0.159	-0.0827	0.0271	-0.0319	-0.0444	0.0448	-0.0299	0.0465	-0.0309
51	1972 3	0.1306	-0.752	-0.2915	0.0135	-0.0126	-0.0543	0.0835	-0.1298	0.0780	-0.0700
52	1972 4	0.2295	-1.015	-0.5545	0.0006	0.0344	-0.0883	0.0201	-0.2078	0.1957	-0.2285
53	1973 1	0.2370	-1.631	-0.9095*	0.0688	0.2753*	0.1943	-0.3222*	-0.6555*	0.0205	-0.2216
54	1973 2	0.1867	-0.620	-0.2973	0.1363	-0.0103	-0.0609	-0.0412	-0.1704	0.0416	-0.0154
55	1973 3	0.7292*	0.016	0.0270	-0.0038	-0.0068	-0.0029	0.0179	0.0082	-0.0008	0.0038
56	1973 4	0.3996*	2.230*	1.8195*	-0.6453*	0.3770*	0.3431*	-0.2433	0.1289	0.3178*	-0.8235*

*Exceeds cutoff values: $h_i = 0.25$; RSTUDENT = 2.0; DFFITS = 0.71; DFBETAS = 0.27.

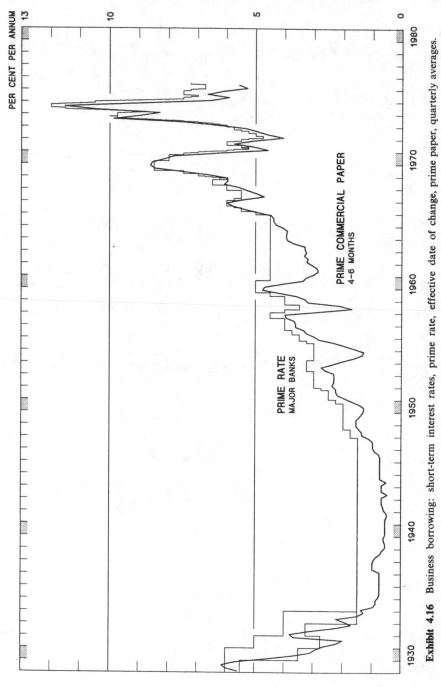

Exhibit 4.16 Business borrowing: short-term interest rates, prime rate, effective date of change, prime paper, quarterly averages.

222

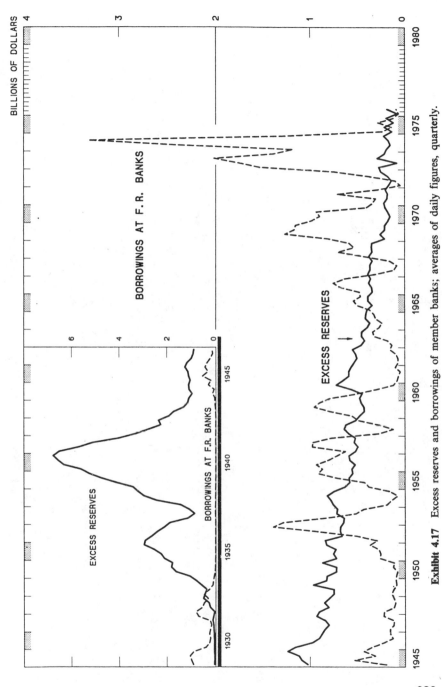

Exhibit 4.17 Excess reserves and borrowings of member banks; averages of daily figures, quarterly.

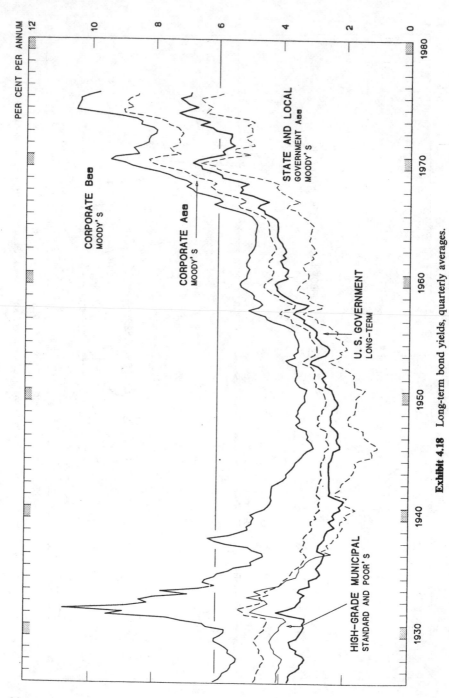

Exhibit 4.18 Long-term bond yields, quarterly averages.

224

Thus, we have chosen ridge regression as a second-best means for improving conditioning, and this (see Section 4.1 relating ridge regression to mixed estimation) in effect gives equal prior weight to each of the (scaled) variates. Conceptually, this is an empirical analogue to obtaining more identifying information since, in the limiting case of one or more exact linear dependencies, the parameters are unidentified. Thus, resorting to ridge regression can be thought of as logically prior to estimation and exploration of data-model interactions.

Equation (4.15) shows the new estimates with the ridge parameter $k = 0.001$. This relatively small perturbation of the (scaled) $\mathbf{X}$ matrix reduces the condition number to 65, still an indication of potential collinearity problems, but to a less severe extent. Coefficient signs are unchanged, while magnitudes change moderately. The choice of a relatively small k value has provided roughly comparable regression estimates so that we have not drastically altered the nature of the principal results by using ridge regression.[14]

$$HCB(T) = 0.02737\,HAFA(T) \cdot MR(T) + 0.000420\,HFA(T-1) \cdot MR(T)$$
$$\qquad\quad (0.00349) \qquad\qquad\qquad\qquad (0.000053)$$

$$\qquad\quad -0.01145\,PER(T) \cdot HAFA(T)$$
$$\qquad\qquad (0.0059)$$

$$\qquad\quad -0.01800\,CPR(T) \cdot HAFA(T) + 0.87228\,HCB(T-1)$$
$$\qquad\qquad (0.00344) \qquad\qquad\qquad (0.01287)$$

$$\qquad\quad +0.00541\,HLA(T-1)$$
$$\qquad\qquad (0.00149)$$

$$\qquad\quad -0.00534\,HE(T-1). \qquad\qquad\qquad\qquad\qquad\qquad (4.15)$$
$$\qquad\qquad (0.00094)$$

$$R^2 = .99 \quad \kappa(\mathbf{X}) = 65 \quad \text{Ridge } k = 0.001$$

Column 1 of Exhibit 4.19, presents the diagonals of the hat matrix. Comparison with Exhibit 4.15, which records equivalent values for the original regression, shows that the leverage effects for the original data

[14]The calculation of the ridge estimator and the corresponding condition indexes have here been accomplished by setting $\mathbf{A} = \mathbf{I}$ in the expression surrounding (4.5). Deletion of a "ridge row" is thereby associated with a column of $\mathbf{X}$ (a specific variate), as is desirable in this case. Other choices of $\mathbf{A}$ are conceivably more suitable to other diagnostic needs. Interesting and unresolved questions arise concerning the calculation of the standard errors of the regression and the coefficients, since ridge is an inherently biased estimator. The reported standard errors for the coefficients have been calculated according to Obenchain (1977).

Exhibit 4.19 Some major diagnostics using ridge regression: household corporate-bond demand data

Index	Year	(1) h_i	DFBETAS						
			(2) b_1 HAFA(T)·MR(T)	(3) b_2 HFA(T−1)·MR(T)	(4) b_3 PER(T)·HAFA(T)	(5) b_4 CPR(T)·HAFA(T)	(6) b_5 HCB(T−1)	(7) b_6 HLA(T−1)	(8) b_7 HE(T−1)
1	1960 1	0.0210	−0.0289	−0.0144	0.0147	0.0102	0.0315	−0.0173	0.0363
2	1960 2	0.0177	−0.0002	−0.0021	0.0000	−0.0006	0.0001	0.0019	−0.0008
3	1960 3	0.0280	0.0011	−0.0102	−0.0038	−0.0003	0.0046	0.0080	−0.0020
4	1960 4	0.0417	0.0171	−0.0215	−0.0176	−0.0047	−0.0094	0.0338	−0.0261
5	1961 1	0.0457	−0.0687	0.1153	0.0865	−0.0464	−0.0287	−0.0808	0.0104
6	1961 2	0.0339	0.0682	−0.1225	−0.0808	0.0461	0.0540	0.0356	0.0681
7	1961 3	0.0415	−0.0183	0.0334	0.0238	−0.0166	−0.0171	−0.0106	−0.0160
8	1961 4	0.0476	−0.0178	0.0360	0.0243	−0.0215	−0.0200	−0.0080	−0.0213
9	1962 1	0.0596	0.0021	−0.0113	−0.0058	0.0088	0.0113	−0.0025	0.0138
10	1962 2	0.0800	0.0767	−0.2521*	−0.1764	0.1844	0.1886	0.0672	0.1359
11	1962 3	0.0764	0.0264	−0.0141	−0.0406	0.0638	0.0259	−0.0267	0.0544
12	1962 4	0.0594	−0.0008	0.0155	−0.0054	0.0319	0.0192	−0.0419	0.0473
13	1963 1	0.0410	−0.0010	−0.0063	0.0018	−0.0060	−0.0011	0.0075	−0.0056
14	1963 2	0.0377	−0.0060	−0.0120	0.0099	−0.0187	−0.0031	0.0123	−0.0076
15	1963 3	0.0356	0.0101	0.0220	−0.0084	0.0064	−0.0075	−0.0082	−0.0058
16	1963 4	0.0347	0.0002	−0.0083	−0.0013	0.0020	0.0012	0.0035	0.0023
17	1964 1	0.0365	−0.0010	0.0574	0.0081	−0.0179	−0.0077	−0.0192	−0.0237
18	1964 2	0.0439	0.0117	0.0069	−0.0145	0.0050	0.0015	0.0075	−0.0168
19	1964 3	0.0448	−0.0013	−0.0012	0.0018	−0.0007	−0.0006	−0.0005	0.0020
20	1964 4	0.0431	0.0153	−0.0641	−0.0241	0.0286	−0.0068	0.0282	0.0244
21	1965 1	0.0616	−0.0503	0.1176	0.0817	−0.0799	−0.0126	−0.0681	−0.0171
22	1965 2	0.0494	0.0264	−0.1034	−0.0527	0.0551	−0.0129	0.0661	0.0126
23	1965 3	0.0779	−0.0623	0.0343	0.0876	−0.0581	−0.0505	−0.0251	0.0019
24	1965 4	0.0721	0.0140	−0.0051	−0.0166	0.0064	0.0029	0.0115	−0.0116
25	1966 1	0.0590	−0.0464	0.0216	0.0494	−0.0135	−0.0185	−0.0382	0.0426
26	1966 2	0.0406	0.0298	−0.0134	−0.0246	0.0088	0.0356	0.0025	−0.0074
27	1966 3	0.0606	−0.0126	0.0365	0.0175	−0.0435	−0.0790	0.0338	−0.0557
28	1966 4	0.1291	0.2577*	−0.3231*	−0.3561*	0.3332	0.3734*	−0.0100	0.2742*
29	1967 1	0.0907	−0.0577	0.0483	0.0793	−0.0519	−0.0888	0.0048	−0.0486
30	1967 2	0.0434	−0.0139	−0.0746	−0.0429	0.1288	0.2141	−0.0486	0.0689
31	1967 3	0.0592	−0.0261	−0.0278	−0.0078	0.0365	0.0994	−0.0124	0.0184

32	1967 4	0.0856	0.0182	0.0608	0.0254	-0.0431	-0.1098	-0.0294	0.0205
33	1968 1	0.0524	-0.0453	-0.0372	0.0724	-0.0071	0.1291	-0.0411	-0.0139
34	1968 2	0.1256	-0.0751	-0.0729	-0.0227	0.0073	0.2634*	0.0167	0.0039
35	1968 3	0.0759	-0.0254	0.0169	0.0446	-0.0914	0.0138	0.0488	-0.1214
36	1968 4	0.1121	0.0069	0.0556	0.0024	0.0613	-0.0513	-0.1305	0.1912
37	1969 1	0.1314	0.0003	0.0082	-0.0185	0.0303	0.0061	-0.0227	0.0463
38	1969 2	0.1141	0.0140	-0.0693	0.1132	-0.2186	-0.0057	0.1101	-0.2202
39	1969 3	0.1226	-0.0167	0.0787	-0.0476	0.0796	-0.0520	-0.0206	0.0336
40	1969 4	0.1311	-0.0107	-0.1169	0.0311	-0.0689	0.1325	0.0281	-0.0051
41	1970 1	0.1168	0.0001	0.0149	-0.0066	0.0079	-0.0133	0.0010	-0.0046
42	1970 2	0.2861*	0.0921	-0.1996	-0.0685	0.1286	0.0822	-0.0004	0.1122
43	1970 3	0.2991*	-0.1949	-0.0594	0.0640	-0.0077	0.3704*	-0.1484	0.2500*
44	1970 4	0.1802	0.3500*	-0.0148	-0.2103	-0.0527	-0.3311*	0.2511*	-0.2874*
45	1971 1	0.1171	0.0866	-0.0152	-0.0322	-0.0645	-0.0345	0.0485	-0.0540
46	1971 2	0.0902	0.0599	-0.0030	-0.0361	-0.0358	-0.0202	0.0111	-0.0029
47	1971 3	0.0787	0.2969*	-0.1440	-0.2204	0.0439	0.1618	0.0602	0.0269
48	1971 4	0.1019	0.0079	-0.0034	0.0151	-0.0356	0.0495	-0.0206	0.0172
49	1972 1	0.1959	0.0009	0.0018	0.0031	-0.0079	0.0015	-0.0014	-0.0004
50	1972 2	0.1679	0.0019	-0.0099	-0.0182	0.0321	-0.0198	0.0183	-0.0116
51	1972 3	0.1237	-0.0141	0.0084	-0.0136	0.0483	-0.0717	0.0232	-0.0224
52	1972 4	0.2094	-0.0661	0.0701	-0.0235	0.0145	-0.1328	0.0896	-0.1211
53	1973 1	0.2203*	0.0534	0.1014	0.0703	-0.1217	-0.3100*	0.0124	-0.0854
54	1973 2	0.1761	0.0353	0.0004	-0.0102	-0.0179	-0.0528	0.0030	0.0025
55	1973 3	0.6916*	-0.0465	-0.0564	-0.0196	0.1746	0.0709	-0.0077	0.0264
56	1973 4	0.3593*	-0.6890*	0.3055*	0.3061*	-0.1086	0.3186*	0.2619*	-0.6975*
57	Ridge β_1	0.0902	-0.4189*	0.2301	0.3145*	-0.0607	0.1610	-0.2330	0.0980
58	Ridge β_2	0.1340	0.3540*	-0.7097*	-0.4885*	0.5250*	0.3668*	0.2718*	0.1430
59	Ridge β_3	0.2110	-0.2018	0.2037	0.2434	-0.1957	-0.0144	-0.1443	0.0119
60	Ridge β_4	0.0474	0.0191	-0.1076	-0.0962	0.1976	0.0796	0.0139	0.0655
61	Ridge β_5	0.0378	0.5167*	0.7653*	0.0723	-0.8105*	-2.1221*	0.5731*	-1.0894*
62	Ridge β_6	0.3569*	-0.5639*	0.4276*	0.5443*	-0.1070	0.4321*	-0.8144*	0.7330*
63	Ridge β_7	0.1416	-0.1130	-0.1072	0.0215	0.2395	0.3914*	-0.3493*	0.5610*

*Exceeds cutoff values: $h_i = 0.22$; DFBETAS = 0.25.

Exhibit 4.20 Extremes of individual DFBETAS dervied from the OLS and ridge* estimates of the household corporate-bond equation

Coefficient	OLS		Ridge	
	Smallest	Largest	Smallest	Largest
b_1	−0.65	0.67	−0.68	0.35
b_2	−0.50	0.38	−0.32	0.30
b_3	−0.53	0.34	−0.36	0.31
b_4	−0.32	0.48	−0.22	0.33
b_5	−0.66	0.50	−0.33	0.37
b_6	−0.44	0.63	−0.15	0.26
b_7	−0.82	0.50	−0.70	0.27

*Ridge extremes based on only the first 56 rows of Exhibit 4.19.

(observations 1–56) are very little affected by the ridge augmentation, while the last seven ridge-related h_i show moderately sizable leverage for only one entry, that associated with 62.

Columns 2–8 of Exhibit 4.19 record the DFBETAS for the ridge-augmented data.[15] The highly intriguing result emerges that, despite leverage having changed only moderately, the number of DFBETAS above the size-adjusted cutoff has dropped dramatically for the 56 original observations, almost by one half. Thus, coefficient changes that had previously appeared may have been the consequence of collinearity. Nevertheless, periods of tight money remain relatively influential and their continued presence cannot be readily attributed to ill conditioning alone. Other interesting information can be gleaned from this exhibit by looking at row deletions for the ridge parameter (rows 57–63). In particular, deletion of the ridge parameter k associated with the lagged dependent variable (row 61) has a noticeably large impact on all coefficients except b_3. Row 62 for liquid assets, HLA($T-1$), also has a pervasive influence. The largest and smallest values for each DFBETAS are summarized in Exhibit 4.20. Clearly, individual coefficient sensitivity has been markedly reduced by ridge regression.

Summary

In the preceding analysis of a household corporate-bond equation, using the influential-data diagnostics, we discovered the presence of two periods

---

[15]In light of the previous footnote, we have calculated the ridge DFBETAS by scaling the ridge DFBETA by the OLS-estimated standard errors of the coefficients from (4.12).

of high influence associated with periods of tight money. This suggests the possibility that the model may be incompletely specified and may not deal adequately with periods of tight money. At the same time, the high condition number of the original data suggests that ill conditioning may also be a problem leading to parameter sensitivity. Ridge regression was therefore employed, using a very mild value for $k = .001$ as a means for improving data conditioning. The condition number was thereby roughly halved, and the periods of relatively high influence were reduced in duration and severity, but not removed altogether. Thus, while ill conditioning appears to be one major source of parameter sensitivity in this case, we must still consider the possibility that the model is not fully adequate to deal with periods of tight money.

4.4   ROBUST ESTIMATION OF A HEDONIC HOUSING-PRICE EQUATION

The final illustrative analysis in this chapter is based on a paper by Harrison and Rubinfeld (1978), in which a hedonic price index for housing is estimated for use in a subsequent estimation of the marginal-willingness-to-pay for clean air. Hedonic price indexes were introduced into the recent literature by Griliches (1968) and, in essence, are based on the fitted values of a regression of price on various explanatory variables used to represent its qualitative determinants. Harrison and Rubinfeld are principally interested in examining the impact of air pollution (as measured by the square of nitrogen oxide concentration (NOXSQ)) on the price of owner-occupied homes and include NOXSQ and thirteen other explanatory variables as indicators of qualities that affect the price variable relevant to this analysis.

The basic data, listed in Appendix 4A, are a sample of 506 observations on census tracts in the Boston Standard Metropolitan Statistical Area (SMSA) in 1970. Tracts containing no housing units, or those composed entirely of institutions, have been excluded. To facilitate later interpretation, the various census tracts are associated with their towns in Exhibit 4.21.

This application of the influential-data diagnostics complements the previous examples in several significant ways. First, it includes many sociodemographic variables at a relatively disaggregated level and, as such, is typical of many current studies of urban markets, voting patterns, wage patterns, and the like. While it is *prima facie* absurd to generalize from this one sample to all others with similar data bases, there may nevertheless be some suggestive insights of more than purely methodological interest.

Exhibit 4.21 Census tracts in the Boston SMSA in 1970

Observation	Town	Observation	Town
1	Nahant	275–279	Needham
2–3	Swampscott	280–283	Wellesley
4–6	Marblehead	284	Dover
7–13	Salem	285	Medfield
14–35	Lynn	286	Millis
36–39	Saugus	287	Norfolk
40–41	Lynnfield	288–290	Walpole
42–50	Peabody	291–293	Westwood
51–54	Danvers	294–298	Norwood
55	Middleton	299–301	Sharon
56	Topsfield	302–304	Canton
57	Hamilton	305–308	Milton
58	Wenham	309–320	Quincy
59–64	Beverly	321–328	Braintree
65	Manchester	329–331	Randolph
66–67	North Reading	332–333	Holbrook
68–70	Wilmington	334–341	Weymouth
71–74	Burlington	342	Cohasset
75–80	Woburn	343	Hull
81–84	Reading	344–345	Hingham
85–88	Wakefield	346–347	Rockland
89–92	Melrose	348	Hanover
93–95	Stoneham	349	Norwell
96–100	Winchester	350–351	Scituate
101–111	Medford	352–353	Marshfield
112–120	Malden	354	Duxbury
121–127	Everett	355–356	Pembroke
128–142	Somerville	357–488	Boston
143–172	Cambridge	357–364	Allston-Brighton
173–179	Arlington	365–370	Back Bay
180–187	Belmont	371–373	Beacon Hill
188–193	Lexington	374–375	North End
194–195	Bedford	376–382	Charlestown
196	Lincoln	383–393	East Boston
197–199	Concord	394–406	South Boston
200–201	Sudbury	407–414	Downtown
202–203	Wayland		(South Bay)
204–205	Weston	415–433	Roxbury
206–216	Waltham	434–456	Savin Hill
217–220	Watertown	457–467	Dorchester
221–238	Newton	468–473	Mattapan
239–244	Natick	474–480	Forest Hills
245–254	Framingham	481–484	West Roxbury
255–256	Ashland	484–488	Hyde Park
257	Sherborn	489–493	Chelsea
258–269	Brookline	494–501	Revere
270–274	Dedham	502–506	Winthrop

Second, the sample size is much larger than any of the others that were examined. We have a chance, then, to see whether problems of a different nature arise on this account. Third, the residual distribution is found to possess much heavier tails than the Gaussian (normal) distribution, unlike the previous cases. It becomes natural therefore to employ estimators which are more robust than OLS to departures from normality in the error structure.

The Model

The hedonic housing-price model used by Harrison and Rubinfeld is

$$LMV = \beta_1 + \beta_2 CRIM + \beta_3 ZN + \beta_4 INDUS + \beta_5 CHAS + \beta_6 NOXSQ + \beta_7 RM$$

$$+ \beta_8 AGE + \beta_9 DIS + \beta_{10} RAD + \beta_{11} TAX + \beta_{12} PTRATIO + \beta_{13} B$$

$$+ \beta_{14} LSTAT + \varepsilon. \tag{4.16}$$

A brief description of each variable is given in Exhibit 4.22. Further details may be found in Harrison and Rubinfeld (1978).

Exhibit 4.22 Definition of model variables

Symbol	Definition
LMV	logarithm of the median value of owner-occupied homes
CRIM	per capita crime rate by town
ZN	proportion of a town's residential land zoned for lots greater than 25,000 square feet
INDUS	proportion of nonretail business acres per town
CHAS	Charles River dummy variable with value 1 if tract bounds on the Charles River
NOXSQ	nitrogen oxide concentration (parts per hundred million) squared
RM	average number of rooms squared
AGE	proportion of owner-occupied units built prior to 1940
DIS	logarithm of the weighted distances to five employment centers in the Boston region
RAD	logarithm of index of accessibility to radial highways
TAX	full-value property-tax rate (per $10,000)
PTRATIO	pupil-teacher ratio by town
B	$(Bk - 0.63)^2$ where Bk is the proportion of blacks in the population
LSTAT	logarithm of the proportion of the population that is lower status

Exhibit 4.23   OLS estimates: Housing-price equation

Variable	Coefficient Estimate	Standard Error	t-Statistic
INTERCEPT	9.758	0.150	65.23
CRIM	-0.0119	0.00124	-9.53
ZN	7.94×10^{-5}	5.06×10^{-4}	0.16
INDUS	2.36×10^{-4}	2.36×10^{-3}	0.10
CHAS	0.0914	0.0332	2.75
NOXSQ	-0.00639	0.00113	-5.64
RM	0.00633	0.00131	4.82
AGE	8.86×10^{-5}	5.26×10^{-4}	0.17
DIS	-0.191	0.0334	-5.73
RAD	0.0957	0.0191	5.00
TAX	-4.20×10^{-4}	1.23×10^{-4}	-3.42
PTRATIO	-0.0311	0.00501	-6.21
B	0.364	0.103	3.53
LSTAT	-0.371	0.0250	-14.83

$$R^2 = .806 \qquad SER = 0.182 \qquad \kappa(X) = 66$$

Exhibit 4.23 reports least-squares estimates of (4.16). The overall fit reflected in R^2 is relatively good for cross-sections, and the condition number $\kappa(X)$ of 66, while large, is not deemed large enough in this context to be worth pursuing. While several coefficient estimates have low t-statistics, we note that the NOXSQ term representing pollution has the correct sign and a t-statistic of -5.64.

As a first diagnostic step we look at the normal probability plot in Exhibit 4.24. It is clear from visual inspection of this exhibit that there are very substantial departures from normality.

Robust Estimation

One appropriate means for estimating a model with an error structure that is not Gaussian is to use the maximum-likelihood estimator relevant to the "correct" error structure. Since, in the case of the Harrison-Rubinfeld model, the "correct" error structure is unknown, a reasonable alternative strategy is to explore structures in a neighborhood of the Gaussian to see how sensitive the estimated coefficients are to such changes. An analysis of this sort should be based on an estimator that is reasonably efficient both when the errors are Gaussian and when they are nearly so.

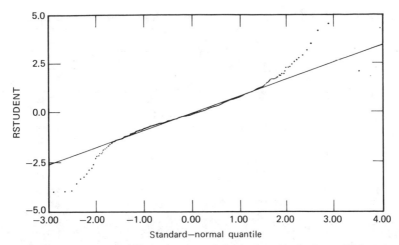

Exhibit 4.24 Normal probability plot for studentized residuals from OLS estimates; housing-price equation.

Recall from Chapter 2 [see (2.23)] that Huber has proposed just such an estimator, with a criterion function given by

$$\rho_c(t) = \begin{cases} \dfrac{t^2}{2} & |t| \leqslant c \\[2mm] c|t| - \dfrac{c^2}{2} & |t| > c. \end{cases} \qquad (4.17)$$

The parameters (including scale) are estimated by minimizing

$$\sum_{i=1}^{n} \sigma \rho_c\left(\frac{y_i - x_i \beta}{\sigma}\right) + d\sigma. \qquad (4.18)$$

As noted in Chapter 2, when $c = \infty$ (4.18) reduces to least squares, and when $c \rightarrow 0$ it is equivalent to least-absolute residuals. For this analysis we chose an intermediate value of $c = 1.345$ along with $d = 0.3591(n-p)$, which results in a Huber estimator having 95% efficiency relative to OLS when the error structure is in fact Gaussian.[16]

Exhibit 4.25 compares the Huber estimates and the OLS estimates for the Harrison-Rubinfeld model, and it reveals a number of sources of sensitivity, especially LSTAT and RM, which change by more than three OLS standard errors.[17]

[16]These results were computed using the ROSEPACK (1980) program.
[17]Since robust estimates of scale have a less secure theoretical basis than those for OLS estimates, they have not been reported here.

Exhibit 4.25   Huber and OLS estimates: housing-price equation

| Variable | Coefficient Estimates | | $\dfrac{b_{LS} - b_H}{s_{b_{LS}}}$ |
	Huber	OLS	
INTERCEPT	9.630	9.758	0.854
CRIM	-0.0110	-0.0119	-0.700
ZN	3.65×10^{-5}	7.94×10^{-5}	0.085
INDUS	1.21×10^{-3}	2.36×10^{-4}	-0.413
CHAS	0.0768	0.0914	0.441
NOXSQ	-0.00505	-0.00639	-1.182
RM	0.0115	0.00633	-3.943
AGE	-6.60×10^{-4}	8.86×10^{-5}	1.423
DIS	-0.164	-0.191	-0.810
RAD	0.0705	0.0957	1.319
TAX	3.61×10^{-4}	-4.20×10^{-4}	-0.480
PTRATIO	-0.0290	-0.0311	-0.436
B	0.551	0.364	-1.813
LSTAT	-0.281	-0.371	-3.587

Although Harrison and Rubinfeld did not examine the distributional form of the residuals, they were careful to explore a number of alternative specifications that led to coefficient changes of at least the amount produced by the Huber estimates and to report their final results by bracketing the ranges of the NOXSQ effect for alternative model structures and correspondingly different coefficient estimates.

Since, apart from the intercept, LSTAT had the largest t-statistic, it is somewhat disturbing for it to have shifted so much. The coefficient of the variable of principal interest, NOXSQ, also changed by one standard error.

It is interesting to compare studentized residuals from OLS (Exhibit 4.24) with those from the Huber estimates which appear in Exhibit 4.26. The latter, while not having normality imposed, have had the more extreme residuals downweighted to such an extent that the sample distribution of the weighted studentized residuals[18] is now broadly consistent with an underlying normal distribution. Since it is characteristic of the Huber

---

[18]The Huber estimates are computed using iteratively reweighted least squares [Holland and Welsch (1978)]. The weights W (a diagonal weight matrix) at the final iteration are used to form $W^{1/2}X$ and $W^{1/2}y$, which replace X and y in the derivation of the studentized residuals in (2.26).

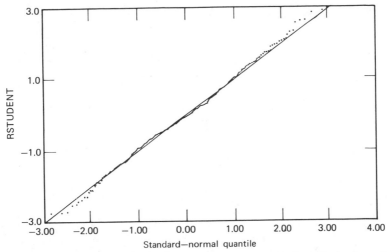

Exhibit 4.26 Normal probability plot of weighted studentized residuals from Huber estimates: housing-price equation.

estimator to downweight large residuals, this result is hardly surprising. It is even more informative to make a direct comparison of OLS residuals with unweighted residuals obtained from the data and the Huber-estimated coefficients. Because the choice of scale is a problem with robust estimates, the cleanest contrast is between the two sets of raw residuals rather than studentized ones. These are shown on a scatter diagram in Exhibit 4.27 with the Huber residuals on the vertical axis and the OLS residuals on the horizontal axis. A line with unit slope has been added. It is significant to note that the shape in the middle of the bivariate scatter is different from that of both ends. For the more sizable residuals at either end, the unweighted Huber residuals are larger in absolute value than the OLS residuals. When a robust fit (Tukey's resistant line [Velleman and Hoaglin (1980)]) is fitted to these data, it has a slope of 0.91, so that the Huber residuals tend to be smaller than the OLS residuals for the main body of the data but larger in the tails. This behavior, which is consistent with the underlying differences in the two estimators, is useful in the interpretation of DFFITS below.

Partial Plots

The partial-regression leverage plots, of which only two are shown here, reveal useful information of both a positive and a negative nature. Exhibit 4.28 for NOXSQ, the variable of principal concern, reveals a scatter which

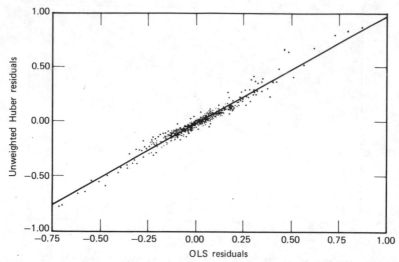

Exhibit 4.27 Scatter plot of unweighted Huber residuals versus OLS residuals: housing-price equation.

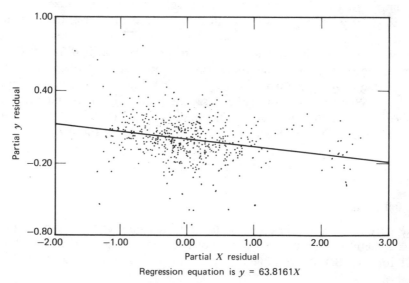

Regression equation is $y = 63.8161X$

Exhibit 4.28 Partial-regression leverage plot for b_6 (NOXSQ), S.E. = 0.00113: housing-price equation.

236

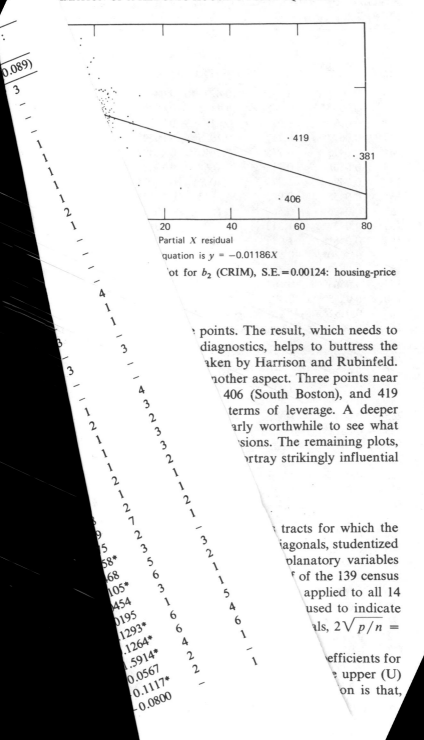

· 419

· 381

· 406

20   40   60   80

Partial X residual

quation is $y = -0.01186X$

ot for b_2 (CRIM), S.E. = 0.00124: housing-price

points. The result, which needs to
diagnostics, helps to buttress the
aken by Harrison and Rubinfeld.
other aspect. Three points near
406 (South Boston), and 419
terms of leverage. A deeper
arly worthwhile to see what
sions. The remaining plots,
ortray strikingly influential

tracts for which the
agonals, studentized
planatory variables
of the 139 census
applied to all 14
used to indicate
als, $2\sqrt{p/n}$ =

efficients for
upper (U)
on is that,

Exhibit 4.30 Hat-matrix diagonals, studentized residuals, DFFIT,
DFBETAS for selected census tracts: housing-price data*

Census Tract	h_i	RSTUDENT	DFFITS	DFBETAS NOXSQ	CRIM	Row Summar DFBETAS L (-0.089) U (
8	0.0338	2.286*	0.4276*	−0.0464	0.0196	1
124	0.0585*	0.443	0.1103	−0.0230	0.0028	−
127	0.0599*	0.071	0.0180	−0.0043	0.0001	−
143	0.0672*	−0.717	−0.1925	−0.0933*	−0.0034	2
144	0.0475	0.723	0.1614	0.1049*	0.0046	−
148	0.0508	0.718	0.1662	0.1032*	−0.0055	−
149	0.0485	1.773	0.4006*	0.2609*	−0.0144	−
151	0.0456	0.980	0.2142	0.1612*	0.0026	−
152	0.0496	0.666	0.1520	0.1051*	−0.0014	−
153	0.0737*	−1.312	−0.3701*	−0.2013*	−0.0028	2
154	0.0461	1.075	0.2364	0.1664*	−0.0061	−
155	0.0630*	−0.538	−0.1395	−0.0842	−0.0032	−
156	0.0840*	−0.427	−0.1292	−0.0650	0.0007	−
157	0.0666*	−0.468	−0.1251	−0.0707	0.0083	−
160	0.0555*	0.007	0.0017	0.0013	0.0001	−
161	0.0552*	−1.525	−0.3688*	0.1152*	−0.0003	2
162	0.0679*	−0.528	−0.1424	0.0166	−0.0107	−
163	0.0767*	−0.975	−0.2809	0.0406	−0.0325	
164	0.0676*	−0.105	−0.0283	0.0042	−0.0028	
215	0.0579*	2.910*	0.7214*	−0.0776	−0.0751	
258	0.0572*	0.437	0.1076	0.0122	−0.0057	
284	0.0690*	0.495	0.1348	−0.0059	0.0159	
285	0.0443	1.607	0.3461*	−0.0005	0.0207	
343	0.0823*	−1.014	−0.3036	−0.0406	−0.0323	
358	0.0493	1.002	0.2282	0.0994*	−0.0185	
359	0.0520	1.250	0.2930	0.1364*	−0.0056	
360	0.0246	1.790	0.2841	0.2000*	−0.0419	
361	0.0316	1.277	0.2306	0.1502*	−0.0190	
362	0.0204	1.203	0.1737	0.1192*	−0.0406	
363	0.0329	0.969	0.1787	0.0906*	−0.031	
365	0.0891*	−2.747*	−0.8590*	−0.1780*	0.064	
366	0.0773*	1.960	0.5671*	0.0549	−0.08	
367	0.0223	1.718	0.2597	0.0427	−0.09	
368	0.0607*	2.764*	0.7026*	−0.1952*	−0.0	
369	0.0982*	2.663*	0.8791*	−0.1256*	−0.	
370	0.0656*	1.723	0.4566*	−0.0808	−0.	
371	0.0731*	0.965	0.2712	−0.0467	−0	
372	0.0242	4.512*	0.7110*	−0.2709*	−0	
373	0.0532	4.160*	0.9856*	−0.1751*	−(	
381	0.2949*	2.559*	1.6551*	0.0975*		
386	0.0183	−2.564*	−0.3499*	0.0113	−	
388	0.0231	−1.779	−0.2738	−0.0077	−	
92	0.0126	2.334*	0.2632	0.0731	−	

Census Tract	h_i	RSTUDENT	DFFITS	DFBETAS NOXSQ	DFBETAS CRIM	Row Summary: DFBETAS L (−0.089)	Row Summary: DFBETAS U (0.089)
398	0.0125	−3.212*	−0.3617*	−0.0025	0.0928*	1	2
399	0.0458	−3.301*	−0.7235*	−0.0168	−0.5774*	3	−
400	0.0213	−3.936*	−0.5808*	0.0132	0.1176*	4	4
401	0.0225	−3.954*	−0.5995*	−0.0242	−0.3424*	4	−
402	0.0141	−3.988*	0.4766*	−0.0204	−0.0636	3	1
404	0.0212	−2.030*	−0.2989	−0.0258	−0.1833*	2	−
406	0.1533*	2.141*	−0.9112*	−0.0552	−0.8699*	3	1
408	0.0198	2.422*	0.3446*	−0.0925*	−0.0271	4	−
410	0.0252	3.162*	0.5079*	−0.2685*	−0.0308	3	3
411	0.1116*	1.688	0.5983*	−0.1278*	0.4130*	4	2
412	0.0394	1.084	0.2195	−0.1008*	−0.0268	2	−
413	0.0477	3.520*	0.7878*	−0.3782*	−0.0060	5	2
414	0.0307	1.947	0.3470*	−0.1554*	0.1783*	2	2
415	0.0674*	0.588	0.1581	−0.0042	0.1190*	−	1
416	0.0329	−2.178*	−0.4019*	0.0409	−0.0251	3	1
417	0.0387	−2.852*	−0.5724*	0.0473	0.1098*	3	3
419	0.1843*	2.316*	1.1009*	0.0352	1.0041*	5	3
420	0.0377	−2.292*	−0.4536*	−0.0606	0.0619	2	2
427	0.0410	−1.956	−0.4012*	0.1416*	0.0582	1	4
467	0.0373	1.803	0.3549*	−0.0155	−0.1191*	2	−
474	0.0189	2.060*	0.2860	−0.0250	−0.0655	1	1
490	0.0514	−3.534*	−0.8225*	0.2957*	0.1797*	3	4
491	0.0527	−2.019*	−0.4763*	0.1760*	0.1107*	2	3
506	0.0357	−3.070*	−0.5906*	−0.1193*	−0.0547	3	3

*Cutoff values: $h_i = 0.055$; RSTUDENT = 2.0; DFFITS = 0.333; DFBETAS = 0.089.

while Boston comprises 131 census tracts of a total of 506 in the sample, it accounts for 40 of the 67 observations that surfaced. While we did not explore the point further, one might speculate from this that central-city behavior differs systematically from that of the surrounding towns. A second general characteristic is that adjacent areas often have similar diagnostic magnitudes. This is particularly striking with respect to the studentized residuals. Here we see that observations 398–406 (in South Boston) all have negative values, only two of which fall short of 3 in absolute value. Similar patterns hold for most of the DFBETAS, which the reader should see for himself. Thus there appear to be potentially significant neighborhood effects on housing prices that have not been fully captured by this model.

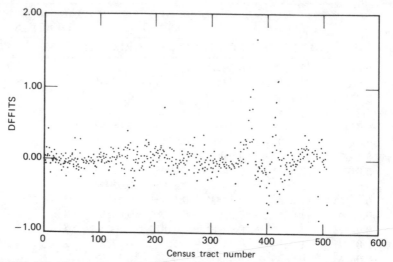

Exhibit 4.31 Scatter plots of DFFITS versus census tract number: housing-price data.

The portmanteau diagnostic DFFITS is tabulated in Exhibit 4.30 and plotted against census tract number in Exhibit 4.31. This latter graph is most helpful in conveying impressions both of the relative magnitudes of the DFFITS and notions of neighborhood clustering. Indeed, in returning to Exhibit 4.30, we note that 35 census tracts exceed the cutoff, and there is apparent clustering around tracts 365–370 and 371–373 (in the adjacent neighborhoods of Back Bay and Beacon Hill), where large residuals are often accompanied by sizable hat-matrix diagonals. It is of further interest to note that the largest DFFITS has a value of 1.655 in tract 381 (in Charlestown), a result that indicates how powerful 1 of 506 observations can be.[19]

When the Huber estimates are used to calculate unweighted DFFITS[20] (again employing OLS scale to preserve comparability), only 10 census

---

[19]It should be noted, however, that the fitted values, in addition to b_6NOXSQ, play an important role in determining W, the willingness to pay for clean air. Changes in the fit, therefore, can have a serious impact on the willingness-to-pay equations (cf. Harrison and Rubinfeld (1978), (2.3) and Table 1). To assess fully the impact on W, a careful analysis of the off-diagonal change in fit (2.12) corresponding to large |DFFITS| would be needed.

[20]The unweighted DFFITS are calculated as

$$\frac{x_i(b_H - b_H(i))}{s(i)\sqrt{h_i}},$$

where b_H is the Huber-estimated coefficient vector, and the denominator is computed from corresponding rows of OLS estimates.

tracts exceed the cutoff in comparison with 35 for OLS. These include census tracts 365, 366, 368–371, 381, 406, 411, and 419. The rank ordering is reasonably similar for both sets of calculations. Robust estimation is intended to provide more reliable estimates in the presence of heavy-tailed error distributions, an objective that has been only partially attained in this instance. We discuss this point further in Chapter 5.

Returning to the OLS results in Exhibit 4.30, census tract 381, already noted above as having an extreme DFFITS, also has 10 of its 14 coefficients with DFBETAS that exceed the cutoff, the largest of the hat-matrix diagonals (0.295) and a large studentized residual. Four other census tracts—365, 369 (in Back Bay); 372, and 373 (in Beacon Hill)—have nine or more coefficients with DFBETAS that exceed the cutoff. All of these tracts have large residuals, while the first two also have hat-matrix diagonals larger than the cutoff.

Turning to the NOXSQ column, there does not appear to be a powerful association of large DFBETAS with large h_i or studentized residuals, although once again there is clustering: eight of the Cambridge tracts, 143–153; all of the Allston-Brighton tracts, 358–362; and most of the downtown Boston tracts, 410–414. However, the Back Bay tracts, 365–370, and the Beacon Hill tracts, 371–373, are clearly linked to hat-matrix diagonals and studentized residuals. Broadly speaking, it is reasonable to conclude that, apart from some neighborhood effects previously noted, the NOXSQ coefficient does not pose severe problems from the viewpoint of single-row deletion diagnostics, but the question arises: could deleting two or more observations cause trouble for NOXSQ?[21] We comment on this shortly.

An examination of the CRIM variable shows it to have the largest DFBETAS, three of which (381, 406, and 419) are larger than 0.8 and the largest of which is -1.59. These are associated with the three largest leverage points as well as sizable, though far from the largest, residuals. The other large DFBETAS for CRIM are more closely linked to large residuals than to leverage.

Multiple-Row Diagnostics

Since some influential observations may be overlooked by single-row diagnostics, we turn now to multiple-row techniques. Various multiple-row methods are discussed in Chapter 2, many of which prove quite costly. We resort here to the least expensive sequential procedure (see (2.62) and

---

[21]However, the problems indicated in footnote 19 in determining W, the willingness to pay, arise here as well.

following) that has nevertheless proven to be quite effective for large data sets.

This procedure, we recall, is based on the principle that the largest changes in fit should occur for those discrepant observations not used in the estimation of the coefficients. Different starting sets $D_m^{(0)}$ can lead to different final sets. Rather than a drawback, we have found this characteristic to be an advantage and often use both of the starting procedures described in Chapter 2.

For the Harrison-Rubinfeld data both starting methods converged to tracts 381, 419, 406, and 411, with 415 a possibility. The multiple-row analysis therefore has not revealed any masked observations, except perhaps 415, which might have been overlooked earlier had we only examined DFFITS.

We have listed the results of deleting these five observations in Exhibit 4.32. There is little change in NOXSQ from the least-squares results, but a substantial change in the CRIM coefficient and the fitted values (which we have not displayed). While a similar outcome arises when observations are deleted individually (Exhibit 4.30), this will not always be the case and there can be some situations in which multiple-row techniques will affect a coefficient when single-row methods do not.

Exhibit 4.32   OLS estimates: housing-price equation with census tracts 381, 419, 406, 411, and 415 deleted

Variable	Coefficient Estimate	Standard Error	t-Statistic
INTERCEPT	9.788	0.147	66.42
CRIM	-0.0191	0.00219	-8.73
ZN	3.12×10^{-4}	4.97×10^{-4}	0.63
INDUS	-7.31×10^{-4}	2.32×10^{-3}	-0.31
CHAS	0.0873	0.0325	2.69
NOXSQ	-0.00637	0.00111	-5.73
RM	0.00626	0.00130	4.82
AGE	-8.08×10^{-5}	5.17×10^{-4}	-0.16
DIS	-0.217	0.0333	-6.51
RAD	0.111	0.0191	5.80
TAX	-3.48×10^{-4}	1.21×10^{-4}	-2.87
PTRATIO	-0.0305	0.00490	-6.23
B	0.391	0.105	3.73
LSTAT	-0.355	0.0249	-14.27

$R^2 = .80$ SER $= 0.78$ $\kappa(X) = 66$

Summary

Several matters of substantive interest emerge. First, the residual distribution based on ordinary least-squares estimation departs strongly from normality according to the normal probability plot. A robust estimator designed to reduce the influence of large errors changes the value of the NOXSQ coefficient by one OLS standard error and several other coefficients by much more. We believe it is useful from a diagnostic perspective to examine estimates from robust fitting procedures in these circumstances and contrast them with OLS estimates. Second, about 10% of the observations are shown to be influential by single-row deletion diagnostics for the elements on which we have concentrated here and more than twice that number for the entire problem. It should be emphasized that the large proportion of observations indicated in this example by the diagnostics as being worthy of further attention is due not only to a large proportion of leverage points but also to a severely non-Gaussian error distribution. Third, a multiple-row deletion procedure indicates that five points are strongly influential (a result consistent with single-row procedures) and also points to one census tract whose influence might have been overshadowed, if not masked, in the single-row deletion diagnostics. Fourth, the NOXSQ coefficient seems well determined according to row-deletion diagnostics in the sense that its magnitude is not strongly affected by data perturbations. Fifth, the influential data tend to be quite heavily concentrated in a few neighborhoods and these are, for the most part, in the central city of Boston, which leads us to believe that the housing-price equation is not as well specified as it might be.

The housing-price equation has provided useful information on the applicability of deletion diagnostics to large data sets. We have found only slight differences between the analysis of large and small data sets. One might imagine that costs would be burdensome, but in fact a full set of single-row deletion diagnostics costs less than $10 and the sequential multiple-row procedure about $25. Since a complete set of output for DFBETAS, for instance, will produce 7084 values, it is advisable to rely more extensively on graphics and other summary measures. Thus, the plot of DFFITS against census tract number in Exhibit 4.31 provides useful global information about magnitude and the potential existence of correlated geographical effects, given the way the data matrix was naturally constructed. There may well be other more powerful pattern-recognition algorithms to detect such systematic behavior, a point that calls for further research.

It proves informative to evaluate DFBETAS not only by examining those coefficients that exceed the chosen cutoff, but also by looking at the number of them in a given census tract that exceed the cutoff. When well over one-half do so, we are especially alert to potential difficulties associated with the specific census tract deleted. In quite a few instances deletion of one observation out of the 506 makes a surprisingly large change as measured by external scale; for example, deletion of census tract 419 alone changes the CRIM coefficient by one standard deviation, and the deletion of census tract 381 changes DFFITS by 1.65 standard deviations. Thus, what we conceive to be extremely influential observations are discernible by "absolute" criteria even in a large data set.

. . . .

Census Tract	LMV	CRIM	ZN	INDUS	CHAS	NOXSQ	RM
				Variable			
1	10.0858	0.00632	18.	2.31	0.	28.9444	43.2306
2	9.98045	0.02731	0.	7.07	0.	21.9961	41.2292
3	10.4545	0.0273	0.	7.07	0.	21.9961	51.6242
4	10.4163	0.03237	0.	2.18	0.	20.9764	48.972
5	10.4968	0.06905	0.	2.18	0.	20.9764	51.0796
6	10.2647	0.02985	0.	2.18	0.	20.9764	41.3449
7	10.0389	0.08829	12.5	7.87	0.	27.4576	36.1441
8	10.2073	0.14455	12.5	7.87	0.	27.4576	38.0936
9	9.71112	0.21124	12.5	7.87	0.	27.4576	31.7082
10	9.84692	0.17004	12.5	7.87	0.	27.4576	36.048
11	9.61581	0.22489	12.5	7.87	0.	27.4576	40.6661
12	9.84692	0.11747	12.5	7.87	0.	27.4576	36.1081
13	9.98507	0.09378	12.5	7.87	0.	27.4576	34.6803
14	9.92329	0.62976	0.	8.14	0.	28.9444	35.3906
15	9.80918	0.63796	0.	8.14	0.	28.9444	37.1612
16	9.89848	0.62739	0.	8.14	0.	28.9444	34.0355
17	10.0476	1.05393	0.	8.14	0.	28.9444	35.2242
18	9.76996	0.7842	0.	8.14	0.	28.9444	35.8801
19	9.91344	0.80271	0.	8.14	0.	28.9444	29.7679
20	9.80918	0.7258	0.	8.14	0.	28.9444	32.7985
21	9.51783	1.25179	0.	8.14	0.	28.9444	31.0249
22	9.88329	0.85204	0.	8.14	0.	28.9444	35.5812
23	9.62905	1.23247	0.	8.14	0.	28.9444	37.7242
24	9.5819	0.98843	0.	8.14	0.	28.9444	33.791
25	9.65503	0.75026	0.	8.14	0.	28.9444	35.0937
26	9.53964	0.84054	0.	8.14	0.	28.9444	31.3488
27	9.71716	0.67191	0.	8.14	0.	28.9444	33.791
28	9.60238	0.95578	0.	8.14	0.	28.9444	36.5662
29	9.82011	0.77299	0.	8.14	0.	28.9444	42.185
30	9.95228	1.00245	0.	8.14	0.	28.9444	44.5423
31	9.44936	1.13081	0.	8.14	0.	28.9444	32.6384
32	9.5819	1.35472	0.	8.14	0.	28.9444	36.8692
33	9.48797	1.38799	0.	8.14	0.	28.9444	35.4025
34	9.48037	1.15172	0.	8.14	0.	28.9444	32.5014
35	9.51045	1.61282	0.	8.14	0.	28.9444	37.1612
36	9.84692	0.06417	0.	5.96	0.	24.9001	35.2005
37	9.90349	0.09744	0.	5.96	0.	24.9001	34.1173
38	9.95228	0.08014	0.	5.96	0.	24.9001	34.2225
39	10.1146	0.17505	0.	5.96	0.	24.9001	35.5932
40	10.3353	0.02763	75.	2.95	0.	18.3184	43.494
41	10.4602	0.03359	75.	2.95	0.	18.3184	49.3365
42	10.1887	0.12744	0.	6.91	0.	20.0704	45.8329
43	10.1386	0.1415	0.	6.91	0.	20.0704	38.0565
44	10.1146	0.15936	0.	6.91	0.	20.0704	38.5765

APPENDIX 4A. THE HARRISON AND RUBINFELD HOUSING-PRICE DATA

The following are the data used for the analysis of the Harrison and Rubinfeld (1978) Housing-Price equation treated in Section 4.4. For identification of census tracts, see Exhibit 4.21. We are grateful to David Harrison and Daniel L. Rubinfeld for making these data available.

. . . .

Variable

Census Tract	AGE	DIS	RAD	TAX	PTRATIO	B	LSTAT
1	65.2	1.40854	0.	296.	15.3	0.3969	-3.00074
2	78.9	1.60283	0.69315	242.	17.8	0.3969	-2.39251
3	61.1	1.60283	0.69315	242.	17.8	0.39283	-3.21165
4	45.8	1.80207	1.09861	222.	18.7	0.39464	-3.52744
5	54.2	1.80207	1.09861	222.	18.7	0.3969	-2.93163
6	58.7	1.80207	1.09861	222.	18.7	0.39412	-2.95555
7	66.6	1.71569	1.60944	311.	15.2	0.3956	-2.08482
8	96.1	1.78347	1.60944	311.	15.2	0.3969	-1.65276
9	100.	1.80535	1.60944	311.	15.2	0.38664	-1.20638
10	85.9	1.88587	1.60944	311.	15.2	0.38671	-1.76627
11	94.3	1.84793	1.60944	311.	15.2	0.39251	-1.58733
12	82.9	1.82885	1.60944	311.	15.2	0.3969	-2.01966
13	39.	1.69578	1.60944	311.	15.2	0.3905	-1.85075
14	61.8	1.54916	1.38629	307.	21.	0.3969	-2.49411
15	84.5	1.49557	1.38629	307.	21.	0.38002	-2.27692
16	56.5	1.50377	1.38629	307.	21.	0.39562	-2.46852
17	29.3	1.50377	1.38629	307.	21.	0.38685	-2.72174
18	81.7	1.44878	1.38629	307.	21.	0.38675	-1.91943
19	36.6	1.33408	1.38629	307.	21.	0.28899	-2.14652
20	69.5	1.33408	1.38629	307.	21.	0.39095	-2.18196
21	98.1	1.33445	1.38629	307.	21.	0.37657	-1.55989
22	89.2	1.38936	1.38629	307.	21.	0.39253	-1.97818
23	91.7	1.3805	1.38629	307.	21.	0.3969	-1.67563
24	100.	1.40981	1.38629	307.	21.	0.39453	-1.6153
25	94.1	1.48151	1.38629	307.	21.	0.39433	-1.81376
26	85.7	1.49394	1.38629	307.	21.	0.30342	-1.80108
27	90.3	1.54372	1.38629	307.	21.	0.37689	-1.90987
28	88.8	1.49367	1.38629	307.	21.	0.30638	-1.75585
29	94.4	1.49396	1.38629	307.	21.	0.38794	-2.05557
30	87.3	1.44433	1.38629	307.	21.	0.38023	-2.12218
31	94.1	1.44291	1.38629	307.	21.	0.36017	-1.4874
32	100.	1.42911	1.38629	307.	21.	0.37673	-2.03738
33	82.	1.38379	1.38629	307.	21.	0.2326	-1.28348
34	95.	1.33163	1.38629	307.	21.	0.35877	-1.69549
35	96.9	1.32436	1.38629	307.	21.	0.24831	-1.59263
36	68.2	1.21203	1.60944	279.	19.2	0.3969	-2.33542
37	61.4	1.21725	1.60944	279.	19.2	0.37756	-2.1705
38	41.5	1.36971	1.60944	279.	19.2	0.3969	-2.43361
39	30.2	1.34737	1.60944	279.	19.2	0.39343	-2.29016
40	21.8	1.6866	1.09861	252.	18.3	0.39563	-3.14238
41	15.8	1.6866	1.09861	252.	18.3	0.39562	-3.9246
42	2.9	1.74413	1.09861	233.	17.9	0.38541	-3.02764
43	6.6	1.74413	1.09861	233.	17.9	0.38337	-2.84525
44	6.5	1.74413	1.09861	233.	17.9	0.39446	-2.59776

Census Tract	LMV	CRIM	ZN	INDUS	CHAS	NOXSQ	RM
45	9.96176	0.12269	0.	6.91	0.	20.0704	36.8327
46	9.86786	0.17142	0.	6.91	0.	20.0704	32.2851
47	9.90349	0.18836	0.	6.91	0.	20.0704	33.4778
48	9.71716	0.22927	0.	6.91	0.	20.0704	36.3609
49	9.57498	0.25387	0.	6.91	0.	20.0704	29.1492
50	9.87303	0.21977	0.	6.91	0.	20.0704	31.3824
51	9.88837	0.08873	21.	5.64	0.	19.2721	35.5574
52	9.92818	0.04337	21.	5.64	0.	19.2721	37.3932
53	10.1266	0.0536	21.	5.64	0.	19.2721	42.3931
54	10.0605	0.04981	21.	5.64	0.	19.2721	35.976
55	9.84692	0.0136	75.	4.	0.	16.81	34.6685
56	10.4745	0.01311	90.	1.22	0.	16.2409	52.548
57	10.1146	0.02056	85.	0.74	0.	16.81	40.7427
58	10.3609	0.01432	100.	1.32	0.	16.8921	46.4579
59	10.0562	0.15445	25.	5.13	0.	20.5209	37.761
60	9.88329	0.10328	25.	5.13	0.	20.5209	35.1293
61	9.83628	0.14932	25.	5.13	0.	20.5209	32.9591
62	9.68034	0.17171	25.	5.13	0.	20.5209	35.5932
63	10.0078	0.11027	25.	5.13	0.	20.5209	41.6799
64	10.1266	0.12651	25.	5.13	0.	20.5209	45.7246
65	10.4043	0.01951	17.5	1.38	0.	17.3056	50.4668
66	10.0648	0.03584	80.	3.37	0.	15.8404	39.5641
67	9.87303	0.04379	80.	3.37	0.	15.8404	33.4893
68	9.9988	0.05789	12.5	6.07	0.	16.7281	34.5509
69	9.76423	0.13555	12.5	6.07	0.	16.7281	31.2928
70	9.9475	0.12816	12.5	6.07	0.	16.7281	34.6332
71	10.0941	0.08826	0.	10.81	0.	17.0569	41.1779
72	9.98507	0.15876	0.	10.81	0.	17.0569	35.5335
73	10.0345	0.09164	0.	10.81	0.	17.0569	36.7842
74	10.0605	0.19539	0.	10.81	0.	17.0569	39.
75	10.09	0.07896	0.	12.83	0.	19.0969	39.3505
76	9.97115	0.09512	0.	12.83	0.	19.0969	39.5138
77	9.90349	0.10153	0.	12.83	0.	19.0969	39.4258
78	9.94271	0.08707	0.	12.83	0.	19.0969	37.6996
79	9.96176	0.05646	0.	12.83	0.	19.0969	38.8378
80	9.91838	0.08387	0.	12.83	0.	19.0969	34.5038
81	10.24	0.04113	25.	4.86	0.	18.1476	45.2525
82	10.0816	0.04462	25.	4.86	0.	18.1476	43.8111
83	10.1186	0.03659	25.	4.86	0.	18.1476	39.7152
84	10.0389	0.03551	25.	4.86	0.	18.1476	38.0319
85	10.0816	0.05059	0.	4.49	0.	20.1601	40.8193
86	10.1887	0.05735	0.	4.49	0.	20.1601	43.9569
87	10.0213	0.05188	0.	4.49	0.	20.1601	36.1802
88	10.0078	0.07151	0.	4.49	0.	20.1601	37.4666
89	10.069	0.0566	0.	3.41	0.	23.9121	49.098
90	10.2647	0.05302	0.	3.41	0.	23.9121	50.1122
91	10.0257	0.04684	0.	3.41	0.	23.9121	41.1779
92	9.9988	0.03932	0.	3.41	0.	23.9121	41.024
93	10.0389	0.04203	28.	15.04	0.	21.5296	41.4993
94	10.1266	0.02875	28.	15.04	0.	21.5296	38.5765
95	9.93305	0.04294	28.	15.04	0.	21.5296	39.05
96	10.2541	0.12204	0.	2.89	0.	19.8025	43.8906
97	9.97115	0.11504	0.	2.89	0.	19.8025	37.9826
98	10.5636	0.12083	0.	2.89	0.	19.8025	65.1087
99	10.6874	0.08187	0.	2.89	0.	19.8025	61.1524
100	10.4103	0.0686	0.	2.89	0.	19.8025	54.9971
101	10.2219	0.14866	0.	8.56	0.	27.04	45.2525
102	10.1849	0.11432	0.	8.56	0.	27.04	45.9819
103	9.83092	0.22876	0.	8.56	0.	27.04	41.024
104	9.86786	0.21161	0.	8.56	0.	27.04	37.6628
105	9.90848	0.1396	0.	8.56	0.	27.04	38.0319

Variable

Census Tract	AGE	DIS	RAD	TAX	PTRATIO	B	LSTAT
45	40.	1.74413	1.09861	233.	17.9	0.38939	-2.34832
46	33.8	1.62932	1.09861	233.	17.9	0.3969	-2.28229
47	33.3	1.62932	1.09861	233.	17.9	0.3969	-1.95567
48	85.5	1.7386	1.09861	233.	17.9	0.39274	-1.67147
49	95.3	1.76985	1.09861	233.	17.9	0.3969	-1.17723
50	62.	1.80627	1.09861	233.	17.9	0.3969	-1.81997
51	45.7	1.91908	1.38629	243.	16.8	0.39557	-2.00656
52	63.	1.91908	1.38629	243.	16.8	0.39397	-2.36138
53	21.1	1.91908	1.38629	243.	16.8	0.3969	-2.94181
54	21.4	1.91908	1.38629	243.	16.8	0.3969	-2.47385
55	47.6	1.99057	1.09861	469.	21.1	0.3969	-1.91041
56	21.9	2.16293	1.60944	226.	17.9	0.39593	-3.03426
57	35.7	2.21785	0.69315	313.	17.3	0.3969	-2.85319
58	40.5	2.11924	1.60944	256.	15.1	0.3929	-3.23196
59	29.2	2.05602	2.07944	284.	19.7	0.39067	-2.67946
60	47.2	1.93615	2.07944	284.	19.7	0.3969	-2.38369
61	66.2	1.9776	2.07944	284.	19.7	0.39511	-2.02837
62	93.4	1.91964	2.07944	284.	19.7	0.37808	-1.93538
63	67.8	1.97762	2.07944	284.	19.7	0.3969	-2.69874
64	43.4	2.07705	2.07944	284.	19.7	0.39558	-2.35377
65	59.5	2.22169	1.09861	216.	18.6	0.39324	-2.51912
66	17.8	1.88881	1.38629	337.	16.1	0.3969	-3.0648/
67	31.1	1.88881	1.38629	337.	16.1	0.3969	-2.27926
68	21.4	1.87149	1.38629	345.	18.9	0.39621	-2.5138
69	36.8	1.87149	1.38629	345.	18.9	0.3969	-2.03302
70	33.	1.87149	1.38629	345.	18.9	0.3969	-2.43178
71	6.6	1.66531	1.38629	305.	19.2	0.38373	-2.70053
72	17.5	1.66531	1.38629	305.	19.2	0.37693	-2.31476
73	7.8	1.66531	1.38629	305.	19.2	0.39091	-2.89061
74	6.2	1.66531	1.38629	305.	19.2	0.37717	-2.58535
75	6.	1.44727	1.60944	398.	18.7	0.39492	-2.69119
76	45.	1.50465	1.60944	398.	18.7	0.38324	-2.41463
77	74.5	1.39926	1.60944	398.	18.7	0.37366	-2.12293
78	45.8	1.40867	1.60944	398.	18.7	0.38696	-2.27585
79	53.7	1.61225	1.60944	398.	18.7	0.3864	-2.09216
80	36.6	1.50465	1.60944	398.	18.7	0.39606	-2.39679
81	33.5	1.68653	1.38629	281.	19.	0.3969	-2.9386
82	70.4	1.68653	1.38629	281.	19.	0.39563	-2.62804
83	32.2	1.68653	1.38629	281.	19.	0.3969	-2.70038
84	46.7	1.68653	1.38629	281.	19.	0.39064	-2.58907
85	48.	1.56431	1.09861	247.	18.5	0.3969	-2.34091
86	56.1	1.49014	1.09861	247.	18.5	0.3923	-2.72876
87	45.1	1.48777	1.09861	247.	18.5	0.39599	-2.05136
88	56.8	1.32111	1.09861	247.	18.5	0.39515	-2.47231
89	86.3	1.23014	0.69315	270.	17.8	0.3969	-2.90097
90	63.1	1.22803	0.69315	270.	17.8	0.39606	-2.8647
91	66.1	1.12891	0.69315	270.	17.8	0.39218	-2.4294
92	73.9	1.12885	0.69315	270.	17.8	0.39356	-2.5014
93	53.6	1.29907	1.38629	270.	18.2	0.39501	-2.50556
94	28.9	1.29907	1.38629	270.	18.2	0.39633	-2.77885
95	77.3	1.28509	1.38629	270.	18.2	0.3969	-2.24507
96	57.8	1.25139	0.69315	276.	18.	0.35797	-2.71115
97	69.6	1.25139	0.69315	276.	18.	0.39183	-2.17648
98	76.	1.25139	0.69315	276.	18.	0.3969	-3.16794
99	36.9	1.25139	0.69315	276.	18.	0.39353	-3.33148
100	62.5	1.25139	0.69315	276.	18.	0.3969	-2.78224
101	79.9	1.02166	1.60944	384.	20.9	0.39476	-2.36191
102	71.3	1.04946	1.60944	384.	20.9	0.39558	-2.56733
103	85.4	0.99868	1.60944	384.	20.9	0.0708	-2.24102
104	87.4	0.99868	1.60944	384.	20.9	0.39447	-2.00716
105	90.	0.88418	1.60944	384.	20.9	0.39269	-2.09354

247

Census Tract	LMV	CRIM	ZN	INDUS	CHAS	NOXSQ	RM
106	9.87817	0.13262	0.	8.56	0.	27.04	34.2342
107	9.87817	0.1712	0.	8.56	0.	27.04	34.0589
108	9.92329	0.13116	0.	8.56	0.	27.04	37.5401
109	9.89344	0.12802	0.	8.56	0.	27.04	41.9127
110	9.87303	0.26363	0.	8.56	0.	27.04	38.8004
111	9.98507	0.10793	0.	8.56	0.	27.04	38.378
112	10.0345	0.10084	0.	10.01	0.	29.9209	45.0912
113	9.84161	0.12329	0.	10.01	0.	29.9209	34.9636
114	9.83628	0.22212	0.	10.01	0.	29.9209	37.1124
115	9.82553	0.14231	0.	10.01	0.	29.9209	39.1125
116	9.81466	0.17134	0.	10.01	0.	29.9209	35.1412
117	9.96176	0.13158	0.	10.01	0.	29.9209	38.143
118	9.86267	0.15098	0.	10.01	0.	29.9209	36.2524
119	9.92329	0.13058	0.	10.01	0.	29.9209	34.4804
120	9.86786	0.14476	0.	10.01	0.	29.9209	32.8444
121	9.9988	0.06899	0.	25.65	0.	33.7561	34.4569
122	9.91838	0.07165	0.	25.65	0.	33.7561	36.048
123	9.92818	0.093	0.	25.65	0.	33.7561	35.5335
124	9.75846	0.15038	0.	25.65	0.	33.7561	34.2927
125	9.84161	0.09849	0.	25.65	0.	33.7561	34.5626
126	9.97115	0.16902	0.	25.65	0.	33.7561	35.8322
127	9.66142	0.38735	0.	25.65	0.	33.7561	31.5058
128	9.69277	0.25915	0.	21.89	0.	38.9376	32.4102
129	9.79813	0.32543	0.	21.89	0.	38.9376	41.3578
130	9.56802	0.88125	0.	21.89	0.	38.9376	31.7758
131	9.86267	0.34006	0.	21.89	0.	38.9376	41.7058
132	9.88329	1.19294	0.	21.89	0.	38.9376	40.0183
133	10.0433	0.59005	0.	21.89	0.	38.9376	40.6024
134	9.82011	0.32982	0.	21.89	0.	38.9376	33.8957
135	9.65503	0.97617	0.	21.89	0.	38.9376	33.143
136	9.80367	0.55778	0.	21.89	0.	38.9376	40.1322
137	9.76423	0.32264	0.	21.89	0.	38.9376	35.3073
138	9.74683	0.35233	0.	21.89	0.	38.9376	41.6541
139	9.49552	0.2498	0.	21.89	0.	38.9376	34.3044
140	9.78695	0.54452	0.	21.89	0.	38.9376	37.8348
141	9.54681	0.2909	0.	21.89	0.	38.9376	38.1183
142	9.57498	1.62864	0.	21.89	0.	38.9376	25.1903
143	9.50301	3.32105	0.	19.58	1.	75.864	29.1924
144	9.65503	4.0974	0.	19.58	0.	75.864	29.899
145	9.37586	2.77974	0.	19.58	0.	75.864	24.0394
146	9.53242	2.37934	0.	19.58	0.	75.864	37.5769
147	9.65503	2.15505	0.	19.58	0.	75.864	31.6744
148	9.58878	2.36862	0.	19.58	0.	75.864	24.2655
149	9.78695	2.33099	0.	19.58	0.	75.864	26.8946
150	9.64212	2.73397	0.	19.58	0.	75.864	31.3264
151	9.97581	1.6566	0.	19.58	0.	75.864	37.4789
152	9.88329	1.49632	0.	19.58	0.	75.864	29.2032
153	9.63561	1.12658	0.	19.58	1.	75.864	25.1201
154	9.87303	2.14918	0.	19.58	0.	75.864	32.5927
155	9.74097	1.41385	0.	19.58	1.	75.864	37.5646
156	9.65503	3.53501	0.	19.58	1.	75.864	37.8471
157	9.48037	2.44668	0.	19.58	0.	75.864	27.794
158	10.6286	1.22358	0.	19.58	0.	36.6025	48.2052
159	10.0982	1.34284	0.	19.58	0.	36.6025	36.7964
160	10.0562	1.42502	0.	19.58	0.	75.864	42.3801
161	10.2036	1.27346	0.	19.58	1.	36.6025	39.0625
162	10.8198	1.46336	0.	19.58	0.	36.6025	56.0851
163	10.8198	1.83377	0.	19.58	1.	36.6025	60.8712
164	10.8198	1.51902	0.	19.58	1.	36.6025	70.1406
165	10.0301	2.24236	0.	19.58	0.	36.6025	34.2693
166	10.1266	2.924	0.	19.58	0.	36.6025	37.2222

248

Census Tract	AGE	DIS	RAD	TAX	PTRATIO	B	LSTAT
106	96.7	0.74522	1.60944	384.	20.9	0.39404	-1.80357
107	91.9	0.79344	1.60944	384.	20.9	0.39567	-1.67868
108	85.2	0.75255	1.60944	384.	20.9	0.38769	-1.95956
109	97.1	0.88908	1.60944	384.	20.9	0.39524	-2.09801
110	91.2	0.93417	1.60944	384.	20.9	0.39123	-1.86104
111	54.4	1.02166	1.60944	384.	20.9	0.39349	-2.03991
112	81.6	0.98488	1.79176	432.	17.8	0.39559	-2.28651
113	92.9	0.85586	1.79176	432.	17.8	0.39495	-1.81954
114	95.4	0.93531	1.79176	432.	17.8	0.3969	-1.76656
115	84.2	0.81381	1.79176	432.	17.8	0.38874	-2.25857
116	88.2	0.90142	1.79176	432.	17.8	0.34491	-1.84776
117	72.5	1.00434	1.79176	432.	17.8	0.3933	-2.11694
118	82.6	1.01065	1.79176	432.	17.8	0.39451	-2.27322
119	73.1	0.90725	1.79176	432.	17.8	0.33863	-1.87295
120	65.2	1.01494	1.79176	432.	17.8	0.3915	-1.99429
121	69.7	0.81435	0.69315	188.	19.1	0.38915	-1.93989
122	84.1	0.78727	0.69315	188.	19.1	0.37767	-1.94673
123	92.9	0.73568	0.69315	188.	19.1	0.37809	-1.71881
124	97.	0.66495	0.69315	188.	19.1	0.37031	-1.37003
125	95.8	0.69629	0.69315	188.	19.1	0.37938	-1.73824
126	88.4	0.68959	0.69315	188.	19.1	0.38502	-1.91
127	95.6	0.56372	0.69315	188.	19.1	0.35929	-1.29975
128	96.	0.58126	1.38629	437.	21.2	0.39211	-1.76084
129	98.8	0.59471	1.38629	437.	21.2	0.3969	-1.87171
130	94.7	0.68305	1.38629	437.	21.2	0.3969	-1.69592
131	98.9	0.75071	1.38629	437.	21.2	0.39504	-2.07123
132	97.7	0.82022	1.38629	437.	21.2	0.3969	-2.0985
133	97.9	0.84475	1.38629	437.	21.2	0.38577	-2.19625
134	95.4	0.90418	1.38629	437.	21.2	0.38869	-1.89505
135	98.4	0.85271	1.38629	437.	21.2	0.26276	-1.75365
136	98.2	0.74702	1.38629	437.	21.2	0.39467	-1.77408
137	93.5	0.67646	1.38629	437.	21.2	0.37825	-1.77762
138	98.4	0.61508	1.38629	437.	21.2	0.39408	-1.92504
139	98.2	0.51198	1.38629	437.	21.2	0.39204	-1.54566
140	97.9	0.51204	1.38629	437.	21.2	0.3969	-1.68951
141	93.6	0.47741	1.38629	437.	21.2	0.38808	-1.42047
142	100.	0.36423	1.38629	437.	21.2	0.3969	-1.06691
143	100.	0.27884	1.60944	403.	14.7	0.3969	-1.31606
144	100.	0.34487	1.60944	403.	14.7	0.3969	-1.33105
145	97.8	0.29706	1.60944	403.	14.7	0.3969	-1.22806
146	100.	0.35002	1.60944	403.	14.7	0.17291	-1.28028
147	100.	0.41647	1.60944	403.	14.7	0.16927	-1.79276
148	95.7	0.37898	.1.60944	403.	14.7	0.39171	-1.21983
149	93.8	0.42501	1.60944	403.	14.7	0.35699	-1.26171
150	94.9	0.42245	1.60944	403.	14.7	0.35185	-1.5394
151	97.3	0.48119	1.60944	403.	14.7	0.3728	-1.95914
152	100.	0.46474	1.60944	403.	14.7	0.34161	-2.01906
153	88.	0.47636	1.60944	403.	14.7	0.34328	-2.11015
154	98.5	0.4844	1.60944	403.	14.7	0.26195	-1.84586
155	96.	0.55927	1.60944	403.	14.7	0.32102	-1.88922
156	82.6	0.55704	1.60944	403.	14.7	0.08801	-1.89579
157	94.	0.55181	1.60944	403.	14.7	0.08863	-1.82375
158	97.4	0.62983	1.60944	403.	14.7	0.36343	-3.08173
159	100.	0.56378	1.60944	403.	14.7	0.35389	-2.74373
160	100.	0.56866	1.60944	403.	14.7	0.36431	-2.60531
161	92.6	0.5869	1.60944	403.	14.7	0.33892	-2.8997
162	90.8	0.67849	1.60944	403.	14.7	0.37443	-4.05821
163	98.2	0.71329	1.60944	403.	14.7	0.38961	-3.95337
164	93.9	0.77103	1.60944	403.	14.7	0.38845	-3.40611
165	91.8	0.88459	1.60944	403.	14.7	0.39511	-2.15047
166	93.	0.82566	1.60944	403.	14.7	0.24016	-2.32156

Census Tract	LMV	CRIM	ZN	INDUS	CHAS	NOXSQ	RM
167	10.8198	2.01019	0.	19.58	0.	36.6025	62.869
168	10.0774	1.80028	0.	19.58	0.	36.6025	34.5391
169	10.0774	2.3004	0.	19.58	0.	36.6025	39.9297
170	10.0123	2.44953	0.	19.58	0.	36.6025	40.9856
171	9.76423	1.20742	0.	19.58	0.	36.6025	34.5156
172	9.85744	2.3139	0.	19.58	0.	36.6025	34.5744
173	10.0476	0.13914	0.	4.05	0.	26.01	31.0472
174	10.069	0.09178	0.	4.05	0.	26.01	41.1651
175	10.0257	0.08447	0.	4.05	0.	26.01	34.3279
176	10.2888	0.06664	0.	4.05	0.	26.01	42.8501
177	10.0519	0.07022	0.	4.05	0.	26.01	36.2404
178	10.1105	0.05425	0.	4.05	0.	26.01	39.8792
179	10.3056	0.06642	0.	4.05	0.	26.01	47.0596
180	10.5241	0.0578	0.	2.46	0.	23.8144	48.7204
181	10.5916	0.06588	0.	2.46	0.	23.8144	60.2952
182	10.4968	0.06888	0.	2.46	0.	23.8144	37.7487
183	10.5427	0.09103	0.	2.46	0.	23.8144	51.194
184	10.389	0.10008	0.	2.46	0.	23.8144	43.073
185	10.1811	0.08308	0.	2.46	0.	23.8144	31.4048
186	10.2955	0.06047	0.	2.46	0.	23.8144	37.8594
187	10.8198	0.05602	0.	2.46	0.	23.8144	61.3246
188	10.3735	0.07875	45.	3.44	0.	19.0969	45.9955
189	10.3023	0.12579	45.	3.44	0.	19.0969	42.9811
190	10.4602	0.0837	45.	3.44	0.	19.0969	51.6242
191	10.5187	0.09068	45.	3.44	0.	19.0969	48.3164
192	10.3255	0.06911	45.	3.44	0.	19.0969	45.4141
193	10.5023	0.08664	45.	3.44	0.	19.0969	51.5237
194	10.345	0.02187	60.	2.93	0.	16.0801	46.24
195	10.2785	0.01439	60.	2.93	0.	16.0801	43.6128
196	10.8198	0.01381	80.	0.46	0.	17.8084	62.0156
197	10.4133	0.04011	80.	1.52	0.	16.3216	53.1003
198	10.3189	0.04666	80.	1.52	0.	16.3216	50.5094
199	10.4516	0.03768	80.	1.52	0.	16.3216	52.911
200	10.4602	0.0315	95.	1.47	0.	16.2409	48.6506
201	10.4012	0.01778	95.	1.47	0.	16.2409	50.9082
202	10.09	0.03445	82.5	2.03	0.	17.2225	37.9702
203	10.6525	0.02177	82.5	2.03	0.	17.2225	57.9121
204	10.7893	0.0351	95.	2.68	0.	17.3056	61.6696
205	10.8198	0.02009	95.	2.68	0.	17.3056	64.5451
206	10.0257	0.13642	0.	10.59	0.	23.9121	34.7039
207	10.1023	0.22969	0.	10.59	0.	23.9121	40.0183
208	10.0213	0.25199	0.	10.59	0.	23.9121	33.4431
209	10.1023	0.13587	0.	10.59	1.	23.9121	36.7721
210	9.90349	0.43571	0.	10.59	1.	23.9121	28.5583
211	9.98507	0.17446	0.	10.59	1.	23.9121	35.5216
212	9.86786	0.37578	0.	10.59	1.	23.9121	29.2032
213	10.0168	0.21719	0.	10.59	1.	23.9121	33.7212
214	10.2435	0.14052	0.	10.59	0.	23.9121	40.6406
215	10.0732	0.28955	0.	10.59	0.	23.9121	29.2897
216	10.1266	0.19802	0.	10.59	0.	23.9121	38.2171
217	10.0562	0.0456	0.	13.89	1.	30.25	34.6685
218	10.2647	0.07013	0.	13.89	0.	30.25	44.1161
219	9.97581	0.11069	0.	13.89	1.	30.25	35.4144
220	10.0433	0.11425	0.	13.89	1.	30.25	40.6151
221	10.1924	0.35809	0.	6.2	1.	25.7049	48.3164
222	9.98507	0.40771	0.	6.2	1.	25.7049	37.9949
223	10.2219	0.62356	0.	6.2	1.	25.7049	47.3206
224	10.3123	0.6147	0.	6.2	0.	25.7049	43.7979
225	10.71	0.31533	0.	6.2	0.	25.4016	68.3268
226	10.8198	0.52692	0.	6.2	0.	25.4016	76.1256
227	10.5348	0.38213	0.	6.2	0.	25.4016	64.6416

250

Census Tract	AGE	DIS	RAD	TAX	PTRATIO	B	LSTAT
167	96.2	0.71584	1.60944	403.	14.7	0.3693	-3.29549
168	79.2	0.8862	1.60944	403.	14.7	0.22761	-2.10875
169	96.1	0.74194	1.60944	403.	14.7	0.29709	-2.19786
170	95.2	0.81647	1.60944	403.	14.7	0.33004	-2.17878
171	94.6	0.8862	1.60944	403.	14.7	0.29229	-1.936
172	97.3	0.87075	1.60944	403.	14.7	0.34814	-2.11793
173	.88.5	0.95401	1.60944	296.	16.6	0.3969	-1.91773
174	84.1	0.97316	1.60944	296.	16.6	0.3955	-2.40307
175	68.7	0.99395	1.60944	296.	16.6	0.39323	-2.33966
176	33.1	1.14177	1.60944	296.	16.6	0.39096	-2.93163
177	47.2	1.26833	1.60944	296.	16.6	0.39323	-2.29145
178	73.4	1.19921	1.60944	296.	16.6	0.3956	-2.76605
179	74.4	1.06997	1.60944	296.	16.6	0.39128	-2.67133
180	58.4	1.03992	1.09861	193.	17.8	0.3969	-2.98776
181	83.3	1.00832	1.09861	193.	17.8	0.39557	-2.58243
182	62.2	0.9547	1.09861	193.	17.8	0.3969	-2.35863
183	92.2	0.99347	1.09861	193.	17.8	0.39412	-3.03177
184	95.6	1.04626	1.09861	193.	17.8	0.3969	-2.86769
185	89.8	1.09457	1.09861	193.	17.8	0.391	-1.96733
186	68.8	1.18775	1.09861	193.	17.8	0.38711	-2.02844
187	53.6	1.1629	1.09861	193.	17.8	0.39263	-3.11316
188	41.1	1.332	1.60944	398.	15.2	0.39387	-2.7065
189	29.1	1.51879	1.60944	398.	15.2	0.38284	-3.08785
190	38.9	1.51879	1.60944	398.	15.2	0.3969	-2.92025
191	21.5	1.86869	1.60944	398.	15.2	0.37768	-2.97671
192	30.8	1.86869	1.60944	398.	15.2	0.38971	-3.05888
193	26.3	1.86869	1.60944	398.	15.2	0.39049	-3.5526
194	9.9	1.8277	0.	265.	15.6	0.39337	-2.98995
195	18.8	1.8277	0.	265.	15.6	0.3767	-3.12835
196	32.	1.73137	1.38629	255.	14.4	0.39423	-3.51526
197	34.1	1.98911	0.69315	329.	12.6	0.3969	-3.19981
198	36.6	1.98911	0.69315	329.	12.6	0.35431	-2.45236
199	38.3	1.98911	0.69315	329.	12.6	0.3922	-2.71583
200	15.3	2.03515	1.09861	402.	17.	0.3969	-3.08829
201	13.9	2.03515	1.09861	402.	17.	0.3843	-3.11137
202	38.4	1.83578	0.69315	348.	14.7	0.39377	-2.59951
203	15.7	1.83578	0.69315	348.	14.7	0.39538	-3.46958
204	33.2	1.63276	1.38629	224.	14.7	0.39278	-3.26702
205	31.9	1.63276	1.38629	224.	14.7	0.39055	-3.54738
206	22.3	1.37255	1.38629	277.	18.6	0.3969	-2.21962
207	52.5	1.4713	1.38629	277.	18.6	0.39487	-2.20973
208	72.7	1.4713	1.38629	277.	18.6	0.38943	-1.7113
209	59.1	1.44437	1.38629	277.	18.6	0.38132	-1.91984
210	100.	1.35454	1.38629	277.	18.6	0.3969	-1.46577
211	92.1	1.35509	1.38629	277.	18.6	0.39325	-1.75631
212	88.6	1.29883	1.38629	277.	18.6	0.39524	-1.42795
213	53.8	1.29544	1.38629	277.	18.6	0.39094	-1.83083
214	32.3	1.37255	1.38629	277.	18.6	0.38581	-2.3668
215	9.8	1.27745	1.38629	277.	18.6	0.34893	-1.21902
216	42.4	1.37255	1.38629	277.	18.6	0.39363	-2.35662
217	56.	1.1353	1.60944	276.	16.4	0.3928	-2.00144
218	85.1	1.22996	1.60944	276.	16.4	0.39278	-2.33428
219	93.8	1.06101	1.60944	276.	16.4	0.3969	-1.71936
220	92.4	1.21292	1.60944	276.	16.4	0.39374	-2.25351
221	88.5	1.05141	2.07944	307.	17.4	0.3917	-2.33232
222	91.3	1.11448	2.07944	307.	17.4	0.39524	-1.53907
223	77.7	1.18543	2.07944	307.	17.4	0.39039	-2.31011
224	80.8	1.18543	2.07944	307.	17.4	0.3969	-2.57669
225	78.3	1.06278	2.07944	307.	17.4	0.38505	-3.18399
226	83.	1.06278	2.07944	307.	17.4	0.382	-3.07175
227	86.5	1.16804	2.07944	307.	17.4	0.38738	-3.46318

			Variable				
Census Tract	LMV	CRIM	ZN	INDUS	CHAS	NOXSQ	RM
228	10.3609	0.41238	0.	6.2	0.	25.4016	51.3086
229	10.7515	0.29819	0.	6.2	0.	25.4016	59.0746
230	10.3577	0.44178	0.	6.2	0.	25.4016	42.9287
231	10.0982	0.537	0.	6.2	0.	25.4016	35.7724
232	10.3641	0.46296	0.	6.2	0.	25.4016	54.9377
233	10.6383	0.57529	0.	6.2	0.	25.7049	69.5055
234	10.7852	0.33147	0.	6.2	0.	25.7049	68.013
235	10.2751	0.44791	0.	6.2	1.	25.7049	45.2391
236	10.0858	0.33046	0.	6.2	0.	25.7049	37.0394
237	10.1306	0.52058	0.	6.2	1.	25.7049	43.9702
238	10.3577	0.51183	0.	6.2	0.	25.7049	54.1402
239	10.0732	0.08244	30.	4.93	0.	18.3184	42.0034
240	10.0562	0.09252	30.	4.93	0.	18.3184	43.6392
241	9.9988	0.11329	30.	4.93	0.	18.3184	47.5686
242	9.90848	0.10612	30.	4.93	0.	18.3184	37.149
243	10.0078	0.1029	30.	4.93	0.	18.3184	40.4241
244	10.0732	0.12756	30.	4.93	0.	18.3184	40.8704
245	9.77565	0.20608	22.	5.86	0.	18.5761	31.2816
246	9.82553	0.19133	22.	5.86	0.	18.5761	31.416
247	10.0982	0.33983	22.	5.86	0.	18.5761	37.3076
248	9.92818	0.19657	22.	5.86	0.	18.5761	38.7631
249	10.1064	0.16439	22.	5.86	0.	18.5761	41.3835
250	10.1735	0.19073	22.	5.86	0.	18.5761	45.1315
251	10.1023	0.1403	22.	5.86	0.	18.5761	42.0812
252	10.1186	0.21408	22.	5.86	0.	18.5761	41.4478
253	10.2955	0.08221	22.	5.86	0.	18.5761	48.3998
254	10.6643	0.36894	22.	5.86	0.	18.5761	68.2111
255	9.99424	0.04819	80.	3.64	0.	15.3664	37.3076
256	9.9475	0.03548	80.	3.64	0.	15.3664	34.5274
257	10.6919	0.01538	90.	3.75	0.	15.5236	55.5621
258	10.8198	0.61154	20.	3.97	0.	41.8608	75.7596
259	10.4913	0.66351	20.	3.97	0.	41.8608	53.7729
260	10.3123	0.65665	20.	3.97	0.	41.8608	46.8129
261	10.4282	0.54011	20.	3.97	0.	41.8608	51.8832
262	10.6713	0.53412	20.	3.97	0.	41.8608	56.5504
263	10.7955	0.52014	20.	3.97	0.	41.8608	70.5264
264	10.3417	0.82526	20.	3.97	0.	41.8608	53.6849
265	10.5051	0.55007	20.	3.97	0.	41.8608	51.9264
266	10.0345	0.76162	20.	3.97	0.	41.8608	30.9136
267	10.332	0.7857	20.	3.97	0.	41.8608	49.1962
268	10.8198	0.57834	20.	3.97	0.	33.0625	68.8402
269	10.6805	0.5405	20.	3.97	0.	33.0625	55.8009
270	9.93789	0.09065	20.	6.96	1.	21.5296	35.0464
271	9.95703	0.29916	20.	6.96	0.	21.5296	34.2927
272	10.1346	0.16211	20.	6.96	0.	21.5296	38.9376
273	10.1023	0.1146	20.	6.96	0.	21.5296	42.7454
274	10.4688	0.22188	20.	6.96	1.	21.5296	59.1515
275	10.3859	0.05644	40.	6.41	1.	19.9809	45.6706
276	10.3735	0.09604	40.	6.41	0.	19.9809	46.9773
277	10.4103	0.10469	40.	6.41	1.	19.9809	52.8093
278	10.4073	0.06127	40.	6.41	1.	19.9809	46.5943
279	10.2785	0.07977	40.	6.41	0.	19.9809	42.0163
280	10.466	0.21038	20.	3.33	0.	19.6249	46.4033
281	10.7233	0.03578	20.	3.33	0.	19.6249	61.1524
282	10.4745	0.03705	20.	3.33	0.	19.6249	48.553
283	10.7364	0.06129	20.	3.33	1.	19.6249	58.446
284	10.8198	0.01501	90.	1.21	1.	16.0801	62.7739
285	10.3797	0.00906	90.	2.97	0.	16.	50.2397
286	9.9988	0.01096	55.	2.25	0.	15.1321	41.6412
287	9.90848	0.01965	80.	1.76	0.	14.8225	38.8129
288	10.0519	0.03871	52.5	5.32	0.	16.4025	38.5517

Census Tract	AGE	DIS	RAD	TAX	PTRATIO	B	LSTAT
228	79.9	1.16804	2.07944	307.	17.4	0.37208	−2.75514
229	17.	1.21642	2.07944	307.	17.4	0.37751	−3.23882
230	21.4	1.21642	2.07944	307.	17.4	0.38034	−3.28208
231	68.1	1.3006	2.07944	307.	17.4	0.37835	−2.14952
232	76.9	1.3006	2.07944	307.	17.4	0.37614	−2.9477
233	73.3	1.34505	2.07944	307.	17.4	0.38591	−3.70217
234	70.4	1.29525	2.07944	307.	17.4	0.37895	−3.23145
235	66.5	1.29525	2.07944	307.	17.4	0.3602	−2.51925
236	61.5	1.29525	2.07944	307.	17.4	0.37675	−2.21788
237	76.5	1.42263	2.07944	307.	17.4	0.38845	−2.3501
238	71.6	1.42263	2.07944	307.	17.4	0.39008	−3.05209
239	18.5	1.82292	1.79176	300.	16.6	0.37941	−2.75561
240	42.2	1.82292	1.79176	300.	16.6	0.38378	−2.60748
241	54.3	1.84626	1.79176	300.	16.6	0.39125	−2.17296
242	65.1	1.84626	1.79176	300.	16.6	0.39462	−2.08723
243	52.9	1.95097	1.79176	300.	16.6	0.37275	−2.1872
244	7.8	1.95097	1.79176	300.	16.6	0.37472	−2.95882
245	76.5	2.07379	1.94591	330.	19.1	0.37249	−2.07952
246	70.2	2.07379	1.94591	330.	19.1	0.38913	−1.68946
247	34.9	2.08636	1.94591	330.	19.1	0.39018	−2.39011
248	79.2	2.08636	1.94591	330.	19.1	0.37614	−2.2873
249	49.1	2.05752	1.94591	330.	19.1	0.37472	−2.35156
250	17.5	2.05752	1.94591	330.	19.1	0.39374	−2.72372
251	13.	2.00103	1.94591	330.	19.1	0.39628	−2.82988
252	8.9	2.00103	1.94591	330.	19.1	0.37707	−3.3259
253	6.8	2.1868	1.94591	330.	19.1	0.38609	−3.34416
254	8.4	2.1868	1.94591	330.	19.1	0.3969	−3.3402
255	32.	2.22141	0.	315.	16.4	0.39289	−2.7222
256	19.1	2.22141	0.	315.	16.4	0.39518	−2.38098
257	34.2	1.84626	1.09861	244.	15.9	0.38634	−3.4699
258	86.9	0.58834	1.60944	264.	13.	0.3897	−2.97104
259	100.	0.63901	1.60944	264.	13.	0.38328	−2.55207
260	100.	0.69848	1.60944	264.	13.	0.39193	−2.67408
261	81.8	0.74768	1.60944	264.	13.	0.3928	−2.34445
262	89.4	0.76071	1.60944	264.	13.	0.38837	−2.62293
263	91.5	0.8279	1.60944	264.	13.	0.38686	−2.82852
264	94.5	0.73179	1.60944	264.	13.	0.39342	−2.18471
265	91.6	0.65757	1.60944	264.	13.	0.38789	−2.51331
266	62.8	0.68637	1.60944	264.	13.	0.3924	−2.25828
267	84.6	0.75748	1.60944	264.	13.	0.38407	−1.91115
268	67.	0.88443	1.60944	264.	13.	0.38454	−2.59897
269	52.6	1.05501	1.60944	264.	13.	0.3903	−3.45555
270	61.5	1.36545	1.09861	223.	18.6	0.39134	−1.99121
271	42.1	1.48817	1.09861	223.	18.6	0.38865	−2.03991
272	16.3	1.48817	1.09861	223.	18.6	0.3969	−2.71947
273	58.7	1.36545	1.09861	223.	18.6	0.39496	−2.56019
274	51.8	1.47396	1.09861	223.	18.6	0.39077	−2.72098
275	32.9	1.40551	1.38629	254.	17.6	0.3969	−3.34444
276	42.8	1.45098	1.38629	254.	17.6	0.3969	−3.51224
277	49.	1.56594	1.38629	254.	17.6	0.38925	−2.80429
278	27.6	1.58161	1.38629	254.	17.6	0.39346	−3.17941
279	32.1	1.42077	1.38629	254.	17.6	0.3969	−2.6322
280	32.2	1.41116	1.60944	216.	14.9	0.3969	−3.02619
281	64.5	1.54643	1.60944	216.	14.9	0.38731	−3.28208
282	37.2	1.65722	1.60944	216.	14.9	0.39223	−3.08042
283	49.7	1.65094	1.60944	216.	14.9	0.37707	−3.50423
284	24.8	1.77241	0.	198.	13.6	0.39552	−3.4546
285	20.8	1.98887	0.	285.	15.3	0.39472	−2.54427
286	31.9	1.98887	0.	300.	15.3	0.39472	−2.49738
287	31.5	2.20709	0.	241.	18.2	0.34161	−2.04531
288	31.3	1.99023	1.79176	293.	16.6	0.3969	−2.64002

Census Tract	LMV	CRIM	ZN	INDUS	CHAS	NOXSQ	RM
289	10.0123	0.0459	52.5	5.32	0.	16.4025	39.8792
290	10.1186	0.04297	52.5	5.32	0.	16.4025	43.0992
291	10.2577	0.03502	80.	4.95	0.	16.8921	47.0733
292	10.5267	0.07886	80.	4.95	0.	16.8921	51.0939
293	10.2364	0.03615	80.	4.95	0.	16.8921	43.9569
294	10.0816	0.08265	0.	13.92	0.	19.0969	37.5401
295	9.98507	0.08199	0.	13.92	0.	19.0969	36.1081
296	10.2612	0.12932	0.	13.92	0.	19.0969	44.5957
297	10.2073	0.05372	0.	13.92	0.	19.0969	42.8894
298	9.91838	0.14103	0.	13.92	0.	19.0969	33.5241
299	10.0213	0.06466	70.	2.24	0.	16.	40.259
300	10.2751	0.05561	70.	2.24	0.	16.	49.5757
301	10.1186	0.04417	70.	2.24	0.	16.	47.2106
302	9.9988	0.03537	34.	6.09	0.	18.7489	43.4281
303	10.1811	0.09266	34.	6.09	0.	18.7489	42.185
304	10.4073	0.1	34.	6.09	0.	18.7489	48.7483
305	10.494	0.05515	33.	2.18	0.	22.2784	52.3597
306	10.2541	0.0548	33.	2.18	0.	22.2784	43.7715
307	10.4163	0.07503	33.	2.18	0.	22.2784	55.0564
308	10.2471	0.04932	33.	2.18	0.	22.2784	46.9088
309	10.0345	0.49298	0.	9.9	0.	29.5936	44.0232
310	9.91838	0.3494	0.	9.9	0.	29.5936	35.6648
311	9.68657	2.63548	0.	9.9	0.	29.5936	24.7307
312	10.0033	0.79041	0.	9.9	0.	29.5936	37.4789
313	9.87303	0.26169	0.	9.9	0.	29.5936	36.2765
314	9.98045	0.26938	0.	9.9	0.	29.5936	39.2628
315	10.0774	0.3692	0.	9.9	0.	29.5936	43.1255
316	9.69277	0.25356	0.	9.9	0.	29.5936	32.547
317	9.78695	0.31827	0.	9.9	0.	29.5936	34.9754
318	9.89344	0.24522	0.	9.9	0.	29.5936	33.4315
319	10.0476	0.40202	0.	9.9	0.	29.5936	40.7299
320	9.95228	0.47547	0.	9.9	0.	29.5936	37.3688
321	10.0774	0.1676	0.	7.38	0.	24.3049	41.2935
322	10.0476	0.18159	0.	7.38	0.	24.3049	40.6534
323	9.92329	0.35114	0.	7.38	0.	24.3049	36.4937
324	9.82553	0.28392	0.	7.38	0.	24.3049	32.5813
325	10.1266	0.34109	0.	7.38	0.	24.3049	41.1522
326	10.1105	0.19186	0.	7.38	0.	24.3049	41.3578
327	10.0433	0.30347	0.	7.38	0.	24.3049	39.8413
328	10.0078	0.24103	0.	7.38	0.	24.3049	37.0029
329	9.86786	0.06617	0.	3.24	0.	21.16	34.4334
330	10.0257	0.06724	0.	3.24	0.	21.16	40.1069
331	9.89344	0.04544	0.	3.24	0.	21.16	37.7487
332	9.74683	0.05023	35.	6.06	0.	19.1844	32.5584
333	9.87303	0.03466	35.	6.06	0.	19.1844	36.373
334	10.0078	0.05083	0.	5.19	0.	26.5225	39.8919
335	9.93789	0.03738	0.	5.19	0.	26.5225	39.8161
336	9.95703	0.03961	0.	5.19	0.	26.5225	36.4453
337	9.87817	0.03427	0.	5.19	0.	26.5225	34.4451
338	9.82553	0.03041	0.	5.19	0.	26.5225	34.751
339	9.93305	0.03306	0.	5.19	0.	26.5225	36.7115
340	9.85219	0.05497	0.	5.19	0.	26.5225	35.8202
341	9.83628	0.06151	0.	5.19	0.	26.5225	35.617
342	10.3951	0.01301	35.	1.52	0.	19.5364	52.4321
343	9.71112	0.02498	0.	1.89	0.	26.8324	42.7716
344	10.0816	0.02543	55.	3.78	0.	23.4256	44.8364
345	10.3482	0.03049	55.	3.78	0.	23.4256	47.2518
346	9.76996	0.03113	0.	4.39	0.	19.5364	36.1682
347	9.75266	0.06162	0.	4.39	0.	19.5364	34.7864
348	10.0476	0.0187	85.	4.15	0.	18.4041	42.4583
349	10.1064	0.01501	80.	2.01	0.	18.9225	44.0232

254

Census Tract	AGE	DIS	RAD	TAX	PTRATIO	B	LSTAT
289	45.6	1.99023	1.79176	293.	16.6	0.3969	-2.57755
290	22.9	1.99023	1.79176	293.	16.6	0.37172	-2.35325
291	27.9	1.63251	1.38629	245.	19.2	0.3969	-3.4025
292	27.7	1.63251	1.38629	245.	19.2	0.3969	-3.33569
293	23.4	1.63251	1.38629	245.	19.2	0.3969	-3.05676
294	18.4	1.70524	1.38629	289.	16.	0.3969	-2.45527
295	42.3	1.70524	1.38629	289.	16.	0.3969	-2.26375
296	31.1	1.78514	1.38629	289.	16.	0.3969	-2.76939
297	51.	1.78514	1.38629	289.	16.	0.39285	-2.60504
298	58.	1.84372	1.38629	289.	16.	0.3969	-1.84276
299	20.1	2.05768	1.60944	358.	14.8	0.36824	-3.00255
300	10.	2.05768	1.60944	358.	14.8	0.37158	-3.04934
301	47.4	2.05768	1.60944	358.	14.8	0.39086	-2.80198
302	40.4	1.70324	1.94591	329.	16.1	0.39575	-2.3543
303	18.4	1.70324	1.94591	329.	16.1	0.38361	-2.44495
304	17.7	1.70324	1.94591	329.	16.1	0.39042	-3.02331
305	41.1	1.39178	1.94591	222.	18.4	0.39368	-2.66873
306	58.1	1.21491	1.94591	222.	18.4	0.39335	-2.41542
307	71.9	1.13114	1.94591	222.	18.4	0.3969	-2.7383
308	70.3	1.15773	1.94591	222.	18.4	0.3969	-2.58628
309	82.5	1.19921	1.38629	304.	18.4	0.3969	-3.09224
310	76.7	1.13221	1.38629	304.	18.4	0.39624	-2.30519
311	37.8	0.92402	1.38629	304.	18.4	0.35045	-2.06854
312	52.8	0.97089	1.38629	304.	18.4	0.3969	-2.81625
313	90.4	1.04169	1.38629	304.	18.4	0.3963	-2.14404
314	82.8	1.18259	1.38629	304.	18.4	0.39339	-2.53818
315	87.3	1.28157	1.38629	304.	18.4	0.39569	-2.37752
316	77.7	1.37245	1.38629	304.	18.4	0.39642	-2.16291
317	83.2	1.38594	1.38629	304.	18.4	0.3907	-1.6969
318	71.7	1.39419	1.38629	304.	18.4	0.3969	-1.8364
319	67.2	1.262	1.38629	304.	18.4	0.39521	-2.26712
320	58.8	1.38677	1.38629	304.	18.4	0.39623	-2.06152
321	52.3	1.51301	1.60944	287.	19.6	0.3969	-2.63081
322	54.3	1.51301	1.60944	287.	19.6	0.3969	-2.67757
323	49.9	1.55204	1.60944	287.	19.6	0.3969	-2.56434
324	74.3	1.55204	1.60944	287.	19.6	0.39113	-2.14242
325	40.1	1.55204	1.60944	287.	19.6	0.3969	-2.79393
326	14.7	1.68934	1.60944	287.	19.6	0.39368	-2.97946
327	28.9	1.68934	1.60944	287.	19.6	0.3969	-2.78953
328	43.7	1.68934	1.60944	287.	19.6	0.3969	-2.05651
329	25.8	1.65146	1.38629	430.	16.9	0.38244	-2.30559
330	17.2	1.65146	1.38629	430.	16.9	0.37521	-2.61238
331	32.2	1.77047	1.38629	430.	16.9	0.36857	-2.39799
332	28.4	1.89322	0.	304.	16.9	0.39402	-2.08538
333	23.3	1.89322	0.	304.	16.9	0.36225	-2.54708
334	38.1	1.86538	1.60944	224.	20.2	0.38971	-2.86839
335	38.5	1.86538	1.60944	224.	20.2	0.3894	-2.69503
336	34.5	1.78931	1.60944	224.	20.2	0.3969	-2.52498
337	46.3	1.65462	1.60944	224.	20.2	0.3969	-2.32309
338	59.6	1.72544	1.60944	224.	20.2	0.39481	-2.24829
339	37.3	1.57115	1.60944	224.	20.2	0.39614	-2.46393
340	45.4	1.57115	1.60944	224.	20.2	0.3969	-2.32862
341	58.5	1.57115	1.60944	224.	20.2	0.3969	-2.37591
342	49.3	1.95131	0.	284.	15.5	0.39474	-2.90279
343	59.7	1.83528	0.	422.	15.9	0.38996	-2.44738
344	56.4	1.74608	1.60944	370.	17.6	0.3969	-2.63345
345	28.1	1.86646	1.60944	370.	17.6	0.38797	-3.07629
346	48.5	2.08114	1.09861	352.	18.8	0.38564	-2.25123
347	52.3	2.08114	1.09861	352.	18.8	0.36461	-2.06625
348	27.7	2.14421	1.38629	351.	17.9	0.39243	-2.75514
349	29.7	2.12154	1.38629	280.	17.	0.39094	-2.81558

Variable

Census Tract	LMV	CRIM	ZN	INDUS	CHAS	NOXSQ	RM
350	10.1887	0.02899	40.	1.25	0.	18.4041	48.1497
351	10.0389	0.06211	40.	1.25	0.	18.4041	42.1201
352	10.09	0.0795	60.	1.69	0.	16.8921	43.2832
353	9.83092	0.07244	60.	1.69	0.	16.8921	34.6214
354	10.3123	0.01709	90.	2.02	0.	16.81	45.266
355	9.80918	0.04301	80.	1.91	0.	17.0569	32.0696
356	9.93305	0.10659	80.	1.91	0.	17.0569	35.2361
357	9.78695	8.98296	0.	18.1	1.	59.2899	38.5889
358	9.98507	3.8497	0.	18.1	1.	59.2899	40.896
359	10.0301	5.20177	0.	18.1	1.	59.2899	37.5401
360	10.0257	4.26131	0.	18.1	0.	59.2899	37.3565
361	10.1266	4.54192	0.	18.1	0.	59.2899	40.9344
362	9.89848	3.83684	0.	18.1	0.	59.2899	39.075
363	9.94271	3.67822	0.	18.1	0.	59.2899	28.751
364	9.72913	4.22239	0.	18.1	1.	59.2899	33.6748
365	9.99424	3.47428	0.	18.1	1.	51.5523	77.0884
366	10.2219	4.55587	0.	18.1	0.	51.5523	12.6807
367	9.99424	3.69695	0.	18.1	0.	51.5523	24.6314
368	10.0476	13.5222	0.	18.1	0.	39.816	14.9228
369	10.8198	4.89822	0.	18.1	0.	39.816	24.7009
370	10.8198	5.66998	0.	18.1	1.	39.816	44.6625
371	10.8198	6.53876	0.	18.1	1.	39.816	49.2243
372	10.8198	9.2323	0.	18.1	0.	39.816	38.6387
373	10.8198	8.26725	0.	18.1	1.	44.6223	34.5156
374	9.53242	11.1081	0.	18.1	0.	44.6223	24.0688
375	9.53242	18.4982	0.	18.1	0.	44.6223	17.123
376	9.61581	19.6091	0.	18.1	0.	45.024	53.4799
377	9.53964	15.288	0.	18.1	0.	45.024	44.2092
378	9.49552	9.82349	0.	18.1	0.	45.024	46.1584
379	9.48037	23.6482	0.	18.1	0.	45.024	40.7044
380	9.23014	17.8667	0.	18.1	0.	45.024	38.7257
381	9.24956	88.9762	0.	18.1	0.	45.024	48.553
382	9.29652	15.8744	0.	18.1	0.	45.024	42.837
383	9.33256	9.18702	0.	18.1	0.	48.9999	30.6473
384	9.41735	7.99248	0.	18.1	0.	48.9999	30.4704
385	9.08251	20.0849	0.	18.1	0.	48.9999	19.0794
386	8.88184	16.8118	0.	18.1	0.	48.9999	27.8467
387	9.25913	24.3938	0.	18.1	0.	48.9999	21.6411
388	8.90924	22.5971	0.	18.1	0.	48.9999	25.
389	9.23014	14.3337	0.	18.1	0.	48.9999	23.8144
390	9.3501	8.15174	0.	18.1	0.	48.9999	29.0521
391	9.62245	6.96215	0.	18.1	0.	48.9999	32.6384
392	10.0519	5.29305	0.	18.1	0.	48.9999	36.6146
393	9.17988	11.5779	0.	18.1	0.	48.9999	25.3613
394	9.53242	8.64476	0.	18.1	0.	48.0248	38.3532
395	9.44936	13.3598	0.	18.1	0.	48.0248	34.6568
396	9.48037	8.71675	0.	18.1	0.	48.0248	41.8738
397	9.43348	5.87205	0.	18.1	0.	48.0248	41.024
398	9.04782	7.67202	0.	18.1	0.	48.0248	33.028
399	8.51719	38.3518	0.	18.1	0.	48.0248	29.7352
400	8.74831	9.91655	0.	18.1	0.	48.0248	34.2459
401	8.63052	25.0461	0.	18.1	0.	48.0248	35.8442
402	8.88184	14.2362	0.	18.1	0.	48.0248	40.2336
403	9.40096	9.59571	0.	18.1	0.	48.0248	41.0112
404	9.02401	24.8017	0.	18.1	0.	48.0248	28.6118
405	9.04782	41.5292	0.	18.1	0.	48.0248	30.5919
406	8.51719	67.9208	0.	18.1	0.	48.0248	32.2965
407	9.38429	20.7162	0.	18.1	0.	43.4281	17.123
408	10.2364	11.9511	0.	18.1	0.	43.4281	31.4497
409	9.75266	7.40389	0.	18.1	0.	35.6409	31.5507
410	10.2219	14.4383	0.	18.1	0.	35.6409	46.9499

Census Tract	AGE	DIS	RAD	TAX	PTRATIO	B	LSTAT
350	34.5	2.17385	0.	335.	19.7	0.38985	-2.83107
351	44.4	2.17385	0.	335.	19.7	0.3969	-2.81675
352	35.9	2.37121	1.38629	411.	18.3	0.37078	-2.90279
353	18.5	2.37121	1.38629	411.	18.3	0.39233	-2.55297
354	36.1	2.49539	1.60944	187.	17.	0.38446	-3.10198
355	21.9	2.3595	1.38629	334.	22.	0.3828	-2.51999
356	19.5	2.3595	1.38629	334.	22.	0.37604	-2.88813
357	97.4	0.75245	3.17805	666.	20.2	0.37773	-1.7375
358	91.	0.91837	3.17805	666.	20.2	0.39134	-2.01966
359	83.4	1.00162	3.17805	666.	20.2	0.39543	-2.16439
360	81.3	0.91992	3.17805	666.	20.2	0.39074	-2.06577
361	88.	0.92354	3.17805	666.	20.2	0.37456	-2.55194
362	91.1	0.83095	3.17805	666.	20.2	0.35065	-1.95298
363	96.2	0.74365	3.17805	666.	20.2	0.38079	-2.28416
364	89.	0.64432	3.17805	666.	20.2	0.35304	-1.92114
365	82.9	0.64432	3.17805	666.	20.2	0.35455	-2.93973
366	87.9	0.47822	3.17805	666.	20.2	0.3547	-2.6424
367	91.4	0.56093	3.17805	666.	20.2	0.31604	-1.96597
368	100.	0.41251	3.17805	666.	20.2	0.13142	-2.01545
369	100.	0.28706	3.17805	666.	20.2	0.37552	-3.42252
370	96.8	0.30505	3.17805	666.	20.2	0.37533	-3.2893
371	97.5	0.18432	3.17805	666.	20.2	0.39205	-3.52066
372	100.	0.15623	3.17805	666.	20.2	0.36615	-2.35083
373	89.6	0.12186	3.17805	666.	20.2	0.34788	-2.42114
374	100.	0.16059	3.17805	666.	20.2	0.3969	-1.05641
375	100.	0.12839	3.17805	666.	20.2	0.3969	-0.96843
376	97.9	0.27482	3.17805	666.	20.2	0.3969	-2.00723
377	93.3	0.29632	3.17805	666.	20.2	0.36302	-1.45925
378	98.8	0.30601	3.17805	666.	20.2	0.3969	-1.54924
379	96.2	0.32649	3.17805	666.	20.2	0.3969	-1.44016
380	100.	0.32649	3.17805	666.	20.2	0.39374	-1.52422
381	91.9	0.34819	3.17805	666.	20.2	0.3969	-1.75974
382	99.1	0.41818	3.17805	666.	20.2	0.3969	-1.55689
383	100.	0.45768	3.17805	666.	20.2	0.3969	-1.44401
384	100.	0.42729	3.17805	666.	20.2	0.3969	-1.40393
385	91.2	0.36429	3.17805	666.	20.2	0.28583	-1.18312
386	98.1	0.35494	3.17805	666.	20.2	0.3969	-1.17723
387	100.	0.38336	3.17805	666.	20.2	0.3969	-1.26291
388	89.5	0.41766	3.17805	666.	20.2	0.3969	-1.13968
389	100.	0.46342	3.17805	666.	20.2	0.37292	-1.18345
390	98.9	0.54702	3.17805	666.	20.2	0.3969	-1.56758
391	97.	0.6557	3.17805	666.	20.2	0.39443	-1.76562
392	82.5	0.77371	3.17805	666.	20.2	0.37838	-1.67323
393	97.	0.57098	3.17805	666.	20.2	0.3969	-1.35934
394	92.6	0.58288	3.17805	666.	20.2	0.3969	-1.88618
395	94.7	0.57779	3.17805	666.	20.2	0.3969	-1.8107
396	98.8	0.54563	3.17805	666.	20.2	0.39198	-1.7651
397	96.	0.51689	3.17805	666.	20.2	0.3969	-1.64155
398	98.9	0.49066	3.17805	666.	20.2	0.3931	-1.61355
399	100.	0.39851	3.17805	666.	20.2	0.3969	-1.18453
400	77.8	0.40573	3.17805	666.	20.2	0.33817	-1.20481
401	100.	0.46298	3.17805	666.	20.2	0.3969	-1.31793
402	100.	0.45368	3.17805	666.	20.2	0.3969	-1.59347
403	100.	0.49409	3.17805	666.	20.2	0.37611	-1.59416
404	96.	0.53227	3.17805	666.	20.2	0.3969	-1.62085
405	85.4	0.47462	3.17805	666.	20.2	0.32946	-1.29543
406	100.	0.35445	3.17805	666.	20.2	0.38497	-1.47072
407	100.	0.1639	3.17805	666.	20.2	0.37022	-1.45483
408	100.	0.25091	3.17805	666.	20.2	0.33209	-2.10932
409	97.9	0.3748	3.17805	666.	20.2	0.31464	-1.33192
410	100.	0.3822	3.17805	666.	20.2	0.17936	-1.62055

Census Tract	LMV	CRIM	ZN	INDUS	CHAS	NOXSQ	RM
411	9.61581	51.1358	0.	18.1	0.	35.6409	33.143
412	9.75266	14.0507	0.	18.1	0.	35.6409	44.3156
413	9.79256	18.811	0.	18.1	0.	35.6409	21.4184
414	9.69892	28.6558	0.	18.1	0.	35.6409	26.574
415	8.85367	45.7461	0.	18.1	0.	48.0248	20.4213
416	8.88184	18.0846	0.	18.1	0.	46.104	41.3963
417	8.92266	10.8342	0.	18.1	0.	46.104	45.9955
418	9.24956	25.9406	0.	18.1	0.	46.104	28.1324
419	9.08251	73.5341	0.	18.1	0.	46.104	35.4858
420	9.03599	11.8123	0.	18.1	0.	51.5523	46.5669
421	9.72316	11.0874	0.	18.1	0.	51.5523	41.1009
422	9.561	7.02259	0.	18.1	0.	51.5523	36.072
423	9.94271	12.0482	0.	18.1	0.	37.6996	31.8999
424	9.50301	7.05042	0.	18.1	0.	37.6996	37.2466
425	9.36734	8.79212	0.	18.1	0.	34.1056	30.9692
426	9.02401	15.8603	0.	18.1	0.	46.104	34.7628
427	9.23014	12.2472	0.	18.1	0.	34.1056	34.0705
428	9.29652	37.6619	0.	18.1	0.	46.104	38.4648
429	9.30565	7.36711	0.	18.1	0.	46.104	38.3532
430	9.15905	9.33889	0.	18.1	0.	46.104	40.7044
431	9.5819	8.49213	0.	18.1	0.	34.1056	40.2971
432	9.55393	10.0623	0.	18.1	0.	34.1056	46.6899
433	9.68657	6.44405	0.	18.1	0.	34.1056	41.2806
434	9.56802	5.58107	0.	18.1	0.	50.8368	41.4221
435	9.36734	13.9134	0.	18.1	0.	50.8368	38.5393
436	9.50301	11.1604	0.	18.1	0.	54.7599	43.9436
437	9.16952	14.4208	0.	18.1	0.	54.7599	41.7445
438	9.07108	15.1772	0.	18.1	0.	54.7599	37.8471
439	9.03599	13.6781	0.	18.1	0.	54.7599	35.2242
440	9.4572	9.39063	0.	18.1	0.	54.7599	31.6631
441	9.25913	22.0511	0.	18.1	0.	54.7599	33.8491
442	9.74683	9.72418	0.	18.1	0.	54.7599	41.0368
443	9.82011	5.66637	0.	18.1	0.	54.7599	38.6759
444	9.64212	9.96654	0.	18.1	0.	54.7599	42.0552
445	9.2873	12.8023	0.	18.1	0.	54.7599	34.2693
446	9.37586	10.6718	0.	18.1	0.	54.7599	41.7187
447	9.60912	6.28807	0.	18.1	0.	54.7599	40.2083
448	9.44145	9.92485	0.	18.1	0.	54.7599	39.075
449	9.55393	9.32909	0.	18.1	0.	50.8368	38.2542
450	9.4727	7.52601	0.	18.1	0.	50.8368	41.1779
451	9.50301	6.71772	0.	18.1	0.	50.8368	45.549
452	9.62905	5.44114	0.	18.1	0.	50.8368	44.289
453	9.68657	5.09017	0.	18.1	0.	50.8368	39.6522
454	9.78695	8.24809	0.	18.1	0.	50.8368	54.6564
455	9.60912	9.51363	0.	18.1	0.	50.8368	45.266
456	9.55393	4.75237	0.	18.1	0.	50.8368	42.5756
457	9.44936	4.66883	0.	18.1	0.	50.8368	35.7126
458	9.51045	8.20058	0.	18.1	0.	50.8368	35.2361
459	9.60912	7.75223	0.	18.1	0.	50.8368	39.7026
460	9.90349	6.80117	0.	18.1	0.	50.8368	36.9786
461	9.70504	4.81213	0.	18.1	0.	50.8368	44.9034
462	9.78132	3.69311	0.	18.1	0.	50.8368	40.6534
463	9.87817	6.65492	0.	18.1	0.	50.8368	39.9045
464	9.91344	5.82115	0.	18.1	0.	50.8368	42.4192
465	9.97115	7.83932	0.	18.1	0.	42.9024	38.5517
466	9.89848	3.1636	0.	18.1	0.	42.9024	33.1661
467	9.85219	3.77498	0.	18.1	0.	42.9024	35.4263
468	9.85744	4.42228	0.	18.1	0.	34.1056	36.036
469	9.85744	15.5757	0.	18.1	0.	33.64	35.1175
470	9.90848	13.0751	0.	18.1	0.	33.64	32.6384
471	9.89848	4.34879	0.	18.1	0.	33.64	38.0319

Variable

Census Tract	AGE	DIS	RAD	TAX	PTRATIO	B	LSTAT
411	100.	0.34571	3.17805	666.	20.2	0.0026	-2.29125
412	100.	0.42363	3.17805	666.	20.2	0.03505	-1.55027
413	100.	0.44077	3.17805	666.	20.2	0.02879	-1.06804
414	100.	0.46336	3.17805	666.	20.2	0.21097	-1.60555
415	100.	0.50573	3.17805	666.	20.2	0.08827	-0.99474
416	100.	0.60688	3.17805	666.	20.2	0.02724	-1.23619
417	90.8	0.59856	3.17805	666.	20.2	0.02157	-1.35522
418	89.1	0.49926	3.17805	666.	20.2	0.12736	-1.32294
419	100.	0.58923	3.17805	666.	20.2	0.01645	-1.5791
420	76.5	0.58445	3.17805	666.	20.2	0.04845	-1.48109
421	100.	0.61998	3.17805	666.	20.2	0.31875	-1.89585
422	95.3	0.62839	3.17805	666.	20.2	0.31998	-1.85138
423	87.6	0.66844	3.17805	666.	20.2	0.29155	-1.95928
424	85.1	0.70399	3.17805	666.	20.2	0.00252	-1.4571
425	70.6	0.7244	3.17805	666.	20.2	0.00366	-1.76253
426	95.4	0.64689	3.17805	666.	20.2	0.00768	-1.41096
427	59.7	0.69195	3.17805	666.	20.2	0.02465	-1.85246
428	78.7	0.62213	3.17805	666.	20.2	0.01882	-1.92971
429	78.1	0.66042	3.17805	666.	20.2	0.09673	-1.53633
430	95.6	0.67712	3.17805	666.	20.2	0.06072	-1.42366
431	86.1	0.71916	3.17805	666.	20.2	0.08345	-1.73506
432	94.3	0.7363	3.17805	666.	20.2	0.08133	-1.62521
433	74.8	0.78864	3.17805	666.	20.2	0.09795	-2.11818
434	87.9	0.83975	3.17805	666.	20.2	0.10019	-1.81862
435	95.	0.7985	3.17805	666.	20.2	0.10063	-1.88618
436	94.6	0.75363	3.17805	666.	20.2	0.10985	-1.45813
437	93.3	0.69445	3.17805	666.	20.2	0.02749	-1.71197
438	100.	0.6493	3.17805	666.	20.2	0.00932	-1.3298
439	87.9	0.59917	3.17805	666.	20.2	0.06895	-1.07816
440	93.9	0.5973	3.17805	666.	20.2	0.3969	-1.47482
441	92.4	0.6239	3.17805	666.	20.2	0.39145	-1.50905
442	97.2	0.72518	3.17805	666.	20.2	0.38596	-1.63363
443	100.	0.69554	3.17805	666.	20.2	0.39569	-1.79613
444	100.	0.68229	3.17805	666.	20.2	0.38673	-1.66871
445	96.6	0.63953	3.17805	666.	20.2	0.24052	-1.43586
446	94.8	0.68708	3.17805	666.	20.2	0.04306	-1.42807
447	96.4	0.72851	3.17805	666.	20.2	0.31801	-1.72653
448	96.6	0.78755	3.17805	666.	20.2	0.38852	-1.80515
449	98.7	0.81607	3.17805	666.	20.2	0.3969	-1.70738
450	98.3	0.78162	3.17805	666.	20.2	0.30421	-1.6447
451	92.6	0.84312	3.17805	666.	20.2	0.00032	-1.74635
452	98.2	0.85662	3.17805	666.	20.2	0.35529	-1.72969
453	91.8	0.86213	3.17805	666.	20.2	0.38509	-1.75637
454	99.3	0.89719	3.17805	666.	20.2	0.37587	-1.76731
455	94.1	0.91473	3.17805	666.	20.2	0.00668	-1.67617
456	86.5	0.89027	3.17805	666.	20.2	0.05092	-1.70766
457	87.9	0.94802	3.17805	666.	20.2	0.01048	-1.66026
458	80.3	1.02216	3.17805	666.	20.2	0.0035	-1.77555
459	83.7	1.02356	3.17805	666.	20.2	0.27221	-1.81849
460	84.4	0.99971	3.17805	666.	20.2	0.3969	-1.91725
461	90.	0.95455	3.17805	666.	20.2	0.25523	-1.80643
462	88.4	0.94278	3.17805	666.	20.2	0.39143	-1.92046
463	83.	1.00591	3.17805	666.	20.2	0.3969	-1.96697
464	89.9	1.03019	3.17805	666.	20.2	0.39382	-2.27429
465	65.4	1.08634	3.17805	666.	20.2	0.3969	-2.02374
466	48.2	1.12054	3.17805	666.	20.2	0.3344	-1.95715
467	84.7	1.05483	3.17805	666.	20.2	0.02201	-1.76311
468	94.5	0.93228	3.17805	666.	20.2	0.33129	-1.54571
469	71.	1.0676	3.17805	666.	20.2	0.36874	-1.70744
470	56.7	1.03805	3.17805	666.	20.2	0.3969	-1.91298
471	84.	1.10968	3.17805	666.	20.2	0.3969	-1.8145

259

Census Tract	LMV	CRIM	ZN	INDUS	CHAS	NOXSQ	RM
472	9.88329	4.03841	0.	18.1	0.	28.3024	38.8004
473	10.0519	3.56868	0.	18.1	0.	33.64	41.435
474	10.3023	4.64689	0.	18.1	0.	37.6996	48.7204
475	9.53242	8.05579	0.	18.1	0.	34.1056	29.4523
476	9.49552	6.39312	0.	18.1	0.	34.1056	37.9702
477	9.72316	4.87141	0.	18.1	0.	37.6996	42.0422
478	9.39266	15.0234	0.	18.1	0.	37.6996	28.1324
479	9.58878	10.233	0.	18.1	0.	37.6996	38.2542
480	9.97115	14.3337	0.	18.1	0.	37.6996	38.8004
481	10.0433	5.82401	0.	18.1	0.	28.3024	38.9625
482	10.0732	5.70818	0.	18.1	0.	28.3024	45.5625
483	10.1266	5.73116	0.	18.1	0.	28.3024	49.8577
484	9.98967	2.81838	0.	18.1	0.	28.3024	33.2006
485	9.93305	2.37857	0.	18.1	0.	33.9889	34.4686
486	9.96176	3.67367	0.	18.1	0.	33.9889	39.8413
487	9.85744	5.69175	0.	18.1	0.	33.9889	37.381
488	9.93305	4.83567	0.	18.1	0.	33.9889	34.869
489	9.62905	0.15086	0.	27.74	0.	37.0881	29.7461
490	8.85367	0.18337	0.	27.74	0.	37.0881	29.3114
491	8.99962	0.20746	0.	27.74	0.	37.0881	25.9386
492	9.51783	0.10574	0.	27.74	0.	37.0881	35.7963
493	9.90848	0.11132	0.	27.74	0.	37.0881	35.7963
494	9.98967	0.17331	0.	9.69	0.	34.2225	32.5698
495	10.1064	0.27957	0.	9.69	0.	34.2225	35.1175
496	10.0476	0.17899	0.	9.69	0.	34.2225	32.1489
497	9.88837	0.2896	0.	9.69	0.	34.2225	29.0521
498	9.81466	0.26838	0.	9.69	0.	34.2225	33.5704
499	9.96176	0.23912	0.	9.69	0.	34.2225	36.2283
500	9.76996	0.17783	0.	9.69	0.	34.2225	31.0137
501	9.72913	0.22438	0.	9.69	0.	34.2225	36.3247
502	10.0168	0.06263	0.	11.93	0.	32.8329	43.4677
503	9.93305	0.04527	0.	11.93	0.	32.8329	37.4544
504	10.0816	0.06076	0.	11.93	0.	32.8329	48.6604
505	9.9988	0.10959	0.	11.93	0.	32.8329	46.1584
506	9.38429	0.04741	0.	11.93	0.	32.8329	36.3609

Census Tract	AGE	DIS	RAD	TAX	PTRATIO	B	LSTAT
472	90.7	1.13118	3.17805	666.	20.2	0.39533	-2.04996
473	75.	1.0635	3.17805	666.	20.2	0.39337	-1.94058
474	67.6	0.92936	3.17805	666.	20.2	0.37468	-2.14926
475	95.4	0.88781	3.17805	666.	20.2	0.35257	-1.70705
476	97.4	0.79118	3.17805	666.	20.2	0.30276	-1.42308
477	93.6	0.83521	3.17805	666.	20.2	0.39621	-1.67777
478	97.3	0.74227	3.17805	666.	20.2	0.34948	-1.38982
479	96.7	0.77496	3.17805	666.	20.2	0.3797	-1.71335
480	88.	0.66844	3.17805	666.	20.2	0.38332	-2.03202
481	64.7	1.23087	3.17805	666.	20.2	0.3969	-2.2312
482	74.9	1.20348	3.17805	666.	20.2	0.39307	-2.55903
483	77.	1.22689	3.17805	666.	20.2	0.39528	-2.65726
484	40.3	1.41057	3.17805	666.	20.2	0.39292	-2.26135
485	41.9	1.3148	3.17805	666.	20.2	0.37073	-2.01463
486	51.9	1.38422	3.17805	666.	20.2	0.38862	-2.2463
487	79.8	1.26579	3.17805	666.	20.2	0.39268	-1.89872
488	53.2	1.14813	3.17805	666.	20.2	0.38822	-2.16727
489	92.7	0.59933	1.38629	711.	20.1	0.39509	-1.71125
490	98.3	0.5627	1.38629	711.	20.1	0.34405	-1.42828
491	98.	0.60026	1.38629	711.	20.1	0.31843	-1.2146
492	98.8	0.62492	1.38629	711.	20.1	0.39011	-1.71097
493	83.5	0.74664	1.38629	711.	20.1	0.3969	-2.01328
494	54.	0.86781	1.79176	391.	19.2	0.3969	-2.11943
495	42.6	0.86781	1.79176	391.	19.2	0.3969	-1.99584
496	28.8	1.02912	1.79176	391.	19.2	0.39329	-1.73733
497	72.9	1.02912	1.79176	391.	19.2	0.3969	-1.55377
498	70.6	1.06219	1.79176	391.	19.2	0.3969	-1.95899
499	65.3	0.87925	1.79176	391.	19.2	0.3969	-2.04624
500	73.5	0.87543	1.79176	391.	19.2	0.39577	-1.89061
501	79.7	0.91557	1.79176	391.	19.2	0.3969	-1.94302
502	69.1	0.90769	0.	273.	21.	0.39199	-2.33604
503	76.7	0.82746	0.	273.	21.	0.3969	-2.3991
504	91.	0.77357	0.	273.	21.	0.3969	-2.87582
505	89.3	0.87083	0.	273.	21.	0.39346	-2.7363
506	80.8	0.91829	0.	273.	21.	0.3969	-2.54033

CHAPTER 5

Research Issues and Directions for Extensions

In this monograph we have presented new diagnostic methods that help to assess the suitability of a given data set for estimating a specific linear regression model by least squares. As such, these techniques go beyond the usual statistical analysis associated with ordinary least squares, an analysis which does not provide means for exploring the interaction between the estimation process itself and the specific data sample employed. These diagnostics, then, complement standard statistical tools. The approach used in developing these diagnostic measures has necessarily had a large empirical content, and the experiments and case studies that have been included in the previous chapters provide meaningful guidelines for their use. It is anticipated, however, that refinements will naturally arise from their more widespread application in a variety of statistical and econometric contexts.

We have limited the scope of this current research to ordinary least squares. This is by far the most widely employed of statistical models, and yet its inability to deal with many practical problems is evident from the variety of "tricks of the trade" that have evolved, mostly in an oral tradition, to help fill the void that arises between theory and practice when the data are less than ideal. It makes sense, then, to concentrate initial systematic efforts to deal with these problems in this most rudimentary but important context.

At the same time we look forward to many interesting elaborations of this work. Section 5.2 contains some preliminary thoughts on the extension of these diagnostics to more complex estimation contexts, including estimation of nonlinear models and systems of simultaneous equations.

Section 5.1 is devoted to some general notions regarding the application and interpretation of the diagnostic procedures that have emerged from

262

our experience with their use. Before proceeding, however, it is instructive to present a framework within which the various diagnostic techniques can be viewed. This is in terms of the two major outputs of the linear regression model that are also the objects toward which the diagnostic procedures are directed: the estimated regression coefficients and the fitted and/or predicted values.

The regression coefficients are, of course, a fundamental element in any structural analysis of a regression equation. Their estimated values, as well as the precision, or the reliability, with which they are estimated, are of central importance. Thus, those diagnostics, such as the DFBETAS or the collinearity analysis, that point to characteristics of the data to which the coefficient estimates or their estimated standard errors are particularly sensitive, are especially useful for examining the suitability of the data for structural estimation.

A different perspective is helpful for assessing the suitability of the data for predictive purposes. In prediction, and in the estimation of the prediction error, less interest is often attached to the estimates of individual coefficients than to their combined effect. Diagnostics such as DFFITS are aimed at pinpointing sources of such overall sensitivity. Some of the diagnostic measures, such as the hat-matrix diagonals and the partial-regression leverage plots, appear to be useful in both the predictive and structural contexts.

5.1 ISSUES IN RESEARCH STRATEGY

1. The principal use of the collinearity diagnostics (Chapter 3) is in evaluating the suitability of the data for structural estimation. This diagnostic provides information on those individual coefficients whose estimates are degraded by the presence of ill-conditioned data. If the investigator finds evidence of such degradation in the estimates of parameters that are of interest to his research, corrective action is indicated. In the absence of new, well-conditioned data, a Bayes-type approach (see Chapter 4) is appealing when the necessary prior information is available. Mixed-estimation provides a simple and flexible means for introducing this prior information, requiring software not much more demanding than a standard regression package. The case is similar for ridge regression.

A commonly employed corrective measure, that of deleting one or more of the collinear variates, is not a correct procedure in the context of structural estimation. This is because the ill-conditioning itself causes the tests of significance on the "collinear variates" to lack power, and hence the investigator stands a high risk of removing a variate that properly belongs in the regression equation. As is well known, this action can severely bias the remaining parameter estimates.

2. The collinearity diagnostics are not devoid of interest in the context of prediction. Even when individual coefficients are unreliably estimated, linear combinations of them often are not, enabling the combined set of coefficients to produce acceptable predictions. In order for this to be the case, one must have reason to believe that the nature of the collinearity of the data on which the estimated parameters are based will continue into the prediction period. The collinearity diagnostics help the investigator to display the various collinear relations that exist in the sample data so that their stability into the prediction period can be investigated.

3. The diagnostics for influential data points (Chapter 2) provide a wholly different point of view for assessing the suitability of the data for structural estimation and prediction, based as they are on an analysis of the effects of row perturbations rather than the column-effects that are observed in the collinearity diagnostics. A quick review of the diagnostic procedures presented in Chapter 2 will convince the reader of the importance of the roles played by the studentized residuals, e_i^*, and the hat-matrix diagonals, h_i. These two elements can often play off against one another in determining the sensitivity of coefficient estimates, DFBETAS, or fitted values, DFFITS, to the deletion of a given row. One sees from (2.7) and (2.11), for example, that when h_i is small, a large coefficient change is often associated with a sizable (absolute) studentized residual, $|e_i^*|$, whereas this effect may be considerably muted with regard to the fitted value. Such a situation, then, can be more critical to structural estimation than to prediction. A row with high leverage, however, can be the root of sensitivity in both coefficient estimates and predicted values. Of course, this need not be the case, for large h_i can be counteracted by a small residual $|e_i^*|$. Such a situation can become a desirable source of increased efficiency in estimation: an important observation that is in close agreement with the regression plane.

4. We have found that the partial-regression leverage plots contain much of the qualitatively interesting information about disparate data. Potential leverage points and large residuals show up clearly, even though the quantitative measure of influence contained in the row-deletion diagnostics is absent from the plots. These plots also reveal convenient information about masking. This phenomenon arises when two or more anomalous data points evade detection through the deletion of a single row (see Exhibit 2.1f). These multiple outliers can also often be detected by combinatorial approaches. The approach described by (2.62) and its

ensuing equations seems to combine an acceptably low cost with effectiveness in detection.

5. The effects that particular data rows may have on the efficiency of estimation are most effectively conveyed by COVRATIO, a ratio of determinants of the estimated variance-covariance matrices of the parameters, the numerator having been row-deleted. While COVRATIO can also be thought of as a summary measure containing information from both explanatory and response variables about possible anomalous rows, we have found the more direct and restrictive interpretation related to precision of estimation to be the most instructive. FVARATIO similarly provides a useful summary of the changes that occur in the precision of the fit.

6. For prediction, as we have noted above, DFFITS is an effective measure of the combined influence of a specific row on all coefficients taken together. DFFITS joins the two key diagnostics, e_i^* and h_i, in an economical way that serves to flag circumstances where either $|e_i^*|$ or h_i is large, or both are moderate but act jointly, and hence points to those rows that have the largest effect on predictions.

7. The residuals play a joint role with the hat-matrix diagonals in the diagnostics described above. But the residuals also play an important role alone. The normal probability plots, for example, convey valuable qualitative information about the extent and types of departures from normality.

8. If there is evidence of influential data, several corrective actions are possible. Influential data should first be checked for accuracy, and any questions relating to model specification for the circumstances surrounding the generation of the influential observations should be cleared up. After this is done, these potentially most interesting data may be set aside or downweighted to assess their impact on the model. Or, an estimation procedure such as bounded-influence regression [Krasker and Welsch (1979)], which is specifically designed to limit the influence of anomalous data, may be used. At the very least, excessively influential data should be mentioned in any discussion of the model fitting and estimation process.

9. Finally, we have found that one cannot be indifferent to the order in which collinearity (columns) and anomalous data (rows) are treated. Collinearity presents a problem akin to the problem of identification. If the X matrix were perfectly collinear, some parameters would be estimable only up to linear combinations. Likewise, the more ill-conditioned the data, the less effectively can the separate effects of the individual parameters be inferred. It is therefore usually advisable to introduce all conditioning information prior to estimation and prior to any additional

diagnosis that makes use of the response data, y. This conclusion, however, cannot be taken categorically. Exhibit 3.1*f* depicts how potential ill conditioning can possibly be masked by one or more strategically placed observations.[1] Should such leverage points also prove to be influential (relative to the response data), and it were deemed reasonable to set them aside, collinearity problems would arise that might not have been foreseen from diagnosing the full data set for ill conditioning. Thus, while we typically suggest employing the collinearity diagnostics first, a careful analysis of ill conditioning would, in the next step, determine whether any influential observations should be set aside, and if so, examine the conditioning of the remaining data.[2] Likewise, should it be decided in the process of diagnosing leverage points to set aside any influential observations, the conditioning of the remaining data set could profitably be examined prior to any further diagnosis for influential data.

5.2   EXTENSIONS OF THE DIAGNOSTICS

The diagnostic techniques developed in Chapters 2 and 3 are designed to apply to the single-equation, linear regression model. In this section we examine some preliminary aspects of extending one or both of these diagnostic methods to systems of simultaneous equations and to nonlinear models. These extensions are by no means complete, and further research is warranted for each.

Extensions to Systems of Simultaneous Equations

Influential-Data Diagnostics.   The basic diagnostic approach of Chapter 2, that for influential data, can be extended to systems of simultaneous linear equations,

$$Y\Gamma + XB + E = 0, \tag{5.1}$$

where Y is an $n \times g$ matrix of n observations on g jointly dependent variables, X is an $n \times p$ matrix of n observations on p predetermined variables, and E is an $n \times g$ matrix of error terms whose rows are independently distributed with a constant mean vector zero and variance-covariance matrix Σ. The matrices Γ and B are, respectively, $g \times g$

---

[1]An extreme case would occur for an observation i when $h_i = 1$ and thus det $(X^T(i)X(i)) = 0$.
[2]This suggests the usefulness of deletion diagnostics based on eigenvalues or condition indexes, but these turn out to be computationally expensive and possibly impractical. In a related vein, recent work by Weisberg (1977) on properties of the hat matrix may offer additional insight into this problem.

and $p \times g$ arrays of structural parameters. We suppose interest centers on estimating the first of these equations, which, after normalization and inclusion of all prior identifying zero restrictions, we write as

$$y_1 = Y_1\gamma_1 + X_1\beta_1 + \varepsilon_1, \tag{5.2}$$

where y_1 is an n-vector of observations on the jointly dependent variable whose coefficient has been normalized to -1, Y_1 is the $n \times g_1$ matrix of other included jointly dependent variables, X_1 is the $n \times p_1$ matrix of included predetermined variables, and γ_1 and β_1 are, respectively, the corresponding elements of the first columns of Γ and $\mathbf{B}$ not known a priori to be zero. The vector ε_1 is the first column of $\mathbf{E}$. We assume (5.2) is identified and, hence, that $p \geqslant g_1 + p_1$. Now let

$$\mathbf{Z} \equiv [\mathbf{Y}_1\ \mathbf{X}_1] \quad \text{and} \quad \delta \equiv \begin{bmatrix} \gamma_1 \\ \beta_1 \end{bmatrix}, \tag{5.3}$$

so that (5.2) becomes

$$y_1 = \mathbf{Z}\delta + \varepsilon_1. \tag{5.4}$$

As is well known, the two-stage least-squares estimator of δ in (5.4) is simply

$$\mathbf{d} = (\mathbf{Z}^T\mathbf{H}_X\mathbf{Z})^{-1}\mathbf{Z}^T\mathbf{H}_X\mathbf{y}_1, \tag{5.5}$$

where $\mathbf{H}_X = \mathbf{X}(\mathbf{X}^T\mathbf{X})^{-1}\mathbf{X}^T$ and $\mathbf{X}$ is the $n \times p$ matrix of full rank containing observations on all predetermined variables of the system. In this case the fit is determined by

$$\hat{\mathbf{y}}_1 \equiv \mathbf{Z}\mathbf{d} = \mathbf{T}\mathbf{y}_1, \tag{5.6}$$

where $\mathbf{T} \equiv \mathbf{Z}(\mathbf{Z}^T\mathbf{H}_X\mathbf{Z})^{-1}\mathbf{Z}^T\mathbf{H}_X$.

In an attempt to parallel the results of Chapter 2, it is natural to focus on the matrix $\mathbf{T}$. The matrix $\mathbf{T}$ is idempotent with trace $g_1 + p_1$, but it is not symmetric and therefore not a projection in the usual sense. No bounds can be placed on the individual diagonal elements of $\mathbf{T}$. If we let $\hat{\mathbf{Z}} \equiv \mathbf{H}_X\mathbf{Z}$, then, since $\mathbf{H}_X$ is a projection matrix,

$$\mathbf{d} = (\hat{\mathbf{Z}}^T\hat{\mathbf{Z}})^{-1}\hat{\mathbf{Z}}^T\mathbf{y}_1, \tag{5.7}$$

and $\hat{\mathbf{H}} \equiv \hat{\mathbf{Z}}(\hat{\mathbf{Z}}^T\hat{\mathbf{Z}})^{-1}\hat{\mathbf{Z}}^T$ is also a projection matrix. While $\hat{\mathbf{H}}$ lacks the direct correspondence to $\hat{\mathbf{y}}_1$ that the matrix $\mathbf{T}$ has, it is nevertheless a projection relevant to the second-stage estimation and can be used to assess the leverage of individual observations.

Naturally, we also want to look at the $\mathbf{d} - \mathbf{d}(i)$, just as in the single-equation case. Phillips (1977, p. 69) has derived the necessary formulas, so these quantities can be computed. These results can also be extended to DFFITS; in this case the relevant statistic becomes

$$\frac{\mathbf{z}_i[\mathbf{d} - \mathbf{d}(i)]}{s(i)\sqrt{\mathbf{z}_i(\hat{\mathbf{Z}}^T\hat{\mathbf{Z}})^{-1}\mathbf{z}_i^T}}, \tag{5.8}$$

where

$$s^2(i) = \frac{1}{n}\sum_{\substack{k=1 \\ k \neq i}}^{n}\left[y_k - \mathbf{z}_k\mathbf{d}(i)\right]^2. \tag{5.9}$$

Phillips (1977, p. 70) also provides a formula for computing $s^2(i)$. An example of the use of these diagnostics is contained in Kuh and Welsch (1979).

If the single-equation diagnostics are all that are available, we suggest putting $\hat{\mathbf{Z}}$ and $\mathbf{y}_1$ in place of $\mathbf{X}$ and $\mathbf{y}$ in Chapter 2 and using the output as an approximation to $\mathbf{d} - \mathbf{d}(i)$ and other diagnostics. In doing so, the effects of the ith observation have only been partially deleted, but useful insights often result.

Collinearity Diagnostics.   The extension of the collinearity diagnostics to some aspects of simultaneous-equations estimation proves to be quite straightforward. Other elements are considerably less obvious and provide fertile areas of research. There is no reason why the collinearity diagnostics cannot be applied directly to the "second stage" of two-stage least squares in order to assess the conditioning of the data for estimation of the given parameters. Here the data matrix whose conditioning is being assessed is the $\hat{\mathbf{Z}} \equiv (\hat{\mathbf{Y}}_1 \mathbf{X}_1)$ matrix, where $\hat{\mathbf{Y}}_1 = \mathbf{H}_X\mathbf{y}_1$ is the set of reduced-form predicted values for the included jointly dependent variables and $\mathbf{X}_1$ is the set of included predetermined variables. Similar considerations apply to three-stage least squares. An interesting research question is whether the conditioning and diagnostics can be obtained directly from an analysis of the raw $\mathbf{Y}$ and $\mathbf{X}$ data. Questions also arise as to an appropriate measure of conditioning and appropriate diagnostics for instrumental variables estimators and for maximum-likelihood estimators.

Another aspect of two-stage least-squares estimation provides a context in which the collinearity diagnostics may have useful application. In the estimation of large models, it often happens that the number of exogenous variates in the model exceeds the number of observations. In this "undersized sample" case, the first stage of two-stage least squares cannot take place. Various solutions have been offered, many of which include

od of $\hat{\theta}$.
expand $f(\theta)$

,   (5.13)

e Hessian of
hout the ith
(i) would be

(5.14)

optimization
roximated by

(5.15)

to (2.7), some
rovided by the

e is desired, all
at

$i)\big] \equiv d_i^2$,   (5.16)

, and then (5.16)

(5.17)

essary to compute

(5.18)

hen a slightly different

the exogenous variates to use in place of the
diagnostics pinpoint those variates that are
hainder through linear relations, they can
set that provides the greatest amount of
relation of this procedure to the use of

gnostics to nonlinear models has some
elements requiring further research. In
rations on a single-equation nonlinear

$\langle, \theta) = \mathbf{u},$ (5.10)

inear stochastic function in the p
turbance term, $\mathbf{y}$ is an n-vector of
ble, and $\mathbf{X}$ is an $n \times k$ matrix of
ables. Since $\mathbf{y}$ and $\mathbf{X}$ often do not
$(\mathbf{y}, \mathbf{X}, \theta)$ more simply as $\phi(\theta)$. In the
rmined that minimizes some specific

$)\big] \equiv f(\theta),$ (5.11)

$(\theta).$ (5.12)

distance) estimator, for example,

r that an examination of the
d be cumbersome indeed. The
irst for the whole data set to
vation deleted to obtain $\hat{\theta}(i)$.
me instances [Duncan (1978)],
tained at much less cost by

also Mitchell and Fisher (1970) and
is not taken.

employing a quadratic approximation to $f(\theta)$ in the neighbor

One approach, extending work by Chambers (1973), is to
about $\hat{\theta}$ as

$$q(\theta) \equiv f(\hat{\theta}) + (\theta - \hat{\theta})^T g(\hat{\theta}) + \frac{1}{2}(\theta - \hat{\theta})^T G(\hat{\theta})(\theta - \hat{\theta})$$

where $g(\theta) \equiv \nabla f(\theta)$, the gradient of $f(\theta)$, and $G(\theta) \equiv \nabla^2 f(\theta)$, th
$f(\theta)$. Often, f, g, and G can be easily computed at $\hat{\theta}$ wit
observation, and a reasonable local approximation to $\hat{\theta} - \hat{\theta}$
obtained from the single Newton step

$$G_{(i)}^{-1}(\hat{\theta}) g_{(i)}(\hat{\theta}).$$

This avoids having to iterate to a new solution. In many
algorithms, such as quasi-Newton methods, $G(\hat{\theta})$ is app
another matrix, $A(\hat{\theta})$, and in this case we could employ

$$A^{-1}(\hat{\theta}) g_{(i)}(\hat{\theta})$$

as an approximation to (5.14). Of course, analogous
appropriate scaling is required, and this can often be p
estimated covariance matrix, $V(\hat{\theta})$.

If there are many parameters and a summary measur
linear combinations can be considered by using the fact th

$$\sup_{\lambda} \frac{\left\{\lambda^T [\hat{\theta} - \hat{\theta}(i)]\right\}^2}{\lambda^T V(\hat{\theta}) \lambda} = [\hat{\theta} - \hat{\theta}(i)]^T V^{-1}(\hat{\theta}) [\hat{\theta} - \hat{\theta}($$

where λ is a p-vector. Often $V(\hat{\theta})$ is taken to be $A^{-1}(\hat{\theta})$
can be approximated by[4]

$$\tilde{d}_i^2 = g_{(i)}^T(\hat{\theta}) A^{-1}(\hat{\theta}) g_{(i)}(\hat{\theta}).$$

To examine changes in fit more directly, it would be ne

$$\phi_i(\hat{\theta}) - \phi_i[\hat{\theta}(i)]$$

with an appropriate scaling.

[4]If $V(\hat{\theta})$ is approximated by $G^{-1}(\hat{\theta})$ and formula (5.14) is used,
version of (5.17) will be obtained.

Often $G_{(i)}(\hat{\theta})$ cannot be computed because a quasi-Newton algorithm is being used. In nonlinear least-squares problems, however, several special approximations to $G_{(i)}(\hat{\theta})$ are available.

For Gauss-Newton algorithms, $A(\hat{\theta})$ is generally $J^T(\hat{\theta})J(\hat{\theta})$, where J is the $n \times p$ Jacobian matrix for $\phi(\hat{\theta})$. Clearly $A_{(i)}(\hat{\theta})$ is then given by $J_{(i)}^T(\hat{\theta})J_{(i)}(\hat{\theta})$, and some algebra shows that

$$A_{(i)}^{-1}(\hat{\theta})g_{(i)}(\hat{\theta}) = \frac{\left[J^T(\hat{\theta})J(\hat{\theta})\right]^{-1}J_i^T(\hat{\theta})\phi_i(\hat{\theta})}{1 - h_i(\hat{\theta})}, \qquad (5.19)$$

where $J_i(\hat{\theta})$ is the ith row of $J(\hat{\theta})$ and $h_i(\hat{\theta})$ is the ith diagonal element of

$$H(\hat{\theta}) = J(\hat{\theta})\left(J^T(\hat{\theta})J(\hat{\theta})\right)^{-1}J(\hat{\theta})^T. \qquad (5.20)$$

Note that (5.19) is exactly $\hat{\theta} - \hat{\theta}(i)$ when, in the linear case, $\phi_i(\theta) = y_i - x_i\theta$ and $J = X$. By analogy, then, we can measure leverage in the neighborhood of $\hat{\theta}$ by the diagonal elements of $H(\hat{\theta})$. Of course, special problems occur when $J(\hat{\theta})$ is singular or has a high condition number.

Specialized quasi-Newton methods for nonlinear least-squares have been developed by Dennis and Welsch (1978) and are described in greater detail in Dennis, Gay, and Welsch (1979). In these algorithms the Hessian is approximated by $J^TJ + S$, and S is updated by rank-two update formulas. It is then possible to construct an approximation to $S_{(i)}(\hat{\theta})$ as well, so that $A_{(i)}(\hat{\theta}) = J_{(i)}^T(\hat{\theta})J_{(i)}(\hat{\theta}) + S_{(i)}(\hat{\theta})$. In large-residual least-squares problems the use of S is quite important and, although the simple relationship of (5.19) to the linear case is lost, the use of S is generally beneficial in computing approximations to $\hat{\theta} - \hat{\theta}(i)$ as well. This procedure is more fully described in Welsch (1977b), and much of the special structure can be applied to the generalized linear models discussed by Nelder and Wedderburn (1972) and to the choice models in McFadden (1976). The recent thesis by Pregibon (1979) provides an in-depth study of a number of these problems.

We recall that the approximation in (5.14) involves just one Newton (or quasi-Newton) step. Naturally, this process could be continued until convergence is achieved. Although more research is necessary, this additional computational burden appears to provide little new information.

Approaches analogous to those discussed in this section can be developed using derivatives with respect to weighted observations rather than deletion. Since we emphasize deletion in the earlier parts of this book, we do not go into further detail here.

There is one fundamental problem with nonlinear regression diagnostics calculated only at the solution: data that are influential during specific iterations can cause the minimization algorithm to find a local minimum different from the one that would have been found had the influential data been modified or set aside. It is, therefore, advisable to monitor influential data as an algorithm proceeds. A simple approach is to compute $\tilde{d}_i^2(k)$ at each iteration, k, and record how often it exceeds a certain cutoff (or is in the top 10% of $\tilde{d}_i^2(k)$ for each k). When specific observations appear overly influential, the nonlinear problem can be rerun from the original starting guess with these data points downweighted or omitted. This technique has worked well in practice, and our early fears that every point would turn up as an influential point at some iteration have proved to be unfounded.

There are special cases where algorithms can be monitored quite easily using the diagnostics developed in Chapter 2, as, for example, when iteratively reweighted least squares is used for robust estimation [Holland and Welsch (1978)]. Once the weights have been determined at each iteration, weighted least squares is used to find the next approximate solution. Since the weights are now fixed, many of the linear regression diagnostics can be used by replacing X by TX and y by Ty, where T is the diagonal matrix with elements consisting of the square root of the weights.

Collinearity Diagnostics.   The extension of the collinearity diagnostics to the estimation of models nonlinear in parameters and/or variables affords many interesting research topics. Alternative model specifications, involving nonlinearities in the variates, are a frequent object of applied statistical research. Economists, for example, are often interested in knowing whether a given explanatory variable should enter the model in the form of x or x^2 or both. Similarly, specifications of models linear in the given variates vie with those linear in the logs, that is, $y = \alpha + \beta x + \varepsilon$ versus $ln\,y = \alpha + \beta\,ln\,x + \eta$. Both of these situations arise from nonlinear transformations of the original data series and result in models that remain linear in the parameters. The collinearity diagnostics would seem to be a useful tool for analyzing directly the effect such nonlinear transformations of the data have on the conditioning of the data for estimation of the model.

Models nonlinear in the parameters present equally interesting problems. There appear to be at least two approaches to extending the collinearity diagnostics to such models. The first applies the existing diagnostics to a linearized approximation to the model in (5.10). Here we assume the components of u to be independently distributed with mean zero and variance σ^2. Several means exist for estimating models of this sort.[5] For example, a least-squares (or minimum-distance) approach would be to find the $\hat{\theta}$ that, given $\mathbf{y}$ and $\mathbf{X}$, minimizes $\mathbf{u}^T\mathbf{u} = \phi^T(\theta)\phi(\theta)$. Once such an estimate has been obtained, the first two terms of a Taylor expansion may be used to form a linear approximation to the model about $\hat{\theta}$ as

$$\mathbf{z} = \mathbf{J}(\hat{\theta})\theta - \mathbf{u},$$

where $\mathbf{z} \equiv \mathbf{J}(\hat{\theta})\hat{\theta} - \phi(\hat{\theta})$ and $\mathbf{J}(\hat{\theta})$ is the $n \times p$ Jacobian matrix defined above. It seems reasonable to assess the conditioning of the data relative to the estimation of θ by examining the conditioning of the $\mathbf{J}(\hat{\theta})$ matrix.

The second possible extension of the diagnostics to models nonlinear in the parameters is a more direct, but less easily effected, approach in need of much more research. We note that the collinearity diagnostics of Chapter 3 are relevant to estimation of the linear regression model by ordinary least squares because they directly exploit properties of that model. In particular, the diagnostics pertain to the data matrix $\mathbf{X}$ in the form in which it enters the model, $\mathbf{y} = \mathbf{X}\beta + \varepsilon$, and the condition indexes of the potential degradation of the estimates are related to a decomposition of the variance-covariance matrix $\sigma^2(\mathbf{X}^T\mathbf{X})^{-1}$ of the least-squares estimator. Of course, no such direct link exists between the diagnostics of Chapter 3 and estimators for models nonlinear in the parameters. Indeed a notion of "conditioning" relevant to such models has yet to be defined.[6] If a measure of conditioning can be defined for such models, it can be exploited as a diagnostic tool only if it can be directly incorporated into specific nonlinear estimation procedures, in the manner, for example, that the μ_k entered the variance decomposition in the linear model.

The collinearity diagnostics and their related concepts of conditioning are also usefully employed as a monitor of the numerical stability of the various optimization algorithms at each iteration leading up to the final parameter estimates. In the Gauss-Newton method, for example, the $\mathbf{J}(\theta)$ matrix is computed at each iteration and enters into the computation of the next step. Clearly, such algorithms should be monitored for the conditioning of $\mathbf{J}(\theta)$, flagging problems that arise. Likewise, in estimators using Newton-like algorithms, knowing that the Hessian is becoming ill conditioned signals problems in determining the next step. Furthermore, in maximum-likelihood estimation, an ill-conditioned Hessian near the solution signals potential problems in inverting the Hessian as an estimator of the variance-covariance matrix.

---

[5]For further discussion of these and related matters, see Malinvaud (1970), Belsley (1974), and Bard (1974).
[6]The $\mathbf{J}(\theta)$ matrix for the linearization obtained above may, however, be a good start.

Additional Topics

Bounded-Influence Regression. In Section 4.4 we examined briefly the use of the regression diagnostics with a robust estimator, and an interesting research issue arises in this context. In particular, we recall that influential observations can result either from high leverage (h_i) or from residuals with large magnitudes. Robust estimation, of course, is designed to lessen the influence of scaled residuals with large magnitudes, and hence helps to mitigate problems arising from this source, but robust regression typically will not downweight influential observations caused by high leverage (h_i) alone. Notice, in Section 4.4, that even after using robust regression, we still have several large values of DFFITS. Thus, a word of warning is in order: since robust regression is not designed to deal with leverage, the weights associated with robust regression (when implemented as a weighted least-squares estimator) are not an adequate diagnostic for influential data from all causes.

A modification of robust regression called bounded-influence regression was proposed by Welsch (1977a) and has been modified and extended in Krasker and Welsch (1979). In this approach the influence of an observation, as measured by generalizations of DFBETAS or DFFITS, is bounded and therefore both leverage *and* residuals are involved. Related alternatives to robust regression have been suggested by Mallows (1975), Hinkley (1977), Hill (1977), Handschin et al. (1975), and Hampel (1978). Much theoretical and computational work remains to be done in this area.

Bounded-influence estimators should provide a powerful, low-cost diagnostic tool. They combine the influence philosophy developed for single-row deletion with the capacity to consider (in a structured way) more than one observation at a time. Thus, the weights obtained from bounded-influence regression can potentially provide a wealth of diagnostic information. In addition, an alternate set of coefficient estimates would be available for comparison with least-squares, robust, and deleted-data regression.

Multiple-Row Procedures. Our discussion in Section 2.1 of multiple-row procedures leaves many fertile areas for further research, especially computational costs, methods of inference, and graphical techniques.

Computational costs rise rapidly when deletion of more than one observation at a time is considered. For this reason, a number of attractive row-diagnostic procedures, based on changes in eigenvalues and condition indexes, have not been given further attention. However, there is some hope that inequalities, branch-and-bound methods, and quadratic-programming techniques can be employed to reduce computational costs.

Multiple-row techiques also lack formal inference procedures, even when a Gaussian sampling distribution is assumed. Andrews and Pregibon (1978) have taken a first step toward providing such procedures, and their techniques can probably be used with other diagnostics. Unfortunately, the calculation of the critical levels required by these tests depends on the specific **X** data employed in a manner that adds greatly to computational costs. Generally applicable tables can not be made available, and it becomes necessary to provide a conditional sampling distribution for each **X**, a formidable task indeed. Perhaps recent work by Efron (1979) on the bootstrap method will be of assistance here.

While formal inference procedures may aid in the analysis of the large volume of information generated by multiple-row methods and other diagnostics, the most likely means for quickly digesting this information is through graphical data analysis. Fortunately, computer graphics technology has advanced to the point where research in this area is feasible, and we think it holds great promise.

Transformations. The influential-data diagnostics of Chapter 2, are based on an examination of the effects that result when some aspect of the overall model has been perturbed, as, for example, the data or the error structure, and includes the possibility that the structural form of the model equations can be perturbed as well. A particular case of this would occur if a response or explanatory variable were to be transformed. For example, we may wish to determine how the regression output is affected by a change in λ, the parameter of the Box and Cox (1964) family of transformations,

$$X^* = \frac{X^\lambda - 1}{\lambda}.$$

A solution to this exercise, consistent with the spirit of Chapter 2, would be to differentiate the least-squares estimates, providing a measure of local sensitivity to λ. (More than one λ could be involved for transformations of different data series.) Recent suggestions by Hinkley (1978) provide interesting ideas on the effect of transformations on response variables.

Data transformations also affect the collinearity diagnostics of Chapter 3. The impact of linear transformations has already been discussed in Appendix 3B, but, as noted above, the effect of nonlinear transformations offers an area of research.

Time Series and Lags.   It is noted in Section 4.2 that the simultaneous presence of several different lags for a given data series can change our interpretation of the deletion diagnostics, for the deletion of a row can no longer be interpreted as the deletion of an observation. An observation can, of course, be deleted by removing all rows where data from that observation period occur. Such a procedure is quite cumbersome when there are higher-order lags, and we have not found it to be particularly effective. Nevertheless, we consider the methods of Chapter 2 only to be approximations to a more complete theory of influential data in the context of time-series. A promising approach to such a theory may be the generalized robust time-series estimation methods being developed in Martin (1979) and (1980). While our research on regression diagnostics has led to the development of bounded-influence regression, it may be that its counterpart in time-series analysis, generalized robust estimation, will lead to the development of better diagnostics for time-series models.

Bibliography

Allen, D. M. (1971), "Mean-Square Error of Prediction as a Criterion for Selecting Variables," *Technometrics*, **13**, pp. 469–475.

Allen, D. M. (1974), "The Relationship between Variable Selection and Data Augmentation and a Method for Prediction," *Technometrics*, **16**, pp. 125–127.

Andrews, D. F. (1971), "Significance Tests Based on Residuals," *Biometrika*, **58**, pp. 139–148.

Andrews, D. F. and D. Pregibon (1978), "Finding the Outliers that Matter," *Journal of the Royal Statistical Society*, Series B, **40**, pp. 85–93.

Anscombe, F. J. (1960), "Rejection of Outliers," *Technometrics*, **2**, pp. 123–147.

Anscombe, F. J. (1961), "Examination of Residuals," *Proceedings of Fourth Berkeley Symposium in Mathematical Statistics and Probability*, **1**, pp. 1–36.

Anscombe, F. J. and J. W. Tukey (1963), "The Estimation and Analysis of Residuals," *Technometrics*, **5**, pp. 141–160.

Bard, Yonathan (1974), *Nonlinear Parameter Estimation*, Academic Press: New York.

Bauer, F. L. (1963), "Optimally Scaled Matrices," *Numerische Mathematik*, **5**, pp. 73–87.

Bauer, F. L. (1971), "Elimination with Weighted Row Combinations for Solving Linear Equations and Least Squares Problems," in *Handbook for Automatic Computation, Vol. II: Linear Algebra*, J. H. Wilkinson, and C. Reisch, Eds., Springer-Verlag: New York, pp. 119–133.

Beaton, A. E., D. B. Rubin, and J. L. Barone (1976), "The Acceptability of Regression Solutions: Another Look at Computational Accuracy," *Journal of the American Statistical Association*, **71**, pp. 158–168.

Becker, R., N. Kaden, and V. Klema (1974), "The Singular Value Analysis in Matrix Computation," Working paper 46, Computer Research Center, National Bureau of Economic Research, Cambridge, Mass. (*See also* Laub and Klema (1980)).

Beckman, R. J. and H. J. Trussell (1974), "The Distribution of an Arbitrary Studentized Residual and the Effects of Updating in Multiple-Regressions," *Journal of the American Statistical Association*, **69**, pp. 199–201.

Behnken, D. W. and N. R. Draper (1972), "Residuals and their Variance Patterns," *Technometrics*, **14**, pp. 101–111.

Belsley, David A. (1969), *Industry Production Behavior: The Order-Stock Distinction*, North-Holland Publishing Co.: Amsterdam.

Belsley, David A. (1974), "Estimation of Systems of Simultaneous Equations, and Computational Specifications of GREMLIN," *Annals of Economic and Social Measurement*, 3, pp. 551–614.

Belsley, David A. (1976), "Multicollinearity: Diagnosing Its Presence and Assessing the Potential Damage It Causes Least-Squares Estimation," Working paper #154, Computer Research Center, National Bureau of Economic Research, Cambridge, Mass.

Belsley, David A. and Virginia C. Klema (1974), "Detecting and Assessing the Problems Caused by Multicollinearity: A Use of the Singular-Value Decomposition," Working paper #66, Computer Research Center, National Bureau of Economic Research, Cambridge, Mass.

Bingham, Christopher (1977), "Some Identities Useful in the Analysis of Residuals from Linear Regression," Technical Report 300, School of Statistics, University of Minnesota.

Box, G. E. P. and D. R. Cox (1964), "An Analysis of Transformations," *Journal of the Royal Statistical Society*, Series B, **26**, pp. 211–252.

Box, G. E. P. and N. R. Draper (1975), "Robust Designs," *Biometrika*, **62**, pp. 347–351.

Box, G. E. P. and D. W. Tidwell (1962), "Transformations of the Independent Variable," *Technometrics* **4**, pp. 531–549.

Brown, R. L., J. Durbin, and J. M. Evans (1975), "Techniques for Testing the Constancy of Regression Relationships," *Journal of the Royal Statistical Society*, Series B, **37**, pp. 149–163.

Bushnell, R. C. and D. A. Huettner (1973), "Multicollinearity, Orthogonality and Ridge Regression Analysis," Unpublished mimeo. Presented to the December 1973 meetings, Econometric Society, New York

Businger, P. and G. H. Golub (1965), "Linear Least Squares Solutions by Householder Transformations," *Numerische Matematik*, **7**, pp. 269–276.

Chambers, John M. (1971), "Regression Updating," *Journal of the American Statistical Association*, **66**, pp. 744–748.

Chambers, John M. (1973), "Fitting Nonlinear Models: Numerical Techniques," *Biometrika*, **60**, 1, pp. 1–13.

Chambers, John M. (1977), *Computational Methods for Data Analysis*, John Wiley and Sons: New York.

Chatterjee, S. and B. Price (1977), *Regression Analysis by Example*, John Wiley and Sons: New York.

Chow, Gregory C. (1960), "Tests of Equality between Sets of Coefficients in Two Linear Regressions," *Econometrica*, **28**, pp. 591–605.

Coleman, D. E. (1977), "Finding Leverage Groups," Working paper 195, Computer Research Center, National Bureau of Economic Research, Cambridge, Mass.

Coleman, D. E., P. Holland, N. E. Kaden, V. Klema and S. C. Peters (1980), "A System of Subroutines for Iteratively Reweighted Least Squares Computations," forthcoming *ACM Transactions on Mathematical Software*.

Cook, R. D. (1977), "Detection of Influential Observations in Linear Regression," *Technometrics*, **19**, pp. 15–18.

Cook, R. D. (1979), "Influential Observations in Linear Regression," *Journal of the American Statistical Association*, **74**, pp. 169–174.

Cook, R. D. and S. Weisberg (1979), "Finding Influential Cases in Linear Regression: A Review," Technical Report 338, School of Statistics, University of Minnesota.

Daniel, C. and F. S. Wood (1971), *Fitting Equations to Data*, 2nd edition, John Wiley and Sons: New York.

Davies, R. B. and B. Hutton (1975), "The Effects of Errors in the Independent Variables in Linear Regression," *Biometrika*, **62**, pp. 383–391.

Dempster, A. P., M. Schatzoff, and N. Wermuth (1977), "A Simulation Study of Alternatives to Ordinary Least Squares," *Journal of the American Statistical Association*, **72**, pp. 77–91.

Denby, L. and C. L. Mallows (1977), "Two Diagnostic Displays for Robust Regression Analysis," *Technometrics*, **19**, pp. 1–14.

Denby, L. and R. D. Martin (1979), "Robust Estimation of the First-Order Autoregressive Parameter," Forthcoming *Journal of the American Statistical Association*.

Dennis, J. E. and R. E. Welsch (1978), "Techniques for Nonlinear Least-Squares and Robust Regression," *Communications in Statistics*, B7, pp. 345–359.

Dennis, J. E., D. Gay, and R. E. Welsch (1979), "An Adaptive Nonlinear Least-Squares Algorithm," to appear in *Transactions on Mathematical Software*.

Devlin, S. J., R. Gnanadesikan, and J. R. Kettenring (1975), "Robust Estimation and Outlier Detection with Correlation Coefficients," *Biometrika*, **62**, 3, pp. 531–545.

Draper, N. R. and H. Smith (1966), *Applied Regression Analysis*, John Wiley and Sons, Inc: New York.

Duncan, G. T. (1978), "An Empirical Study of Jackknife-Constructed Confidence Regions in Nonlinear Regression," *Technometrics*, **20**, pp. 123–129.

EISPACK, II (1976), *Matrix Eigensystem Routines—Eispack Guide*, B. T. Smith *et al.*, Eds., Springer-Verlag: New York.

Efron, B. (1979), "Bootstrap Methods: Another Look at the Jackknife," *Annals of Statistics*, **7**, pp. 1–26.

Ezekiel, M. and K. A. Fox (1959), *Methods of Correlation and Regression Analysis*, 3rd ed., John Wiley and Sons, Inc.: New York.

Faddeeva, V. N. (1959), *Computational Methods of Linear Algebra* (trans. C. D. Berster), Dover: New York.

Farrar, D. E. and R. R. Glauber (1967), "Multicollinearity in Regression Analysis: The Problem Revisited," *Review of Economics and Statistics*, **49**, pp. 92–107.

Fisher, F. M. (1965), "Dynamic Structure and Estimation in Economy-wide Econometric Models," J. S. Duesenberry, G. Fromm, L. R. Klein, and E. Kuh, Eds., Chapter 15, *The Brookings Quarterly Econometric Model of the United States*, Rand McNally: Chicago; North-Holland: Amsterdam.

Fisher, F. M. (1970), "Tests of Equality between Sets of Coefficients in Two Linear Regressions: An Expository Note," *Econometrica*, **38**, pp. 361–365.

Foley, D. (1975), "On Two Specifications of Asset Equilibrium in Macroeconomic Models," *Journal of Political Economy*, **83**, pp. 303–324.

Forsythe, George and Cleve B. Moler (1967), *Computer Solution of Linear Algebraic Systems*, Prentice-Hall: Englewood Cliffs, N. J.

Forsythe, G. E. and E. G. Straus (1955), "On Best Conditioned Matrices," *Proceedings of the American Mathematical Association*, **6**, pp. 340–345.

Friedman, Benjamin (1977), "Financial Flow Variables and the Short Run Determination of Long Term Interest Rates," *Journal of Political Economy*, **85**, pp. 661–689.

Friedman, M. (1957), *A Theory of the Consumption Function*, Princeton University Press: New Jersey.

Frisch, R. (1934), *Statistical Confluence Analysis by Means of Complete Regression Systems*, University Institute of Economics, Oslo.

Frisch, R. and F. V. Waugh (1933), "Partial Time Regression as Compared with Individual Trends," *Econometrica*, 1, pp. 221–223.

Furnival, G. M. and R. W. Wilson, Jr. (1974), "Regression by Leaps and Bounds," *Technometrics* 16, pp. 499–511.

Gentleman, J. F. and M. B. Wilk (1975), "Detecting Outliers II. Supplementing the Direct Analysis of Residuals," *Biometrika*, 62, pp. 387–411.

Gnanadesikan, R. and J. R. Kettenring (1972), "Robust Estimates, Residuals and Outlier Detection with Multiresponse Data," *Biometrics* 29, pp. 81–124.

Golub, Gene H. (1969), "Matrix Decompositions and Statistical Calculations," *Statistical Computation*, R. C. Milton and J. A. Nelder, Eds. Academic Press, New York, pp. 365–397.

Golub, G., V. Klema, and S. Peters (1980), "Rules and Software for Detecting Rank Degeneracy," *Journal of Econometrics*, 12, pp. 41–48.

Golub, G., V. Klema, and G. W. Stewart (1976), "Rank Degeneracy and Least-Squares Problems," *Computer Science Technical Report Series*, #TR-456, University of Maryland.

Golub, G. H. and C. Reinsch (1970), "Singular Value Decomposition and Least-Squares Solutions," *Numerische Mathematik*, 14, pp. 403–420.

Golub, G. H. and J. M. Varah (1974), "On a Characterization of the Best l_2-Scalings of a Matrix," *SIAM Journal of Numerical Analysis*, 5, pp. 472–479.

Griliches, Z. (1968), "Hedonic Price Indexes for Automobiles: An Econometric Analysis of Quality Change," in *Readings in Economic Statistics and Econometrics*, A. Zellner, Ed., Little Brown: Boston.

Haitovsky, Yoel (1969), "Multicollinearity in Regression Analysis: Comment," *Review of Economics and Statistics*, 50, pp. 486–489.

Hampel, F. R. (1974), "The Influence Curve and Its Role in Robust Estimation," *Journal of the American Statistical Association*, 69, pp. 383–393.

Hampel, F. R. (1975), "Beyond Location Parameters: Robust Concepts and Methods," International Statistical Institute Invited Paper, *Proceedings of the 40th Session*, 46, Book 1, pp. 375–382, Warsaw.

Hampel, F. R. (1978), "Optimally Bounding the Gross-Error-Sensitivity and the Influence of Position in Factor Space," *1978 Proceedings of the Statistical Computing Section*, American Statistical Association, Washington, D.C., pp. 59–64.

Handschin, E., J. Kohlas, A. Fiechter, and F. Schweppe (1975), "Bad-Data Analysis for Power System State Estimation," *IEEE Transactions on Power Apparatus and Systems*, PAS-94, No. 2, pp. 329–337.

Hanson, Richard J. and Charles L. Lawson (1969), "Extensions and Applications of the Householder Algorithm for Solving Linear Least Squares Problems," *Mathematics of Computation*, 23, pp. 787–812.

Harrison, D. and D. L. Rubinfeld (1978), "Hedonic Prices and the Demand for Clean Air," *Journal of Environmental Economics and Management*, 5, pp. 81–102.

Hawkins, D. M. (1973), "On the Investigation of Alternative Regressions by Principal Component Analysis," *Applied Statistics*, 22, pp. 275–286.

Hill, R. W. (1977), *Robust Regression When There Are Outliers in the Carriers*, unpublished Ph.D. dissertation, Department of Statistics, Harvard University.

Hinkley, David V. (1977), "On Jackknifing in Unbalanced Situations," *Technometrics*, **19**, pp. 285–292.

Hinkley, David V. (1978), "Some Topics in Robust Regression," *1978 Proceedings of the Statistical Computing Section*, American Statistical Association, Washington, D.C., pp. 55–58.

Hoaglin, D. C. and R. E. Welsch (1978), "The Hat Matrix in Regression and ANOVA," *The American Statistician*, **32**, pp. 17–22 and *Corrigenda* 32, p. 146.

Hoerl, A. E. and R. W. Kennard (1970), "Ridge Regression: Biased Estimation for Nonorthogonal Problems," *Technometrics*, **12**, pp 55–68.

Holland, P. W. (1973), "Weighted Ridge Regression: Combining Ridge and Robust Regression Methods," Working paper 11, Computer Research Center, National Bureau of Economic Research, Cambridge, Mass.

Holland, P. W. and R. E. Welsch (1978), "Robust Regression Using Iteratively Reweighted Least Squares," *Communications in Statistics*, **A9**, pp. 813–827.

Huber, P. J. (1973), "Robust Regression: Asymptotics, Conjectures, and Monte Carlo," *Annals of Statistics*, **1**, pp. 799–821.

Huber, P. J. (1975), "Robustness and Designs," in *A Survey of Statistical Design and Linear Models: Proceedings*, J. N. Srivastava, Ed., North-Holland: Amsterdam.

Huber, P. J. (1977), *Robust Statistical Procedures*, *SIAM*: Philadelphia, Penn.

Kadane, Joseph B., James M. Dickey, Robert L. Winkler, Wayne S. Smith, and Stephen C. Peters (1977), "Interactive Elicitation of Opinion for a Normal Linear Model," forthcoming *Journal of the American Statistical Association*.

Kendall, M. G. (1957), *A Course in Multivariate Analysis*, Griffin: London.

Kloek, T. and L. B. M. Mennes (1960), "Simultaneous Equations Estimation Based on Principal Components of Predetermined Variables," *Econometrica*, **28**, pp. 45–61.

Knuth, D. E. (1973), *The Art of Computer Programming, Vol. 1: Fundamental Algorithms*, Addison-Wesley: Massachusetts.

Krasker, W. S. and R. E. Welsch (1979), "Efficient Bounded-Influence Regression Estimation Using Alternative Definitions of Sensitivity," Technical Report #3, Center for Computational Research in Economics and Management Science, Massachusetts Institute of Technology, Cambridge, Mass..

Kuh, E. and R. E. Welsch (1979), "Econometric Models and their Assessment for Policy: Some New Diagnostics Applied to Energy Research in Manufacturing," forthcoming, Proceedings of the Workshop on Validation and Assessment Issues of Energy Models, U.S. Department of Energy and the National Bureau of Standards, Washington, D.C.

Kumar, T. K. (1975a), "The Problem of Multicollinearity: A Survey," Unpublished mimeo, Abt Associates, Inc., Cambridge, Mass..

Kumar, T. K. (1975b), "Multicollinearity in Regression Analysis," *The Review of Economics and Statistics*, **57**, pp. 365–366.

Larson, W. A. and S. J. McCleary (1972), "The Use of Partial Residual Plots in Regression Analysis," *Technometrics* **14**, pp. 781–790.

Laub, A. and V. Klema (1980), "The Singular Value Decomposition: Its Computation and Some Applications," *IEEE Transactions on Automatic Control*, forthcoming.

Lawson, C. R. and R. J. Hanson (1974), *Solving Least-Squares Problems*, Prentice-Hall: Englewood Cliffs, N. J.

Leamer, E. E. (1973), "Multicollinearity: A Bayesian Interpretation," *Review of Economics and Statistics*, **55**, pp. 371–380.

Leamer, E. E. (1978), *Specifications Searches*, John Wiley and Sons: New York.

LINPACK (1979), *Linpack Users Guide* by J. J. Dongarra, J. R. Bunch, C. B. Moler, and G. W. Stewart, Society for Industrial and Applied Mathematics, Philadelphia, Pa.

Longley, J. W. (1967), "An Appraisal of Least Squares Programs for the Electronic Computer from the Point of View of the User," *Journal of the American Statistical Association*, **62**, pp. 819–831.

Lund, R. E. (1975), "Tables for an Approximate Test for Outliers in Linear Models," *Technometrics*, **17**, pp. 473–476.

Malinvaud, E. (1970), *Statistical Methods of Econometrics*, 2nd rev. ed., North-Holland: Amsterdam.

Mallows, C. L. (1973a), "Influence Functions," paper delivered to National Bureau of Economic Research Conference on Robust Regression, Cambridge, Mass.

Mallows, C. L. (1973b), "Some Comments on C_p," *Technometrics*, **15**, pp. 661–675.

Mallows, C. L. (1975), "On Some Topics in Robustness," paper delivered to the Eastern Regional IMS meeting, Rochester, New York.

Martin, R. D. (1979), "Robust Estimation for time Series Autoregressions," in *Robustness in Statistics*, R. L. Launer and G. Wilkins, Eds., Academic Press: New York.

Martin, R. D. (1980), "Robust Estimation of Autoregressive Models," in *Directions in Time Series*, D. Brillinger, et. al., Eds., Institute of Mathematical Statistics, Stanford University, California.

Mayer, L. and T. Willke (1973), "On Biased Estimation in Linear Models," *Technometrics*, **15**, pp. 497–508.

McFadden, D. (1976), "Quantal Choice Analysis: A Survey," *Annals of Economic and Social Measurement*, **5**, pp. 363–391.

McNeil, Donald R. (1977), *Interactive Data Analysis*, John Wiley and Sons: New York.

Mickey, M. R., O. J. Dunn, and V. Clark (1967), "Note on the Use of Stepwise Regression in Detecting Outliers," *Computers and Biomedical Research*, **1**, pp. 105–111.

Miller, R. G. (1964), "A Trustworthy Jackknife," *Annals of Mathematical Statistics*, **35**, 1594–1605.

Miller, R. G. (1966), *Simultaneous Statistical Inference*, McGraw-Hill: New York.

Miller, R. G. (1974a), "The Jackknife: A Review," *Biometrika*, **61**, pp. 1–15.

Miller, R. G. (1974b), "An Unbalanced Jackknife," *The Annals of Statistics*, **2**, pp. 880–891.

Mitchell, B. (1971), "Estimation of Large Econometric Models by Principal Components and Instrumental Variable Methods," *Review of Economics and Statistics*, **53**, 2, pp. 140–146.

Mitchell, B. and F. M. Fisher (1970), "The Choice of Instrumental Variables in the Estimation of Economy-Wide Econometric Models: Some Further Thoughts," *International Economic Review*, **2**, pp. 226–234.

Modigliani, Franco (1975), "The Life Cycle Hypothesis of Saving, Twenty Years Later," in *Contemporary Issues in Economics*, Michael Parkin, Ed., University Press, Manchester.

Mosteller, F. and J. W. Tukey (1977), *Data Analysis and Regression*, Addison-Wesley: Reading, Mass.

Nelder, J. A. and R. W. M. Wedderburn (1972), "Generalized Linear Models," *Journal of the Royal Statistical Society*, Series A, **135**, pp. 370–384.

Obenchain, R. L. (1977), "Classical F-Tests and Confidence Regions for Ridge Regression," *Technometrics*, **19**, pp. 429–439.

O'Brien, R. J. (1975), "The Sensitivity of OLS and Other Econometric Estimators," unpublished manuscript, University of Southampton.

O'Hagan, J. and B. McCabe (1975), "Tests for the Severity of Multicollinearity in Regression Analysis: A comment," *The Review of Economics and Statistics*, **57**, pp. 369–370.

Peters, S. (1976), "An Implementation of Algorithms Which Produce Jackknifed Least-Squares Estimates," unpublished B. S. Thesis, Department of Electrical Engineering and Computer Science, M.I.T.

Phillips, G. D. A. (1977), "Recursions for the Two-Stage Least-Squares Estimators," *Journal of Econometrics*, **6**, pp. 65–77.

Pregibon, D. (1979), "Data Analytic Methods for Generalized Linear Models," Ph.D. Dissertation, Department of Statistics, University of Toronto, Ontario, Canada.

Prescott, P. (1975), "An Approximate Test for Outliers in Linear Models," *Technometrics*, **17**, 1, pp. 129–132.

Quandt, Richard E. (1972), "A New Approach to Estimating Switching Regressions," *Journal of the American Statistical Association*, **67**, pp. 306–310.

Rao, C. R. (1973), *Linear Statistical Inference and Its Applications*, 2nd ed., John Wiley and Sons: New York.

ROSEPACK (1980), "A System of Subroutines for Iteratively Reweighted Least Squares Computations," by Coleman et al., to appear in *ACM Transactions on Mathematical Software*.

Rosner, B. (1975), "On the Detection of Many Outliers," *Technometrics*, **17**, pp. 221–227.

Schweder, Tore (1976), "Optimal Methods to Detect Structural Shift or Outliers in Regression," *Journal of the American Statistical Association*, **71**, pp. 491–501.

Silvey, S. D. (1969), "Multicollinearity and Imprecise Estimation," *Journal of Royal Statistical Society*, Series B, **31**, pp. 539–552.

Stein, C. M. (1956), "Inadmissibility of the Usual Estimator for the Mean of a Multivariate Normal Distribution," *Proceedings of the Third Berkeley Symposium on Mathematical Statistics and Probability*, **1**, pp. 197–206. University of California Press: Berkeley and Los Angeles.

Sterling, Arlie G. (1977), "An Investigation of the Determinants of the Long-Run Savings Ratio," unpublished B.S. Thesis, Massachusetts Institute of Technology, Cambridge, Mass.

Stewart, G. W. (1973), *Introduction to Matrix Computations*, Academic Press: New York.

Stone, M. (1974), "Cross-Validitory Choice and Assessment of Statistical Predictions," *Journal of the Royal Statistical Society*, Series B, 36, pp. 111–147.

Theil, H. (1963), "On the Use of Incomplete Prior Information in Regression Analysis," *Journal of the American Statistical Association*, **58**, pp. 401–414.

Theil, H. (1971), *Principles of Econometrics*, John Wiley and Sons: New York.

Theil, H. and A. S. Goldberger (1961), "On Pure and Mixed Statistical Estimation in Economics," *International Economic Review*, **2**, pp. 65–78.

Tietjen, G. L., R. H. Moore, and R. J. Beckman (1973), "Testing for a Single Outlier in Simple Linear Regression," *Technometrics*, **15**, pp. 717–721.

Tukey, J. W. (1977), *Exploratory Data Analysis*, Addison-Wesley: Reading, Massachusetts.

Van der Sluis, A. (1969), "Condition Numbers and Equilibration of Matrices," *Numerische Mathematik*, **14**, pp. 14–23.

Van der Sluis, A. (1970), "Condition, Equilibration and Pivoting in Linear Algebraic Systems," *Numerische Mathematik*, **15**, pp. 74–88.

Velleman, P. F. and D. C. Hoaglin (1980), *Applications, Basics and Computing for Exploratory Data Analysis*, Duxbury Press: North Scituate, Mass.

Vinod, H. D. (1978), "Survey of Ridge Regression and Related Techniques for Improvements over Ordinary Least Squares," *Review of Economics and Statistics*, **60**, pp. 121–131.

Von Hohenbalken, B. and W. C. Riddell (1978), "Wedge Regression: A Ridge-Equivalent Procedure That Forecasts and Generalizes Differently," unpublished mimeo, Department of Economics, University of Alberta, Canada.

Webster, J. T., R. F. Gunst, and R. L. Mason (1974), "Latent Root Regression Analysis," *Technometrics*, **16**, pp. 513–522.

Webster, J. T., R. F. Gunst, and R. L. Mason (1976), "A Comparison of Least Squares and Latent Root Regression Estimators," *Technometrics*, **18**, pp. 75–83.

Weisberg, S. (1977), "A Statistic for Allocating C_p to Individual Cases," Technical Report No. 296, School of Statistics, University of Minnesota.

Welsch, R. E. (1976), "Graphics for Data Analysis," *Computers and Graphics*, **2**, pp. 31–37.

Welsch, R. E. (1977a), "Regression Sensitivity Analysis and Bounded-Influence Estimation." Presented at the Conference on Criteria for Evaluation of Econometric Models, Dept. of Econ., Ann Arbor, Michigan, June 9–10; forthcoming in *Evaluation of Econometric Models*, Jan Kmenta and James Ramsey, Eds., Academic Press: New York.

Welsch, R. E. (1977b), "Nonlinear Statistical Data Analysis," *Proceedings of the Tenth Interface Symposium on Computer Science and Statistics*, D. Hogben and D. Fife, Eds., National Bureau of Standards, Washington, D. C. pp. 77–86.

Welsch, R. E. and Edwin Kuh (1977), "Linear Regression Diagnostics," Sloan School of Management Working Paper, 923–77, Massachusetts Institute of Technology, Cambridge, Massachusetts.

Wilkinson, J. H. (1965), *The Algebraic Eigenvalue Problem*, Oxford University Press.

Wilks, Samuel S. (1962), *Mathematical Statistics*, John Wiley and Sons: New York.

Wilks, Samuel S. (1963), "Multivariate Statistical Outliers," *Sankhya*, **A25**, pp. 407–426.

Wood, F. S. (1973), "The Use of Individual Effects and Residuals in Fitting Equations to Data," *Technometrics*, **15**, pp. 677–694.

Wonnacott, R. J. and T. H. Wonnacott (1979), *Econometrics*, 2nd edition, John Wiley and Sons: New York.

Zellner, A. (1971), *An Introduction to Bayesian Inference in Econometrics*, John Wiley and Sons: New York.

Author Index

Subject Index